Value of ICTs in Facilitating Educational Development and Reform

Faiz Abdelhafid
https://orcid.org/0000-0003-4758-9142
Al-Istiqlal University, Palestine

Sameh R. Al-Qubaj
Al-Istiqlal University, Palestine

IGI Global
Scientific Publishing
Publishing Tomorrow's Research Today

Vice President of Editorial	Melissa Wagner
Director of Acquisitions	Mikaela Felty
Director of Book Development	Jocelynn Hessler
Production Manager	Mike Brehm
Cover Design	Jose Rosado

Published in the United States of America by
IGI Global Scientific Publishing
701 East Chocolate Avenue
Hershey, PA, 17033, USA
Tel: 717-533-8845 | Fax: 717-533-7115
Website: https://www.igi-global.com E-mail: cust@igi-global.com

Library of Congress Cataloging-in-Publication Data

Names: Abdelhafid, Faiz, 1979- editor
http://id.loc.gov/rwo/agents/n2023070788 | Al-Qubaj, Sameh R., 1968-
editor http://id.loc.gov/rwo/agents/n2025019628
Title: Value of ICTs in facilitating educational development and reform /
[edited by] Faiz Abdelhafid, Sameh R. Al-Qubaj.
Description: Hershey, PA : IGI Global Scientific Publishing, [2026] |
Includes bibliographical references and index. | Summary: "The book aims
to explore the role of ICT in education from an anthropological
perspective, addressing its impact on teaching, learning,
infrastructure, and globalization"-- Provided by publisher.
Identifiers: LCCN 2025009758 (print) | LCCN 2025009759 (ebook) | ISBN
9798337311906 hardback | ISBN 9798337311913 paperback | ISBN
9798337311920 ebook
Subjects: LCSH: Educational technology
http://id.loc.gov/authorities/subjects/sh85041150 | Educational
innovations http://id.loc.gov/authorities/subjects/sh85041127 |
Educational change http://id.loc.gov/authorities/subjects/sh90003113
Classification: LCC LB1028.3 .V35 2025 (print) | LCC LB1028.3 (ebook) |
DDC 371.33/4--dc23/eng/20250510
LC record available at https://lccn.loc.gov/2025009758
LC ebook record available at https://lccn.loc.gov/2025009759

British Cataloguing in Publication Data
A Cataloguing in Publication record for this book is available from the British Library.

Table of Contents

Preface xii

Chapter 1
Humanising Technology: An Anthropological Perspective With Empathy in University Distance Education 1
Maria Fragkaki, Open University of Cyprus, Cyprus
Maria Krasa, Open University of Cyprus, Cyprus

Chapter 2
Anthropologic Perspective of ICT in Education: How Society Shapes and Is Shaped by EdTech 53
Wasswa Shafik, Dig Connectivity Research Laboratory (DCRLab), Uganda

Chapter 3
Embracing Digital Transformation: Impacts on Teaching and Learning Practices 85
Sam Ramaila, University of Johannesburg, South Africa

Chapter 4
The Future of Education in the Post-Democratic Era: Technology, Governance, and Learning 115
Mustafa Kayyali, Maaref University of Applied Sciences, Syria

Chapter 5
Post-Globalization and Education: Redefining Access and Equity in a Technological Era 137
Mustafa Kayyali, Maaref University of Applied Sciences, Syria

Chapter 6
Unveiling AI's Dual Impact: Analyzing Creativity, Application, and Decision-Making Skills in Education 159
Varanasi Rahul, Jain (Deemed to be) University, India
Y. Fathima, Jain (Deemed to be) University, India
L. B. Muralidhar, Jain (Deemed to be) University, India
R. Manju Priya, Jain (Deemed to be) University, India
N. Sathyanarayana, Jain (Deemed to be) University, India
G. Srividhya, Jain (Deemed to be) University, India

Chapter 7
Leveraging Cloud Computing for Enhanced Efficiency and Collaboration in Education Institutions 187
Pooja Dehankar, Ajeenkya D.Y. Patil School of Engineering, India
Kashvi Chaturvedi, Ajeenkya D.Y. Patil University, India
Susanta Das, Ajeenkya D.Y. Patil University, India

Chapter 8
Blending Computational Thinking and Learner-Paced Segment: Content Expert Evaluation of Digital Video Courseware Development 219
Wan Nor Ashiqin Wan Ali, Universiti Malaysia Perlis, Malaysia
Wan Ahmad Jaafar Wan Yahaya, Universiti Sains Malaysia, Malaysia
Syed Zulkarnain Syed Idrus, Universiti Malaysia Perlis, Malaysia
Mohd Noorul Fakhri Yaacob, Universiti Malaysia Perlis, Malaysia

Chapter 9
The Use of ICTs Tools for Students With ADD/ADHD in a Spanish EFL Classroom 243
Isabel María García Conesa, Centro Universitario de la Defensa de San Javier, Spain

Chapter 10
Leveraging ICTs for Transforming English Language Teaching: Facilitating Educational Development and Reform 269
Vivit Rosmayanti, Universitas Negeri Makassar, Indonesia
Muliaty Ibrahim, Universitas Megarezky, Indonesia
Amra Ariyani, Universitas Negeri Makassar, Indonesia

Chapter 11
Designing an Interactive Website Based on "Life Mathematics" to Develop Mathematical Problem-Solving Skills and Reduce the Level of Anxiety 293
Abdelrahman Mohammad Abu Sarah, Ministry of Education and Higher Education, Palestine
Yasmeen Majed Shkokani, Universiti Sains Islam Malaysia, Malaysia
Husam Tewfeeq Herzallah, Al-Quds Open University, Palestine
Riyad Abdalrazzaq Shkokani, Universiti Sains Islam Malaysia, Malaysia
Abedallah Mohammad Abu Sarah, Ministry of Education, Palestine

Compilation of References 301

About the Contributors 345

Index 353

Detailed Table of Contents

Preface.. xii

Chapter 1
Humanising Technology: An Anthropological Perspective With Empathy in University Distance Education.. 1
Maria Fragkaki, Open University of Cyprus, Cyprus
Maria Krasa, Open University of Cyprus, Cyprus

In an era of rapid digital transformation and higher education massification, University Distance Education must transcend functionality to embrace ethical responsibility, emotional resonance, and pedagogical humanity. This chapter introduces the Empathic Anthropological Framework for University Distance Education (EAF@ UDE), a model that frames empathy as both a pedagogical ethic and epistemological imperative. Informed by critical pedagogy, affect theory, digital humanism, and communicative action, the framework explores the political, emotional, and cultural dimensions of digital learning. A qualitative case study at the Open University of Cyprus employed educational ethnography and constructivist grounded theory to examine how empathy is co-constructed via multimodal dialogue, relational presence, emotional scaffolding, and narrative vulnerability. Findings reveal empathy as a driver of transformative learning, civic consciousness, and social cohesion. The chapter offers recommendations for culturally responsive design, human-centered pedagogy, and ethical digital practice.

Chapter 2
Anthropologic Perspective of ICT in Education: How Society Shapes and Is Shaped by EdTech.. 53
Wasswa Shafik, Dig Connectivity Research Laboratory (DCRLab), Uganda

This chapter investigates the anthropologic perspective of Information and Communication Technology (ICT) in education by examining how cultural, social, and contextual factors influence the adoption, integration, and impact of digital technologies in learning environments. It explores the intersection of educational technology and human behavior, highlighting the role of local traditions, beliefs, and values in shaping technological practices. Through a multidisciplinary lens grounded in educational anthropology, this chapter critically analyzes how ICT transforms pedagogical methods, teacher-learner relationships, and knowledge dissemination in diverse cultural settings. Emphasis is placed on ethnographic case studies and theoretical frameworks that elucidate the socio-technical dynamics underpinning digital learning. The chapter aims to contribute to a more culturally responsive approach to ICT implementation in education, offering insights for policymakers, educators, and technologists seeking to bridge the gap between innovation and indigenous knowledge systems.

Chapter 3
Embracing Digital Transformation: Impacts on Teaching and Learning Practices 85
Sam Ramaila, University of Johannesburg, South Africa

The rapid advancement of digital technologies has transformed teaching and learning. This chapter explores the impacts of digital transformation, focusing on technologies like AI, AR/VR, big data, and cloud computing, which reshape pedagogy, personalize learning, and enhance engagement. It also emphasizes the need for educators' professional development, digital literacy, and adaptive strategies. The chapter addresses digital equity, ensuring accessible education for all, and advocates for a strategic approach to technology, driving meaningful educational change.

Chapter 4
The Future of Education in the Post-Democratic Era: Technology, Governance, and Learning 115
Mustafa Kayyali, Maaref University of Applied Sciences, Syria

As democratic values increasingly come under strain globally, education systems are quietly entering a new era—one marked not by open civic participation, but by algorithmic oversight, technocratic governance, and controlled narratives. This chapter explores the contours of education in a post-democratic context, where the promises of digital transformation coexist with subtle forms of surveillance and behavioral standardization. It interrogates the evolving relationship between learners and power, showing how technologies once celebrated for democratizing knowledge may now reinforce hierarchical structures and limit dissent. At the same time, it examines how educational institutions can either reproduce systems of compliance or become sites of critical resistance. Through an interdisciplinary lens that weaves together political theory, technology ethics, and educational praxis, the chapter advocates for a reimagining of ICT-driven education—one that safeguards human agency, nurtures ethical awareness, and resists reductionist narratives.

Chapter 5
Post-Globalization and Education: Redefining Access and Equity in a Technological Era 137
Mustafa Kayyali, Maaref University of Applied Sciences, Syria

In the wake of globalization's waning influence, a new era is emerging—one where educational access and equity must be reconsidered through the lens of technological transformation. This chapter explores the post-globalization landscape and the evolving role of Information and Communication Technologies (ICTs) in shaping the future of education. While digital tools have expanded access to knowledge, they have also introduced new forms of exclusion and inequality, particularly in underrepresented and marginalized communities. By interrogating how access is defined, who benefits from educational technologies, and what forms of equity are truly being pursued, this chapter offers a nuanced analysis of the intersection between technology and justice in education. Drawing on global examples and critical theory, it argues for a human-centered approach that reclaims education as a public good and not merely a digital service. Ultimately, it calls for systemic reform that aligns technological advancement with inclusive, ethical, and equitable educational futures.

Chapter 6
Unveiling AI's Dual Impact: Analyzing Creativity, Application, and Decision-Making Skills in Education 159
Varanasi Rahul, Jain (Deemed to be) University, India
Y. Fathima, Jain (Deemed to be) University, India
L. B. Muralidhar, Jain (Deemed to be) University, India
R. Manju Priya, Jain (Deemed to be) University, India
N. Sathyanarayana, Jain (Deemed to be) University, India
G. Srividhya, Jain (Deemed to be) University, India

Integrating Artificial Intelligence (AI) tools, such as ChatGPT, into education is transforming learning paradigms. This chapter investigates AI's dual impact on students' creativity, application, and decision-making skills through an experimental study involving control and experimental groups. Results highlight that while AI can significantly enhance learning outcomes, improper usage risks cognitive laziness and dependence. This chapter provides an in-depth discussion on fostering responsible AI engagement, ethical considerations, and strategies to harness its potential for education. This study contributes to the discourse on sustainable AI integration in academia by bridging theoretical frameworks with empirical findings.

Chapter 7
Leveraging Cloud Computing for Enhanced Efficiency and Collaboration in Education Institutions 187
Pooja Dehankar, Ajeenkya D.Y. Patil School of Engineering, India
Kashvi Chaturvedi, Ajeenkya D.Y. Patil University, India
Susanta Das, Ajeenkya D.Y. Patil University, India

The desire for increased productivity, teamwork, & creativity is driving a major revolution in the education industry. Cloud computing, which offers institutions an inexpensive, flexible, & scalable platform for delivering excellent instruction, has played a significant role in enabling this shift. The potential of cloud computing to improve communication & collaboration between administrators, professors, & students is one of its main advantages for institutions. Cloud computing can improve educational outcomes by giving students access to top-notch digital resources. However, there are many issues with cloud computing adoption in institutions. The chapter discusses these issues & offers recommendations on how institutions might reduce these risks. It offers a thorough overview of the potential of cloud computing in institutions, highlighting its advantages, difficulties, & best practices. By utilizing cloud computing, institutions can lower costs, improve collaboration & communication, and outcomes—all of which will ultimately increase their competitiveness in the digital age.

Chapter 8
Blending Computational Thinking and Learner-Paced Segment: Content Expert Evaluation of Digital Video Courseware Development 219
Wan Nor Ashiqin Wan Ali, Universiti Malaysia Perlis, Malaysia
Wan Ahmad Jaafar Wan Yahaya, Universiti Sains Malaysia, Malaysia
Syed Zulkarnain Syed Idrus, Universiti Malaysia Perlis, Malaysia
Mohd Noorul Fakhri Yaacob, Universiti Malaysia Perlis, Malaysia

This study explores the development and implementation of digital video courseware (DVC) combining computational thinking (CT) and segmentation techniques to enhance educational outcomes. Segmentation, particularly self-explanation, improves engagement and achievement, while system-paced segmentation reduces cognitive load for novices, boosting retention. The courseware fosters CT skills like algorithmic thinking and supports learner-paced flexibility for better information processing. Content experts evaluated its design, pedagogical relevance, and CT alignment. Results highlight improved understanding, retention, and performance, offering insights for advancing learner-centered education through CT integration and adaptive pacing.

Chapter 9
The Use of ICTs Tools for Students With ADD/ADHD in a Spanish EFL Classroom 243
Isabel María García Conesa, Centro Universitario de la Defensa de San Javier, Spain

This chapter explores the implementation of Information and Communication Technologies (ICT) as teaching support for students with Attention Deficit Disorder (ADD) and Attention Deficit Hyperactivity Disorder (ADHD) in the context of a Spanish EFL classroom in Compulsory Secondary Education. The analysis focuses on a review of studies that use technological tools to enhance attention, concentration, and participation among students with ADD/ADHD in English language teaching. Overall, the chapter highlights the effectiveness of ICT in interventions with the learners studied. A comprehensive view is provided on these tools as a viable and effective solution for addressing the educational needs of students with ADD/ADHD in English language learning. The purpose of this chapter is to conduct a literature review on the use of ICT in the English as a Foreign Language (EFL) classroom for learners with ADD/ADHD.

Chapter 10
Leveraging ICTs for Transforming English Language Teaching: Facilitating Educational Development and Reform 269
Vivit Rosmayanti, Universitas Negeri Makassar, Indonesia
Muliaty Ibrahim, Universitas Megarezky, Indonesia
Amra Ariyani, Universitas Negeri Makassar, Indonesia

The integration of Information and Communication Technologies (ICTs) in English Language Teaching (ELT) has significantly transformed teaching methods and improved the accessibility of education resources. ICTs, including web-enabled resources, mobile devices, Artificial Intelligence, and computer software, have made ELT more engaging and interactive for learners of all levels. They can bridge gaps in quality education, especially in remote and marginalized regions, and promote self-paced learning, instant response, and active participation. ICTs also help overcome the digital divide by providing learners with knowledge about different cultures and peoples, enhancing international cooperation. They contribute to education reform processes in line with global agendas like Sustainable Development Goal 4 (Equitable Quality Education). However, concerns like lack of technology, teacher training, and culture can limit the efficient use of ICT in ELT.

Chapter 11
Designing an Interactive Website Based on "Life Mathematics" to Develop Mathematical Problem-Solving Skills and Reduce the Level of Anxiety 293
Abdelrahman Mohammad Abu Sarah, Ministry of Education and Higher Education, Palestine
Yasmeen Majed Shkokani, Universiti Sains Islam Malaysia, Malaysia
Husam Tewfeeq Herzallah, Al-Quds Open University, Palestine
Riyad Abdalrazzaq Shkokani, Universiti Sains Islam Malaysia, Malaysia
Abedallah Mohammad Abu Sarah, Ministry of Education, Palestine

The aim of the research is to design an "interactive website based on life mathematics" to develop mathematical problem-solving skills and reduce the level of anxiety among secondary students in Palestine. Researchers used the experimental curriculum with a semi-experimental design. The sample consisted of (40) students, from eleventh grade scientific secondary students, who were distributed equally to two groups: Experimental, examined the content of the module (equations and discrepancies) using the interactive website based on life mathematics, and control group who studied the same content in the usual way. The data was collected using the mathematical problem-solving skills test and mathematical anxiety scale. The results showed a statistically different between the two groups' students (experimental and control), in the testing of mathematical problem-solving skills for the students of the experimental group, and the absence of a difference between the students of the two groups in the mathematical anxiety scale.

Compilation of References 301

About the Contributors 345

Index 353

Preface

INTRODUCTION

The accelerating evolution of Information and Communication Technologies (ICTs) continues to redefine the landscape of education, not merely as a matter of technological enhancement but as a profound transformation of pedagogy, access, governance, and human experience. *Value of ICTs in Facilitating Educational Development and Reform* brings together a diverse collection of scholarly contributions that collectively interrogate the promises, tensions, and implications of digital transformation in education across global contexts.

This volume is grounded in the understanding that ICTs are not neutral tools; rather, they are deeply embedded in social, cultural, political, and ethical frameworks that shape—and are shaped by—educational practices. From anthropological and human-centered perspectives to critical examinations of power, equity, and emerging technologies, the chapters in this book offer a multidimensional exploration of how ICTs can both enable and constrain educational development and reform.

A central theme running throughout the volume is the call to humanize technology in education. As digital tools become increasingly integrated into teaching and learning, there is a growing need to foreground empathy, cultural responsiveness, and ethical responsibility. At the same time, the book critically engages with issues such as the digital divide, algorithmic governance, and the risks of technological dependence, emphasizing that meaningful reform requires not only innovation but also reflection and accountability.

The contributions further highlight the transformative potential of emerging technologies—including artificial intelligence, immersive environments, and cloud computing—while addressing their practical, pedagogical, and ethical challenges. Importantly, the volume situates these discussions within broader global shifts, including post-globalization dynamics and evolving conceptions of equity and access in education.

In addition to theoretical and critical perspectives, this book presents empirical studies and applied research that demonstrates how ICTs can enhance learning outcomes, support diverse learners, and foster new forms of engagement. From language education and computational thinking to inclusive practices for students with attention-related challenges, the chapters illustrate how technology can be leveraged to create more adaptive, interactive, and learner-centered educational environments.

CHAPTER OVERVIEWS

Chapter 1 introduces an empathic anthropological framework for university distance education, emphasizing the role of empathy as both a pedagogical and ethical foundation for digital learning. It highlights how emotional connection, cultural awareness, and relational engagement can foster transformative learning experiences.

Chapter 2 examines ICT in education through an anthropological lens, exploring how cultural, social, and contextual factors influence the adoption and use of educational technologies. It underscores the importance of aligning technological innovation with local values and practices.

Chapter 3 investigates the impact of digital transformation on teaching and learning practices, with a focus on technologies such as artificial intelligence, big data, and cloud computing. It highlights the need for educator preparedness, digital literacy, and strategic implementation.

Chapter 4 critically examines the future of education in a post-democratic era, where digital technologies intersect with issues of governance, surveillance, and power. It calls for safeguarding human agency and fostering critical resistance within ICT-driven educational systems.

Chapter 5 addresses the challenges of access and equity in a post-globalization context, analyzing how ICTs both expand and restrict educational opportunities. It advocates for a redefinition of equity that prioritizes inclusivity and social justice.

Chapter 6 explores the dual impact of artificial intelligence on student learning, particularly in relation to creativity, application, and decision-making skills. It highlights both the benefits and risks of AI integration, emphasizing responsible and ethical use.

Chapter 7 focuses on the role of cloud computing in enhancing efficiency, collaboration, and accessibility in educational institutions. It also addresses the challenges associated with adoption and provides recommendations for effective implementation.

Chapter 8 examines the integration of computational thinking and learner-paced segmentation in digital video courseware. It demonstrates how these approaches can improve engagement, retention, and learning outcomes.

Chapter 9 reviews the use of ICT tools to support students with ADD/ADHD in English as a Foreign Language classrooms. It highlights the potential of technology to enhance attention, participation, and overall learning experiences for diverse learners.

Chapter 10 explores the transformative role of ICTs in English Language Teaching, emphasizing their capacity to improve accessibility, engagement, and global connectivity while addressing challenges related to infrastructure and teacher training.

Chapter 11 investigates the design of an interactive website aimed at developing mathematical problem-solving skills and reducing anxiety among secondary students. The findings demonstrate the potential of ICT-based interventions to improve both cognitive and emotional learning outcomes.

CONCLUSION

Together, these chapters offer a comprehensive and critical examination of ICTs as catalysts for educational development and reform. By bridging theory, research, and practice, this volume aims to inform educators, researchers, policymakers, and practitioners seeking to navigate the complexities of digital transformation and to harness its potential for creating more equitable, human-centered, and future-ready education systems.

Faiz Abdelhafid
Al-Istiqlal University, Palestine

Sameh Alqubaj
Al-Istiqlal University, Palestine

Chapter 1
Humanising Technology: An Anthropological Perspective With Empathy in University Distance Education

Maria Fragkaki
Open University of Cyprus, Cyprus

Maria Krasa
Open University of Cyprus, Cyprus

ABSTRACT

In an era of rapid digital transformation and higher education massification, University Distance Education must transcend functionality to embrace ethical responsibility, emotional resonance, and pedagogical humanity. This chapter introduces the Empathic Anthropological Framework for University Distance Education (EAF@UDE), a model that frames empathy as both a pedagogical ethic and epistemological imperative. Informed by critical pedagogy, affect theory, digital humanism, and communicative action, the framework explores the political, emotional, and cultural dimensions of digital learning. A qualitative case study at the Open University of Cyprus employed educational ethnography and constructivist grounded theory to examine how empathy is co-constructed via multimodal dialogue, relational presence, emotional scaffolding, and narrative vulnerability. Findings reveal empathy as a driver of transformative learning, civic consciousness, and social cohesion. The chapter offers recommendations for culturally responsive design, human-centered pedagogy, and ethical digital practice.

DOI: 10.4018/979-8-3373-1190-6.ch001

1. INTRODUCTION

1.1. Chapter Overview

Higher education has entered a transformative epoch defined by the accelerating forces of digitization, algorithmic personalization, and pedagogical automation. As digital infrastructures become central to university systems, educational models are increasingly governed by logic borrowed from industry—efficiency, scalability, and output optimization. This evolution, though promising in its ability to expand access and diversify delivery (Means et al., 2013; Owusu-Boampong & Holmberg, 2015), simultaneously poses complex risks. Chief among these is the erosion of emotional presence and ethical intentionality in pedagogy. The rise of platform-mediated learning risks converting learners into data points, educators into content managers, and pedagogy into transaction (Williamson, 2017; Zembylas, 2018). In such a context, where technology amplifies both possibility and peril, it becomes urgent to ask whether the very ethos of higher education—its relational, dialogic, and emancipatory character—can be preserved, or perhaps reinvented.

This chapter introduces the Empathic Anthropological Framework for Distance Education (EAF@UDE), a model designed to counteract the dehumanizing tendencies of techno-instrumentalist education. Rather than treating empathy as a soft interpersonal skill, EAF@UDE reframes it as a pedagogical ethic, a designable capacity, and an epistemological imperative. It challenges dominant paradigms that marginalize emotion in favor of data and control, instead centering human relationships, emotional resonance, and ethical reflexivity as foundational to digital learning. The EAF@UDE does not position empathy as an innate quality, but as a culturally situated practice that can be scaffolded through intentional pedagogical and technological strategies. It builds on a wide interdisciplinary foundation—Freirean pedagogy, Habermasian communicative rationality, digital humanism, and affect theory—while grounding its application in a real case study of postgraduate distance education.

The primary research question that underpins this work is clear: To what extent, and through what pedagogical and technological mechanisms, can empathy be intentionally cultivated within university-level distance learning environments as a transformative force for inclusion, critical awareness, and humanization? This inquiry arises not merely from theoretical curiosity but from a pressing ethical imperative. As institutions rush to scale online education—accelerated further by the global shift in teaching during the COVID-19 pandemic (Hodges et al., 2020)—there is a risk that values such as emotional presence, dialogic learning, and civic responsibility may be sacrificed on the altar of functionality. The urgency of this question is heightened by the reality that the future of education will likely remain hybrid, remote, and technologically mediated. If this is true, then education must be reimagined not just in form, but in moral function.

1.2. Situating Empathy in the Age of Digital Massification: Reclaiming Humanity in Distance Higher Education

Empathy is increasingly recognized across disciplines as a driver of engagement and transformation. In health education, it enhances patient communication and learning retention (Cartee, 2021). In intercultural and citizenship education, it serves as a bridge across cultural divides (Andreotti, 2014). In narrative therapy and trauma-informed pedagogy, empathy functions as a therapeutic and restorative force (Bubandt & Willerslev, 2015). Yet, despite its growing relevance, empathy remains under-theorized in the domain of online higher education, where its absence is often presumed due to the lack of physical co-presence, asynchronous communication, and content modularity (Rodriguez, 2022). Some have argued that empathy is incompatible with scalable platforms, arguing instead for objectivity and standardization. However, recent research decisively rebuts this assumption. Borup et al. (2012) found that affectively attuned feedback in online settings significantly enhanced student engagement and retention. Murphy et al. (2022) demonstrated that emotionally resonant design elements—such as reflective tasks and instructor presence—lead to improved persistence and learner satisfaction. Similarly, Tsourli, Dalakoura, and Fragkaki (2022) showed that culturally responsive empathy initiatives fostered stronger intercultural awareness and civic discourse in digital classrooms.

These findings collectively suggest that empathy is not only possible in digital education—it is pedagogically essential. However, this possibility depends on intentional design. Empathy must be scaffolded, not assumed. It must be recognized as a complex, situated, and often contested construct that requires both pedagogical vision and structural support. Simply encouraging instructors to "be empathetic" is insufficient; rather, institutional strategies must embed empathy into the architecture of platforms, the pacing of curriculum, the modes of feedback, and the metrics of assessment.

Crucially, empathy must also be situated within the politics of knowledge and identity. Drawing on Zembylas (2018), this framework does not treat empathy as a fixed emotional trait but as a socially constructed practice. It is shaped by discourse, power relations, and cultural norms. Empathy, in this sense, is also epistemic: it determines who is heard, whose pain matters, and whose stories are allowed into the curriculum. Thus, empathy cannot be universalized without critique. It must always be examined in terms of whose emotions are centered, whose are silenced, and what cultural expectations govern emotional expression. As Andreotti (2014) cautions, empathy can function as a "soft empire"—a way of positioning oneself as benevolent while failing to interrogate one's own power. The EAF@UDE therefore advocates for critical empathy, which includes recognition of asymmetries, interrogation of privilege, and commitment to dialogic co-construction.

This political view of empathy aligns with critical theory's vision of education as a site of ideological reproduction and resistance (Freire, 1970; Habermas, 1987). It also converges with digital humanism, which insists that technology be designed in service of dignity, relationship, and civic agency (Young, 2018). Indeed, to embed empathy in distance education is not merely a pedagogical tactic—it is a philosophical stance that resists the commodification of learning and affirms the relational nature of knowledge. Within this view, platforms are not neutral; they are moral environments. Feedback is not simply information; it is recognition. Assessment is not evaluation alone; it is affirmation of identity and care. Thus, the EAF@UDE argues that distance education must not be designed solely for scale or compliance, but for emotional intimacy, ethical presence, and epistemic justice.

This argument becomes particularly urgent when one considers the phenomenon of massification in higher education. While massification expands access, it also tends to reduce learners to anonymous units within massive platforms. The very scalability that allows thousands to enroll often undermines the conditions necessary for meaningful learning: relational trust, instructor presence, and dialogic feedback. As Brookfield (2017) warns, massified education risks extinguishing learner motivation, flattening student identity, and reducing pedagogy to content delivery. Biesta (2021) adds that such systems tend to produce "learnification," where the process of education is emptied of its political and ethical significance and recast as a technical transaction.

Against this backdrop, empathy becomes a necessary resistance. It is the ethical pulse that reanimates pedagogy, insists on seeing the whole learner, and interrupts the mechanistic rhythms of depersonalized learning. But this is not to romanticize emotion. Empathy, within the EAF@UDE, is not sentimentality; it is rigorous emotional engagement, embedded in social context, anchored in equity, and oriented toward critical reflection. It is a demand, not a decoration.

Yet, critics might argue that emotional design lacks rigor or objectivity. Some suggest that affective pedagogy opens the door to therapeutic excess or ideological manipulation. Others claim that digital platforms are inherently unfit for emotional nuance, citing the absence of body language and tonal feedback. These critiques deserve attention. However, emerging research and practice show that empathy can be structured without sentimentality. Garrison, Anderson, and Archer's (1999) Community of Inquiry model emphasizes "social presence" as a core element of cognitive engagement. Isaias et al. (2017) propose "empathic technologies" as a way to scaffold presence through design features such as mood-check tools, video diaries, and reflective prompts. Bickmore et al. (2022) argue that platforms themselves can be co-designed to support emotional regulation and relational scaffolding.

In this light, the absence of physical co-presence is not a categorical barrier to empathy—it is a design challenge. The use of voice threads, personalized video feedback, and storyboarding tools (e.g., Padlet) can create spaces of emotional resonance, even in asynchronous formats. When learners are invited to reflect on their emotions, share their narratives, and engage in reciprocal feedback, they report higher levels of belonging and motivation. Such findings underscore the need for what this chapter calls affective architecture—the intentional design of emotional pathways and rituals within digital learning.

An essential layer of this architecture is cultural reflexivity. Empathy that is universalized risks becoming hegemonic. Not all cultures express or interpret emotion in the same way. In many cases, expectations for self-disclosure or expressive affirmation may align with Western norms but marginalize other affective traditions. Therefore, the EAF@UDE emphasizes situated empathy, grounded in the recognition of cultural codes, epistemic diversity, and intersectional identities. Emotional safety, in this model, is not simply about avoiding harm—it is about ensuring interpretive dignity. Dialogic empathy, in turn, means that understanding is not presumed or projected; it is co-produced through respectful inquiry and narrative negotiation.

The final argument advanced here is that empathy does more than support learning—it transforms it into a civic and ethical act. Empathy enables learners to move beyond their own frames of reference, to engage with the lived experiences of others, and to confront structures of injustice. This is not a hypothetical proposition. Participants in the EPA524 postgraduate module described how empathic engagements—through story sharing, reflective journaling, and peer response—shifted their understanding of migration, marginalization, and digital colonialism. Such moments align with Mezirow's (1991) theory of transformative learning, which hinges on the disruption of meaning perspectives through critical reflection. They also affirm hooks' (1994) vision of education as a practice of freedom—a space where love, critique, and justice intersect.

In conclusion, this introduction lays the conceptual and ethical foundation for a framework that seeks to reclaim the relational and civic soul of distance education. Empathy, when reconceptualized as an epistemic structure, a design goal, and a pedagogical ethic, has the potential to humanize even the most technologically mediated learning environments. It offers not a retreat into sentiment, but a call to rigorously reimagine education as a relational, ethical, and transformative encounter—even, and especially, at a distance.

1.3. Synopsis

This opening chapter establishes the ethical, technological, and pedagogical stakes of the study by interrogating the dual nature of distance education in the digital age. While online platforms offer unprecedented scalability, access, and personalization (Means et al., 2013; Owusu-Boampong & Holmberg, 2015), they also risk emotional detachment, instrumentalization of learning, and deepening socio-technical inequalities (Selwyn, 2016; Zembylas, 2018). Against this backdrop, the chapter introduces the Empathic Anthropological Framework for Distance Education (EAF@UDE)—a critical, human-centered model that positions empathy as a pedagogical and epistemic necessity rather than a sentimental surplus. Through this lens, distance education becomes not merely an administrative system, but a relational, ethical, and cultural practice. This sets the stage for a rethinking of online education as a site of possibility, where emotional engagement and critical consciousness can be intentionally cultivated.

This theoretical and ethical problematisation naturally leads into Unit 2, which sharpens the focus by presenting the research motivations and questions that drive the inquiry into empathy as a transformative force in digital education.

2. RESEARCH MOTIVATION AND GUIDING QUESTION

2.1. Chapter Overview

Unit 2 establishes the ethical and pedagogical foundations of this study by introducing two interlinked research questions that guide the development of the Empathic Anthropological Framework for Distance Education (EAF@UDE). These questions engage directly with the paradox that lies at the heart of contemporary digital higher education: the tension between the demands of technological scalability and the enduring human need for emotional presence, relational authenticity, and civic engagement. As such, this chapter initiates a necessary rethinking of distance education—from a largely technocratic enterprise oriented toward efficiency and control, to an emotionally and ethically attuned form of digital pedagogy grounded in empathy not just as sentiment but as epistemology, ethics, and praxis. In doing so, the chapter offers the foundational rationale for the inquiry and naturally prepares the intellectual terrain for Unit 3, where the conceptual architecture of the EAF@UDE is developed in detail.

2.2. Situating Empathy in the Age of Digital Massification: The Moral Imperative of Humanizing Distance Higher Education

Unit 2 situates the research within the ethical, pedagogical, and political tensions emerging from the massification and digitization of higher education, especially in the wake of the COVID-19 pandemic. The chapter critiques the technocratic logic of platform-based education and calls for a humanizing reorientation of digital pedagogy. Specifically, it advocates for empathy to be reframed not as a soft or incidental attribute but as a foundational pedagogical infrastructure in digital learning environments.

This unit provides the ethical rationale for the study and introduces the guiding questions that inform the development of the Empathic Anthropological Framework for Distance Education (EAF@UDE) It highlights the growing gap between the promises of scalable digital education and its emotional, civic, and ethical shortcomings, proposing empathy as the connective and transformative force needed to realign these systems with democratic and relational values.

The central question of the chapter is how an empathy be reconceptualized and structurally embedded into distance education to restore relationality, emotional depth, and civic purpose in an era of digital massification?

This central question guides the inquiry and frames empathy as a necessary response to the emotional disconnection, cognitive detachment, and civic dislocation often found in large-scale online learning environments. It positions empathy not as sentimentality, but as an epistemological and pedagogical imperative.

The global shift to remote learning during the COVID-19 pandemic did more than disrupt traditional educational systems; it catalyzed a fundamental transformation in how teaching and learning are imagined and enacted at a distance. Faced with institutional shutdowns, universities rapidly deployed digital infrastructures—Learning Management Systems (LMS), video conferencing platforms, and AI-powered assessment tools—to ensure instructional continuity (Hodges et al., 2020). While these responses were functionally necessary, they also revealed deep limitations within the pedagogical models underpinning digital education. Chief among these was the erosion of emotional presence, ethical dialogue, and relational engagement in online classrooms.

This acceleration exposed not only the strengths of digital tools but also their profound affective shortcomings. As Zembylas (2018) observes, the dominance of technocratic rationality in online education often leads to an "affective flattening" of pedagogy, where emotion, identity, and care are relegated to the margins. Selwyn (2016) similarly warns of the growing risk that platform-based education may become a space of surveillance and instrumentalism—where learners are reduced to data points, and pedagogy is reduced to transaction rather than transformation. It is within this ethically fraught landscape that the current study emerges, animated by an urgent imperative: to recover what was lost in the rush to digitize. Specifically, the study argues for rehumanizing education by positioning empathy not as a sentimental excess, but as a necessary pedagogical infrastructure.

At the heart of this inquiry are two fundamental **research questions**:

a) To what extent can empathy be cultivated in university-level distance learning?
b) Through what pedagogical and technological methods can empathy function as a driver of inclusion, critical awareness, and humanization?

These questions are not theoretical abstractions; they arise from the lived contradictions of the digital university. On the one hand, higher education institutions increasingly promise personalized, flexible, and accessible learning experiences through digital platforms. On the other, these same systems often reproduce emotional detachment, cultural erasure, and learner isolation when relational dimensions are neglected or systemically excluded.

Thus, these research questions operate on two levels. They serve as diagnostic instruments that reveal the systemic limitations of current digital education models, while also functioning as visionary tools that point toward an alternative paradigm—a future of online learning that centers empathy, justice, and relational depth. The first question explores the pedagogical feasibility of embedding empathy as a relational and cognitive practice within asynchronous and digitally mediated educational settings. It investigates the pedagogical strategies, emotional conditions, and instructional designs required to support sustained empathic engagement. The second question shifts the focus toward the structural dimension, asking how empathy can be intentionally built into the architectures of digital education—across platform design, assessment, feedback, tutor presence, and peer interaction. It also calls for attention to diversity, cultural reflexivity, and justice-oriented pedagogy.

2.3 Why Empathy? Reframing a Marginalized Concept

Historically, empathy has been marginalized within the discourse of higher education, often viewed as affective surplus rather than as integral to serious intellectual inquiry. It has been treated as relevant to therapeutic, early childhood, or interpersonal domains, but incompatible with the cognitive rigor and abstraction typically valorized in university settings. However, this view is increasingly being challenged by interdisciplinary scholarship. Calloway-Thomas (2018) redefines empathy as a complex act of intercultural and epistemic engagement, while Cartee (2021) identifies it as foundational to reflective professional development and critical thought.

Furthermore, digital learning environments—previously assumed to inhibit emotion—are now shown to support empathic engagement when designed intentionally. Borup et al. (2012) demonstrated that affectively attuned feedback enhances learner persistence and engagement. Murphy et al. (2022) found that emotionally resonant instructional design, when grounded in reflection and dialogue, fosters greater resilience and critical depth. These findings collectively refute the assumption that empathy and online education are incompatible. Rather, they suggest that empathy is essential to the ethical transformation of digital pedagogy. This study therefore frames empathy not as "niceness" or emotional projection, but as a culturally embedded, designable, and teachable strategy that supports both inclusion and critical engagement.

2.4 The Massification of Education and Its Discontents

As distance education continues to expand globally, propelled by demands for access, efficiency, and scale, it also runs the risk of exacerbating the very exclusions it seeks to remedy. Williamson (2017) articulates this dilemma in his critique of the "platform university"—a system in which students are managed algorithmically, educators are deskilled through automated monitoring, and learning is reduced to performance metrics. This massification process gives rise to a triad of discontents. Emotional disconnection arises as learners increasingly report feelings of isolation, invisibility, and lack of support. Cognitive disengagement is observed as students find it difficult to connect with depersonalized, content-heavy curricula that lack relational anchors. Civic dislocation occurs when education is stripped of its social purpose, becoming detached from community, justice, and ethical responsibility. Against this backdrop, this research positions empathy as a bridge—a connective tissue that links emotional presence with cognitive meaning and civic relevance. Empathy, when intentionally cultivated, becomes a transformative force capable of restoring education as a relational and civic practice.

2.5 Reorienting Distance Learning as a Moral Technology

One of the core assertions of this study is that distance education is not a neutral delivery system; it is a moral technology. Every design decision—whether it concerns platform architecture, assessment modality, or instructional tone—embeds ethical assumptions and reproduces normative values. Just as digital systems can facilitate depersonalization and control, they can also be repurposed to support empathy, ethical presence, and justice-oriented pedagogy. The EAF@UDE represents a deliberate attempt to rewire digital pedagogy toward affective and relational co-presence. In this vision, empathy is not a peripheral concern but a structural foundation. It becomes the very condition under which online education can fulfill its democratic promise. When integrated into feedback, platform design, and instructional ethos, empathy transforms online classrooms into spaces of critical awareness, emotional literacy, and civic engagement.

2.6 Implications of the Research Questions

The formulation of the two research questions opens a set of implications that inform the structure of the entire study. The first question addresses the pedagogical feasibility of cultivating empathy. It asks what emotional, cognitive, and relational conditions must be in place for empathy to emerge and be sustained in asynchronous or semi-synchronous learning environments. This inquiry necessitates an examination of tutor presence, emotional design, feedback practices, and student agency. The second question explores the design challenge of embedding empathy structurally into the architecture of distance learning. It considers how empathy can be supported not through one-off gestures but through sustained pedagogical and technological scaffolds that account for cultural difference, emotional safety, and critical reflexivity. Together, these questions require a reconceptualization of assessment norms, instructional roles, and platform affordances. They also demand a deeper exploration of how power, emotion, and positionality operate within the digital university.

2.7 Linking to the Conceptual Framework: From Inquiry to Design

The two research questions articulated in this chapter are not ends in themselves; they serve as a bridge to the next stage of inquiry. Unit 3 directly addresses these questions by constructing the theoretical scaffolding of the EAF@UDE. If Unit 2 asks whether and how empathy can be embedded into digital education, Unit 3 provides the conceptual tools to answer that question. It draws from critical anthropology, affect theory, digital humanism, and dialogic pedagogy to build a model in which empathy functions not as a personality trait, but as a systemic and epistemological feature of educational design.

The transition from Unit 2 to Unit 3 is thus both logical and necessary. To enact the transformation envisioned in this chapter, a robust framework is required—one that conceptualizes empathy as an infrastructural property of pedagogy and technology. Unit 3 begins the task of building that framework, enabling a reimagining of digital education through the lens of emotional, relational, and civic intentionality.

2.8 Synopsis

Unit 2 introduces and justifies the two central research questions that anchor this study, framing them as urgent responses to the pedagogical, ethical, and relational crises exposed by digital massification in higher education. Rather than treating empathy as incidental, the chapter establishes it as the pedagogical infrastructure of a more inclusive, critical, and humanizing digital university. Drawing on emerging scholarship, critical theory, and empirical insights, this unit makes the case for empathy as a designable and teachable element of online learning environments. In doing so, it lays the intellectual groundwork for Unit 3, which constructs the theoretical foundations of the Empathic Anthropological Framework for Distance Education (EAF@UDE) and operationalizes empathy as a central pillar of pedagogical transformation.

3. CONCEPTUAL FRAMEWORK

3.1. Chapter Overview

Unit 3 lays the theoretical scaffolding for the Empathic Anthropological Framework for Distance Education (EAF@UDE), positioning it as a paradigm-shifting response to the emotional, ethical, and relational deficits in contemporary online education. Drawing from critical anthropology, affect theory, transformative pedagogy, and digital humanism, the chapter establishes empathy as the structural core of educational design—not a supplementary virtue but a foundational epistemic and ethical principle.

The chapter begins by situating EAF@UDE against the backdrop of technocratic and depersonalized trends in distance learning—what Zembylas (2018) calls the "affective flattening" of pedagogy. It argues that digital education systems, while efficient, often erase the emotional nuance and ethical deliberation necessary for deep learning. In response, EAF@UDE proposes a framework that treats empathy as a teachable, designable, and culturally embedded mechanism for rehumanizing digital learning.

The framework redefines empathy not as a soft skill but as a relational epistemology, rooted in narrative, reflexivity, and intercultural presence. This conceptual shift transforms every layer of instructional design—from platform infrastructure and peer interaction to assessment and tutor presence. Through interdisciplinary synthesis and empirical grounding, the chapter proposes that the digital space is not inherently dehumanizing but must be reclaimed as an affective and ethical ecology through intentional design.

Importantly, the chapter critiques universalist models of education and integrates anthropological reflexivity to account for the sociocultural specificity of emotional expression and identity formation. Learning, within EAF@UDE, is reconceived as relational identity construction—not mere cognitive transaction. The educator becomes a cultural mediator, and the platform evolves from content repository to emotional infrastructure.

The chapter closes by presenting dialogic education and emotional design as foundational practices, arguing that when empathy is embedded in platform features, dialogic rituals, and relational pedagogy, the digital classroom becomes a site not only of knowledge acquisition but of civic growth, moral imagination, and collective becoming.

3.2. The "Empathic Anthropological Framework of University Distance Education" (EAF@UDE): Reclaiming the Digital as Relational Space

In the contemporary landscape of global higher education, distance learning has transitioned from being an auxiliary mechanism for expanding access to a central modality for disseminating knowledge. Yet, as this mode of education becomes normalized, its foundational assumptions remain rooted in technocratic paradigms—prioritizing control, standardization, and predictive analytics over ethics, human connection, and dialogic engagement. Over the past two decades, the exponential growth of online education has been justified by policy narratives emphasizing efficiency, scalability, and data-driven personalization (Selwyn, 2016; Williamson, 2017). While these justifications carry institutional appeal, they often obscure the deeper cultural and pedagogical ramifications of digital massification. As Zembylas (2018) cautions, technocentric models tend to result in the "affective flattening" of education, whereby emotional presence, narrative depth, and ethical commitment are subordinated to mechanistic metrics and managerial convenience.

In response to this flattening, the Empathic Anthropological Framework for Distance Education -**EAF@UDE** *(figure 1)* emerges as a critical intervention. Rather than treating platform-driven learning environments as morally neutral conduits for knowledge delivery, EAF@UDE reconceptualizes distance education as a relational and ethical ecology. The framework is informed by critical anthropology, human-centered design, affect theory, and emancipatory pedagogy. It contends that digital learning spaces must be cultivated not solely for the transmission of content but for emotional resonance, cultural reflexivity, and civic development. **(EAF@UDE):** does not reject technology. Instead, it demands that technological tools be reimagined as instruments for empathy—facilitators of co-presence, mutual recognition, and transformative dialogue across digital space.

Figure 1. The Empathic Anthropological Framework for Distance Education (EAF@ UDE)

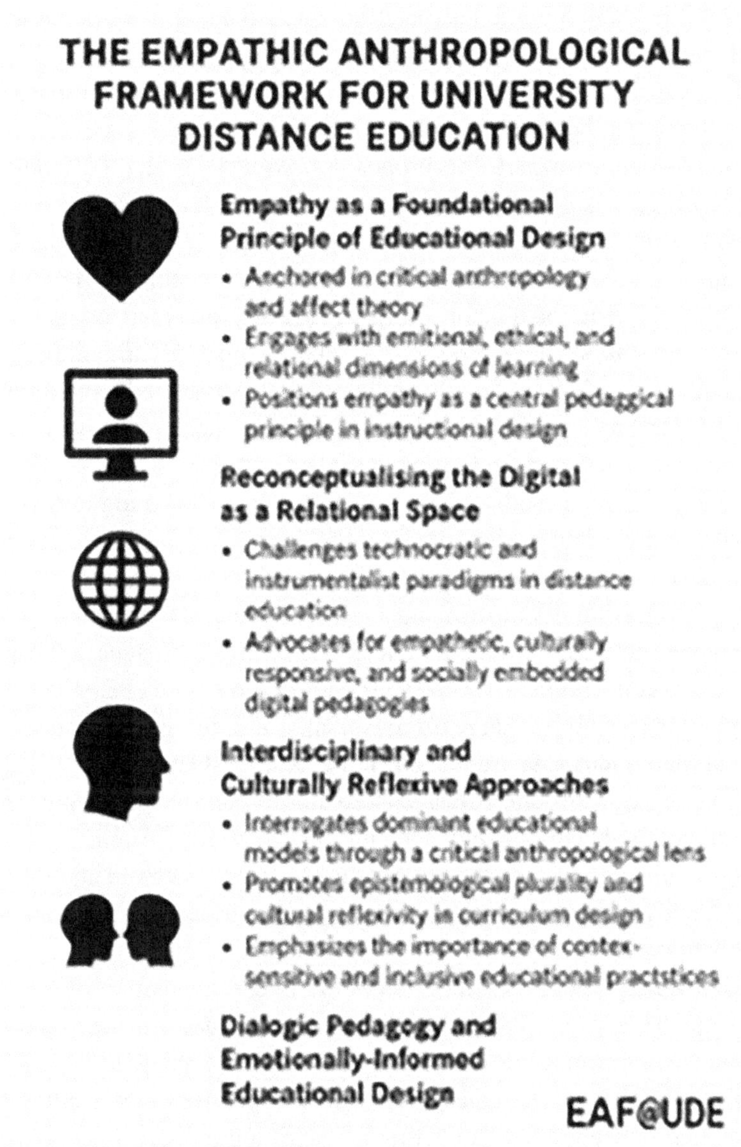

A foundational move in this framework is its redefinition of empathy. Conventional educational discourse often reduces empathy to a soft skill, a passive emotional state, or an interpersonal virtue. This framing, however, elides the epistemological and structural dimensions that empathy can embody. **EAF@UDE):** instead positions empathy as both a pedagogical design principle and an epistemic architecture. Drawing from interdisciplinary research in neuroscience, moral philosophy, and dialogic pedagogy (Calloway-Thomas, 2018; Zembylas, 2018), the framework argues that empathy is not just an emotional disposition but also a cognitive, communicative, and moral capacity. This reframing has material implications: if empathy can be cultivated and structured, then educators and institutions bear a responsibility to integrate it intentionally within digital learning ecosystems.

Empirical evidence supports this proposition. Murphy et al. (2022) demonstrate that emotionally resonant feedback in online learning fosters engagement and introspection, indicating that feedback is both a cognitive and affective encounter. Baran and Correia (2014) show that when tutors practice empathic communication—affirming students' emotional labor, validating identity, and responding to narrative content—such gestures initiate a reciprocal relational pattern among peers. Cartee (2021) extends this insight by showing how rituals of structured vulnerability, such as guided storytelling or reflective sharing, produce more inclusive and trusting learning communities. These findings directly challenge prevailing skepticism about the limitations of screen-mediated emotionality. Critics who argue that empathy cannot thrive in digital education overlook the degree to which careful design can unlock affective depth and intersubjective connection.

However, the design of empathy must transcend one-off interventions or sentimentally framed gestures. It must be embedded within a broader anthropological understanding of education as a cultural and relational process. The EAF@UDE draws on anthropological critiques that resist universalist, cognition-dominated models of learning. Scholars such as Andreotti (2014) and Ndlovu-Gatsheni (2018) argue that prevailing models of instruction are anchored in Eurocentric epistemologies, which often erase emotional heterogeneity and cultural specificity. In contrast, the EAF@UDE foregrounds anthropological reflexivity. It urges educators and designers to interrogate their assumptions, acknowledge their positionalities, and co-construct curricula that respect multiple ways of knowing and feeling.

From this vantage point, learning is not the linear acquisition of disembodied knowledge but the relational formation of identity. Educators are not passive content deliverers but active cultural mediators, and platforms are not neutral tools but emotionally and symbolically charged spaces. This perspective aligns with Indigenous pedagogies that conceptualize education as "learning as becoming," Afrocentric approaches that center communal knowledge production, and Buddhist principles of interbeing that affirm the mutual co-constitution of learning relationships (Nhat Hanh, 1998). These intellectual traditions make clear that empathy is not incidental but central to meaningful education—it is a philosophical and pedagogical practice that affirms the ontological reality of connection.

Given this orientation, the EAF@UDE calls for a radical reconfiguration of instructional design—from content-first models to relationally anchored pedagogical ecologies. Conventional digital education tends to prioritize modularity, text-heavy resources, and outcome-driven assessment strategies. Although these are administratively efficient, they often fail to support affective depth and epistemic diversity. By contrast, the EAF@UDE: advocates for narrative-driven learning arcs, emotionally intelligent feedback loops, and instructional designs that embed affective presence as an educational priority. Work by Norman (2013) in emotional design, as well as Anderson and Dron's (2011) theory of presence in online learning, reinforces the idea that when platforms are attuned to human relational needs, learners report enhanced engagement, autonomy, and intellectual connection.

Concerns that emotionally rich pedagogy undermines academic rigor are not uncommon. However, this objection constructs a false binary between emotion and cognition. The EAF@UDE demonstrates that when learners are invited to reflect critically on emotional experience—through narrative case studies, ethical dilemmas, or reflective writing—they are not distracted from analytical thought but deepened in it. This model resists emotivism and instead affirms that cognitive growth and emotional complexity are mutually reinforcing dimensions of learning. This perspective is consistent with Mezirow's (1991) theory of transformative learning and bell hooks' (1994) vision of teaching as the practice of freedom.

Dialogic pedagogy is at the heart of this framework. The EAF@UDE adopts Paulo Freire's (1970) notion of dialogic praxis, wherein learners and educators collaboratively interrogate the world in pursuit of transformation. Within digital contexts, this dialogic ethic must be intentionally scaffolded. In the absence of shared physical space, empathy does not emerge spontaneously—it must be cultivated through careful ritual and structure. EAF@UDE recommends practices such as moral dilemma prompts, collaborative annotation tools, peer storytelling circles, and reflective discussion forums where learners are encouraged to respond affectively as well as cognitively. Phrases such as "this reminded me of..." or "I felt..." are not superficial; they signal the development of dialogic presence and shared vulnerability.

Nevertheless, scaling dialogic empathy within massive digital systems presents challenges. Most Learning Management Systems (LMS) are optimized for task completion and standardization, not emotional nuance. The EAF@UDE critiques this orientation and proposes the intentional development of emotional infrastructures. Emotional infrastructure refers to platform features that are designed to sustain empathic presence—tools such as mood tracking, presence indicators, reflective journaling spaces, and emotion-aware prompts. The integration of emotionally intelligent bots, which detect affective cues and encourage reflective practice, can support learners without replacing the need for human empathy (Bickmore et al., 2022). These proposals align with emerging work in affective computing and digital humanism, which challenge the assumption that platforms must be affectively sterile.

Importantly, this orientation toward emotional infrastructure does not equate to emotional surveillance or forced vulnerability. Zembylas (2013) offers a critical warning about the potential for emotion to be co-opted into managerial regimes, where feeling becomes another metric for institutional control. The EAF@UDE takes this critique seriously. It insists that empathy must always be invitational, culturally responsive, and contextually grounded. Emotional engagement should never be required or standardized; rather, learners must have agency in how, when, and whether they express affect in ways that resonate with their lived identities and cultural contexts.

In summary, the Empathic Anthropological Framework for Distance Education demands a reorientation of digital pedagogy toward relational depth and ethical intentionality. It calls for a shift from content to connection, from efficiency to care, and from instrumental control to dialogic co-presence. It encourages educators and institutions to reimagine the role of technology—not as a neutral delivery tool, but as a relational infrastructure capable of enabling justice-oriented, emotionally intelligent learning. The chapters that follow expand upon this vision by detailing how empathy is translated into pedagogical strategies, implemented methodologically, and evaluated empirically. In doing so, the EAF@UDE affirms that empathy in distance education is not only possible—it is imperative for cultivating the future of higher learning.

The EAF@UDE is, ultimately, a call to action: to design not just for cognition, but for relation. As digital surveillance intensifies and datafication accelerates, educators are urged to reclaim pedagogy as a deeply human endeavor. Empathy, when intentionally cultivated, offers the architecture necessary to support a more ethical, inclusive, and transformative digital university.

3.3. Synopsis

Unit 3 presents the Empathic Anthropological Framework for University Distance Education (EAF@UDE) as a transformative response to the technocratic limitations of conventional online learning. Rooted in critical anthropology, affective theory, and relational pedagogy, the framework positions empathy as a structural and epistemic imperative in the design of just and humanized distance education. It reclaims the digital not as a space of alienation but as a relational field where presence, emotion, and co-becoming are possible—when designed with care, cultural reflexivity, and ethical intent.

The chapter dismantles the prevailing models of standardization, modularity, and cognitive isolation by proposing a design architecture centered on dialogic learning, emotional infrastructure, and narrative-based interaction. It challenges the assumption that digital environments are incapable of supporting empathy, showing instead how platforms, when ethically constructed, can amplify emotional presence and transformative dialogue.

This conceptual framework links directly to Unit 4, which delves into the philosophical and ethical underpinnings of empathy in education. If Unit 3 outlines what the EAF@UDE *is* and how it operates conceptually, Unit 4 explores *why* it matters philosophically—by grounding the framework in epistemic justice, emancipatory theory, and ethical pedagogy. Together, they form the intellectual backbone of the study's pedagogical and methodological applications.

4. PHILOSOPHICAL AND THEORETICAL FRAMEWORK

4.1. Chapter Overview

This chapter lays the theoretical foundation *(figure 2)* for the Empathic Anthropological Framework for Distance Education (EAF@UDE), proposing a redefinition of empathy not as a private sentiment or interpersonal disposition, but as a critical epistemological, ethical, and pedagogical construct. In an era where distance learning is increasingly dominated by instrumental rationalities—such as automation, algorithmic control, and the depersonalized delivery of content—this chapter argues for an urgent reorientation of education as a moral and relational practice. To build this case, the chapter weaves together insights from several philosophical traditions. It begins with critical epistemology, which challenges the traditional assumption of objective, neutral knowledge and instead affirms that knowledge is situated, embodied, and always shaped by historical, cultural, and political power structures. From this premise, the chapter turns to Habermas's theory of communicative action, drawing from his notion of emancipatory rationality to advocate for dialogic digital spaces that promote inclusivity, mutual understanding, and critical reflection. The discussion then moves to Paulo Freire's pedagogy of liberation, emphasizing praxis—the interplay of reflection and action—as a means of awakening learners to their social conditions. Freire's concept of conscientização underscores empathy as a mode of ethical engagement with injustice, requiring learners to confront complexity and contradiction rather than evade them. bell hooks' feminist pedagogy further enriches this perspective, framing teaching as a practice grounded in love, care, vulnerability, and emotional honesty. Empathy, in this context, becomes a political commitment to the flourishing of others. Jack Mezirow's transformative learning theory adds a cognitive and developmental dimension by presenting empathy as a trigger for disorienting dilemmas—those moments of emotional and intellectual discomfort that precede deep personal and epistemic change. Within a digitally mediated context, these moments can arise through narrative exchanges, collaborative reflection, or emotionally resonant feedback. Finally, the chapter draws from digital humanism to argue that technology must be designed not merely for content delivery but for ethical presence and emotional resonance. Online learning environments should become emotional infrastructures—spaces that facilitate care, reflection, recognition, and solidarity. Rather than treating empathy as a disposable soft skill or branding tactic, the EAF@UDE insists that it be structurally embedded in digital education as an ethical necessity. Together, these philosophical strands construct a cohesive vision of empathy as foundational to transformative and just digital learning. When woven into the theoretical, pedagogical, and technological layers of educational design, empathy enables a reimagining of distance learning not just as a mode of instruction, but as a space of ethical becoming, civic responsibility, and shared humanity.

Figure 2. Philosophical and Theoretical Framework

PHILOSOPHICAL AND THEORETICAL FRAMEWORK (EAF@UDE)

- **Empathy as Critical Construct**
  Defined as epistemological, ethical-pedagogical—not a trivial sentiment
- **Critique of Technocratic Rationality**
  Knowledge is situated, embodied, and shaped by power structures (Kincheloe, Fricker)
- **Critical Epistemology**
  Knowledge is dialogic, and critically reflexive in online learning environments
- **Freire - Pedagogy of Liberation**
  Promotes praxis (as interplay of reflection and action): ethical political engagement, conscientização
- **bell hooks - Feminist Pedagogy**
  Frames teaching as love, care, and vulnerability: emotional justice
- **Mezirow - Transformative Learning**
  Empathy as a catalyst for disorienting dilemmas leading to deep cognitive transformation
- **Digital Humanism**
  Insists that learning fosters ethical presence and emotional resonance—beyond just efficiency
- **Emotional Infrastructure**
  Online learning environments should be structurally embedded in pedagogy, assessment, interaction, and platform design
- **Rejection of Neoliberalism in Education**
  Empathy is not a "soft skill" but a national, ethical stance resisting commodification

EAF@UDE

4.2. Philosophical Foundations of the Empathic Anthropological Framework for Distance Education (EAF@UDE)

The Empathic Anthropological Framework for Distance Education (EAF@UDE) emerges as both a critique and a corrective to prevailing technocratic models of online learning that prioritize efficiency over empathy, data over dialogue, and standardization over situated understanding. Its philosophical foundation is grounded in a constellation of emancipatory, critical, feminist, and humanist theories, each of which converges on a central proposition: education is inherently relational, affective, and ethical. In rejecting the myth of epistemic neutrality, EAF@UDE invites a reimagining of digital learning as a moral and cultural endeavor—where empathy is not a surplus emotion but a core epistemological and pedagogical imperative.

Traditional Western epistemology often positions knowledge as objective, abstract, and disembodied. This positivist legacy continues to shape much of educational design, especially in online environments, where content is modularized, assessments are standardized, and interaction is minimized. Against this background, critical epistemology asserts that all knowledge is socially situated, culturally mediated, and politically contingent (Kincheloe, 2005; Fricker, 2007). Knowing, in this view, is never detached from the knower's context—it is always filtered through histories, identities, and power relations. This epistemic shift has profound implications for pedagogy. It suggests that to teach is not simply to transmit facts, but to engage learners in reflective dialogue that acknowledges their positionalities. It also redefines empathy not as emotional identification alone, but as a cognitive act of ethical recognition—an openness to alternative ways of knowing and being. In this way, empathy becomes a form of situated knowing, grounded in humility and responsive to otherness. EAF@UDE builds on this insight by embedding empathy into the very architecture of distance learning—not as an affective add-on, but as a core design principle that shapes feedback, assessment, interaction, and platform design. The educator's role is transformed: no longer a content authority, the teacher becomes an emotional facilitator, a cultural mediator, and an ethical witness.

Central to the philosophical scaffolding of EAF@UDE is Jürgen Habermas's (1987) concept of emancipatory cognitive interest. Habermas differentiates three types of knowledge: technical (focused on prediction and control), practical (focused on mutual understanding), and emancipatory (oriented toward critical self-reflection and freedom). EAF@UDE draws specifically from the emancipatory domain, viewing education not merely as a process of information transfer, but as a means of self and social transformation. Communicative rationality—Habermas's framework for dialogic engagement—offers a normative model of inclusive, reasoned discourse in which participants seek mutual understanding free from domination. In the digital university, this principle translates into learning environments that are dialogic rather than hierarchical, collaborative rather than transactional. However, this ideal is tempered by critiques such as those by Nancy Fraser (1990), who warns that real-world educational contexts are marked by asymmetries of power and voice. EAF@UDE responds to this critique by integrating intersectionality and cultural reflexivity, thereby grounding communicative rationality in lived inequality rather than assuming idealized equality. Empathy, then, becomes the emotional condition for communicative rationality to flourish. It allows learners to hear one another beyond surface disagreement, to attune themselves to difference, and to sustain ethical dialogue even amid discomfort. The digital classroom, when designed with empathy, can become a microcosm of deliberative democracy—a space where mutual recognition fosters critical consciousness.

No framework advocating empathy in education can ignore Paulo Freire's (1970) enduring influence. His critique of the "banking model" of education—where learners are treated as empty vessels to be filled with static knowledge—remains powerfully relevant in online education, where LMS platforms often replicate mechanistic delivery models. Freire's alternative is rooted in dialogic pedagogy, where learners become co-authors of knowledge and subjects in the transformation of their world. Freire's notion of conscientização, or critical consciousness, underscores the inseparability of education and liberation. Learners must not only understand their social conditions but be empowered to transform them. EAF@UDE operationalizes this principle by foregrounding empathy as a political effect. It is not about comfort or harmony; it is about feeling-with others in ways that expose injustice, disrupt complacency, and catalyse action. In digitally mediated spaces, this means structuring courses not around passive consumption but around active co-construction—where learners engage in narrative inquiry, ethical dilemma debates, and peer dialogue that push them to reflect on their own positionality and social responsibilities. Empathy becomes the bridge between affect and action, between feeling and praxis.

Feminist philosopher and educator Bell Hooks extends Freire's legacy by placing love at the center of emancipatory pedagogy. In Teaching to Transgress (1994), she defines love as a radical, communal force—an ethic of care, mutuality, and vulnerability that makes education transformative. In the emotionally arid landscapes of many digital classrooms, hooks' vision is not only inspiring but urgent. EAF@UDE integrates this feminist ethic by positioning empathy as a practice of emotional justice. It challenges the notion that feelings have no place in "serious" academic inquiry, insisting instead that learning is inseparable from emotional presence. This redefinition alters every facet of the educational process. Feedback becomes not a corrective mechanism but an ethical act of acknowledgment. Presence is not about attendance but about being attuned and available. The educator is no longer a neutral transmitter of information, but an emotional architect shaping the affective contours of learning. Crucially, hooks and Zembylas (2013) caution against the commodification of empathy within neoliberal education systems. When empathy is reduced to a performance metric or a "soft skill" for employability, its radical potential is neutralized. EAF@UDE resists this by framing empathy not as an institutional output but as a relational condition—a co-created ethos of trust, recognition, and moral commitment.

While Freire and hooks emphasize empathy as an ethical and political affect, Jack Mezirow (1991) contributes a psychological dimension by conceptualizing transformation as a learning process initiated by disorienting dilemmas. These are moments when learners' habitual ways of thinking are unsettled, forcing them into reflective revaluation. Empathy, in this schema, becomes a catalyst for such disruption. When learners encounter narratives that contradict their assumptions—stories of injustice, marginalization, or resilience—they are pushed into emotional and cognitive tension. This discomfort, when scaffolded through dialogic structures such as reflective journaling, peer dialogue, and narrative feedback, can lead to genuine perspective transformation. However, transformative learning is not guaranteed. Without emotional scaffolding, relational trust, and cultural sensitivity, dissonance can result in retreat or resistance. EAF@UDE addresses this by embedding empathy across the digital learning ecosystem—from tutor presence and course design to platform features and peer engagement. Emotional safety is treated not as a luxury but as a condition for critical reflection and intellectual risk-taking.

The digital turn in education has often been framed as a matter of efficiency and access. However, digital humanism reclaims technology as an ethical medium—one that must serve dignity, solidarity, and human flourishing (Young, 2018). EAF@UDE takes this insight seriously by treating platforms and interfaces not as neutral conduits but as cultural artifacts with ethical weight. This leads to the concept of emotional infrastructure—the intentional design of technological features that enable relational presence. Presence indicators, affective avatars, reflective prompts, and emotionally intelligent bots (Bickmore et al., 2022) become tools not for surveillance, but for solidarity. These features help build what Garrison et al. (1999) call "emotional presence"—a sense of being seen and felt in digital space. Such infrastructure aligns with the Community of Inquiry framework, which identifies cognitive, teaching, and social presence as essential to meaningful online learning. But EAF@UDE extends this by asserting that without emotional resonance, these other forms of presence remain thin. Emotional design is not peripheral—it is central to inclusive and engaged digital learning.

Underlying much of EAF@UDE's critique is a rejection of neoliberalism's influence on education. In treating students as consumers, learning as a product, and education as a transaction, neoliberal paradigms hollow out the civic and ethical purposes of pedagogy (Apple, 2006; Biesta, 2021). Empathy, within such a system, risks being instrumentalized—marketed as a skill, measured as a competency, and managed through affective labor. EAF@UDE resists this by reframing empathy as a terrain of ethical encounter—a relational, situated, and political practice that cannot be reduced to metrics. It asserts that empathy must be designed, taught, and institutionalized as a pedagogical commitment—not to placate learners but to awaken them. Not to improve customer satisfaction, but to cultivate moral imagination and civic agency. By centering empathy, EAF@UDE calls for a redefinition of digital education: from platform delivery to dialogic presence, from cognitive acquisition to ethical engagement, from isolated performance to collective becoming. It insists that we reimagine not only how we teach, but why—and for whom.

4.3. Synopsis

Unit 4 constructs the theoretical and philosophical scaffolding that underpins the Empathic Anthropological Framework for Distance Education (EAF@UDE). It positions empathy as a critical, teachable, and culturally situated mode of knowing, rather than a sentimental or interpersonal virtue. By integrating critical epistemology, Freire's pedagogy of liberation, bell hooks' affective feminism, and digital humanism, the chapter reframes distance education as an ethical and relational practice. This theoretical repositioning directly challenges dominant technocratic logics that frame learning as value-free, disembodied, or strictly cognitive. Instead, it argues that education is always already ethical—and that online learning must be deliberately designed to support emotional presence, intercultural listening, and moral imagination. Educators, under this paradigm, are called not to manage students but to engage with them as co-becoming subjects navigating difference, vulnerability, and possibility. Crucially, this unit serves as the bridge between conceptual imagination and pedagogical implementation. It concludes by asserting that if empathy is epistemic, then pedagogy must be redesigned around it. This argument sets the stage for Unit 5, which operationalizes these philosophical commitments into concrete pedagogical practices. In Unit 5, the ethical imperative of empathy becomes a design principle—guiding how feedback, assessment, peer engagement, and platform tools are shaped to foster relational, inclusive, and transformative distance education.

5. PEDAGOGICAL FRAMEWORK

5.1 Chapter Overview

Unit 5 represents a pivotal transition in the overall study with the pedagogical framework (figure 3): it moves from the theoretical underpinnings of empathy (developed in Unit 4) to their pedagogical realization within the digital classroom. Grounded in the philosophical conviction that empathy is epistemic and ethical, this chapter operationalizes the Empathic Anthropological Framework for Distance Education (EAF@UDE) into tangible, teachable, and designable pedagogical principles. This unit reimagines pedagogy in digital distance education by rejecting industrial-age models that view instruction as content delivery, learners as passive receivers, and assessment as compliance (Selwyn, 2016; Williamson, 2017). Instead, the EAF@UDE champions a model where learning is emotionally scaffolded, culturally responsive, and rooted in dialogic, affective presence. It centers empathy as both a design objective and a pedagogical method—structuring learning environments where students feel seen, heard, and empowered to engage deeply and ethically. Drawing from affect theory, critical pedagogy, and recent studies on emotional presence (e.g., Zembylas, 2018; Murphy et al., 2022), this chapter articulates a series of interconnected pedagogical principles. These include the educator's role as an emotional architect, the necessity of peer-to-peer empathy scaffolding, the reframing of feedback as dialogic recognition, the transformation of assessment into ethical inquiry, and the integration of multimodal and culturally inclusive design. By implementing these strategies, the digital classroom becomes not just a site of knowledge delivery, but a relational space where emotional safety, vulnerability, and co-presence can flourish. Importantly, Unit 5 does not romanticize empathy—it anticipates critique. It acknowledges the risks of emotional overreach, cultural misreading, and performative engagement, advocating instead for empathic design that is invitational, flexible, and context-sensitive. The chapter concludes by situating these pedagogical strategies within broader civic and ethical aims, affirming that empathetic learning is not a "soft" add-on but a transformative driver of critical engagement, inclusion, and learner empowerment. This prepares the groundwork for Unit 6, where these pedagogical choices are examined within the methodological architecture of the research itself. In that next chapter, empathy moves from classroom practice to research praxis—revealing how the same principles of care, reflexivity, and dialogic presence guide the study's design and data collection.

Figure 3. The Pedagogical Framework

PEDAGOGICAL FRAMEWORK (EAF@UDE)

- **Empathy as Design Principle**
  Structurally embedded in pedagogy, assessment, feedback, and interaction.
- **Educator as Emotional Architect**
  Shapes emotional climate through presence, honesty, and responsiveness.
- **Emotion as Epistemic**
  Emotions are integral to knowing, meaning-making, and interpretation.
- **Peer-to-Peer Empathy Structures**
  Scaffolded activities foster mutual recognition and dialogic co-creation.
- **Transformative Feedback**
  Feedback as dialogic recognition, not correction - affirms emotional insight.
- **Assessment as Ethical Inquiry**
  Narrative, reflective, and civically anchored - evaluates emotional and cultural reasoning.
- **Multimodal & Culturally Responsive Design**
  Encourages diverse expression (e.g. images, podcasts, metaphor) across platforms.
- **Digital Tools as Ethical Spaces**
  Platforms must support emotional presence and relational connection.

EAF@UDE

5.2. Culturally Situated Empathy and Multimodal Expression

Distance education, as practiced in many global contexts, continues to reflect pedagogical models born in the industrial era. These models prioritize efficiency, standardization, and the delivery of quantifiable outcomes. Within this framework, learning is treated as the transmission of content from instructor to student, instruction is reduced to delivery logistics, and assessment becomes a mechanism for validating mastery through discrete, measurable outputs. As Selwyn (2016) notes, such models render students as passive recipients and educators as content managers. Williamson (2017) further critiques this trend by warning of the growing datafication of education, where learners are converted into data points, and education loses its affective, cultural, and ethical depth.

The Empathic Anthropological Framework for Distance Education (EAF@UDE) responds to this reductive landscape with a transformative proposition: shift distance learning from transactional content delivery toward relational, ethical, and emotionally intelligent praxis. Central to this shift is the repositioning of empathy—not as an optional affective flourish, but as a guiding design principle. Within the EAF@UDE, learning is redefined as a relational, embodied, and transformative experience. Drawing on affect theory, critical pedagogy, and digital humanism, the framework envisions online learning as a holistic ecology in which empathy permeates every layer—from instructor presence to platform infrastructure, from assessment to feedback, from peer engagement to identity affirmation.

At the foundation of this pedagogical reorientation lie five core assumptions that inform the EAF@UDE. First, learning is inherently relational. Knowledge does not emerge in isolation but is co-constructed through emotionally resonant and dialogic encounters. This stands in contrast to individualistic and abstract models that dominate traditional education. Second, emotion is epistemic. As Zembylas (2018) argues, emotions are not disruptions but integral to the process of knowing. They influence how learners interpret information, relate to others, and construct meaning. Third, empathy is teachable. Much like critical thinking or civic literacy, empathy can be intentionally cultivated through scaffolded activities like reflective journaling, storytelling, and structured peer dialogue. Fourth, digital tools are not neutral—they are ethical environments. Their design can either suppress or foster affective presence, cultural voice, and human connection. Finally, educators are emotional architects. Their presence, tone, and transparency actively shape the emotional climate of learning, determining how students engage, belong, and persist.

This reframing necessitates a pedagogical design that explicitly centers emotional presence. Building on the Community of Inquiry model (Garrison et al., 1999), which highlights social presence as a prerequisite for cognitive engagement, the EAF@UDE extends the notion by foregrounding emotional presence as an intentional and designable element. It is not enough to merely allow emotional expression; it must be scaffolded through pedagogical and technological practices. These include the use of personalized video messages to humanize instruction, weekly emotional check-ins via mood maps or anonymous surveys, and multimodal assignments that invite students to articulate feelings through imagery, music, or metaphor. Even simple interventions—such as tone indicators, emojis, and sentiment cues in asynchronous forums—can alleviate the textual flatness of online communication and restore emotional nuance. Murphy et al. (2022) demonstrate that these strategies enhance instructor-learner bonds and increase intrinsic motivation. Contrary to concerns that emotional design undermines academic seriousness, such practices foster psychological safety and relational trust, both of which are essential for intellectual risk-taking.

The reconceptualized role of the educator in this model is not merely that of a subject expert or guide, but of an emotional and ethical presence. Drawing on Brookfield's (2017) concept of critically reflective teaching, EAF@UDE underscores the importance of emotional honesty, responsiveness, and vulnerability in pedagogy. The educator, through practices such as naming emotional tensions, sharing personal narratives, and validating emotional labor, becomes a catalyst for a culture of mutual recognition and co-creation. Feedback, too, is transformed in this model—from a mechanistic error-correction process to a dialogic act of witnessing, affirmation, and ethical invitation. Rather than simply stating what is missing, empathic feedback might say, "Your interpretation reveals a powerful personal connection—can you explore that further?" This acknowledges both the intellectual and affective dimensions of the learner, deepening engagement and clarity.

Peer-to-peer structures are another key dimension of the EAF@UDE. Traditional online education often treats peer interaction as formulaic—thread replies, perfunctory critiques, or disconnected group tasks. The EAF@UDE intervenes here by making empathy a design goal for peer activities. Baran and Correia (2014) found that peer relationships become more meaningful and academically productive when empathy is scaffolded. Techniques such as structured empathy dialogues—where students reflect on prompts like "What moved you?"—or narrative remixing, where a peer's story is reinterpreted through another modality (such as audio or infographic), promote deeper listening and emotional reciprocity. Tools like collaborative annotation platforms (e.g., Hypothesis) allow for multi-voiced engagement with readings, embedding emotional resonance and cultural reflection into academic discourse. Importantly, these activities are reflexive, optional, and culturally responsive, preventing the risk of performative or coercive empathy.

Assessment, too, is radically reimagined. The EAF@UDE positions assessment not as a neutral tool of measurement, but as a relational and ethical act. Traditional assessment models often privilege standardization and decontextualized correctness, leaving little room for diversity of expression or emotional insight. By contrast, EAF@UDE assessments are narrative-based, reflective, and civically anchored. They ask students not just what they know, but how knowledge intersects with their identity, values, and lived experience. For instance, a conventional essay on surveillance might be replaced with a reflective narrative about the student's own experiences with digital privacy. Assessment rubrics evaluate not only conceptual clarity but also emotional insight, cultural reflexivity, and ethical reasoning. This aligns with Mezirow's (1991) transformative learning theory, which posits that learning involves shifts in perspective through emotional and cognitive disruption.

Critics may argue that such assessments are hard to standardize, but the EAF@UDE maintains that standardization must never override epistemic justice. Assessments must empower, not exclude.

The cultural dimension of empathy is crucial. Expression, vulnerability, and emotionality are culturally situated. What counts as authentic or appropriate in one context may be misunderstood or dismissed in another. Thus, empathy must be designed with cultural responsiveness and multimodality. Scholars like Gay (2018) and Noddings (2013) remind us that education devoid of cultural context often perpetuates hegemony. The EAF@UDE addresses this by encouraging assignments that allow expression through diverse forms—photo essays, podcasts, spoken word, or visual storyboards. Platforms are selected or adapted to allow metaphor, image, and color, not just text. Cultural autobiographies and positionality statements help students locate their emotional expressions within broader socio-political narratives. In doing so, the EAF@UDE affirms plural emotional literacies and rejects monolithic norms of engagement.

Ultimately, the EAF@UDE does not treat empathy as decorative or supplementary. It is structural. It is the connective tissue between cognition and care, between critique and compassion. From platform design to assessment, from feedback to peer interaction, empathy functions as the guiding principle that restores emotional depth and relational meaning to distance education. In a time when online education is expanding rapidly, but often without attention to humanity, the EAF@UDE offers a radical alternative—an invitation to reimagine digital pedagogy as a site not only of learning but of ethical presence, cultural recognition, and mutual transformation.

5.3. Synopsis

Unit 5 translates the philosophical commitments of the EAF@UDE into a cohesive and evidence-based pedagogical model for online higher education. It calls for a shift from transactional, cognitively isolated modes of instruction toward emotionally intelligent, ethically grounded, and relationally rich learning environments. At its core is the belief that empathy must be designed into every layer of the course: into how teachers interact, how peers connect, how feedback is given, and how learning is assessed.

The chapter introduces five foundational pedagogical assumptions of the EAF@UDE: that learning is inherently relational; that emotion is epistemic; that empathy is teachable; that technology is an ethical space; and that educators shape emotional climates. These assumptions support the design of emotionally resonant learning strategies, including personalized video messages, affectively tuned feedback, story-sharing circles, narrative remixing, and flexible, multimodal assessments that honor cultural and emotional diversity.

Crucially, Unit 5 does not treat empathy as a universal or uncritical virtue. It insists on cultural reflexivity and ethical boundaries, proposing safeguards to ensure that empathy is invitational, not coercive. The pedagogical model is framed as civic, dialogic, and transformative—one that encourages critical consciousness alongside emotional connection.

The unit closes by reaffirming that these pedagogical strategies are not simply about improving engagement; they are about redesigning education around the principles of care, justice, and shared humanity. As such, it provides the bridge to Unit 6, which shows how the EAF@UDE principles were methodologically enacted in the research study itself—through constructivist inquiry, reflexive practice, and empathically attuned data collection.

6. METHODOLOGICAL FRAMEWORK

6.1. Chapter Overview

Unit 6 deepens the Empathic Anthropological Framework for Distance Education (EAF@UDE) by transitioning from pedagogical application to methodological embodiment. The methodological framework of this chapter *(figure 4)* chapter asserts that in ethically oriented, emotionally attuned education, research itself must also reflect the values it seeks to investigate. Thus, empathy is not only the object of study—it is the methodology. Anchored in constructivist and emancipatory paradigms, the methodological framework outlined in this chapter refuses positivist detachment. Instead, it embraces research as an ethical and relational act, where knowledge is co-constructed through dialogic encounters, emotional reflexivity, and narrative interpretation. Building on the pedagogical commitments detailed in Unit 5, this chapter enacts them within the logic of inquiry—demonstrating that research, like teaching, must be designed with care, vulnerability, and justice at its core. This unit introduces a qualitative case study, enriched by educational ethnography, as the methodological vehicle for exploring empathy in a real-world distance learning context: EPA524, a postgraduate course designed to cultivate empathic, reflective, and socially engaged learners. Drawing from Charmaz's (2006) constructivist grounded theory and Kincheloe and McLaren's (2005) critical research traditions, the chapter details how the study privileges depth, contextual resonance, and emotional complexity over generalization or prediction. Data collection followed a triangulated design: in-depth semi-structured interviews, learner artifacts (e.g., reflective journals, Padlet posts, peer feedback), an Emotional Presence Index questionnaire, and a researcher reflexive journal. Each method served to illuminate how empathy is experienced, enacted, and perceived within the course's digital landscape. Analytical rigor was ensured through coding cycles that identified emotional markers and relational dynamics, culminating in theoretical categories aligned with the EAF@UDE's core pillars: emotional presence, dialogic pedagogy, cultural reflexivity, and ethical co-creation. Importantly, Unit 6 does not shy away from the ethical risks of emotionally embedded research. It discusses researcher positionality, dual roles (as tutor and investigator), and the challenges of ensuring emotional safety and consent. Reflexivity is not treated as an afterthought, but as a method in itself—as both moral stance and analytic tool. This aligns the chapter with Pillow's (2003) notion of "uncomfortable reflexivity" that exposes the ethical tensions of inquiry and calls for accountability through transparency. The chapter concludes by affirming that methodological empathy is not about therapeutic intimacy or emotional indulgence. Rather, it is about research that honors participant voice, centers relational context, and is committed to justice through care. As we move to Unit 7, which presents the results and discussion, we see how this methodological architecture yields rich, transformative insights into the mechanics and meaning of empathy in digital learning.

Figure 4. The Methodological Framework

METHODOLOGICAL FRAMEWORK (EAF@UDE)

- **Empathy as Methodology**
  Research is relational, ethically attuned, and not detached or neutral
- **Inquiry as Relational Praxis**
  Empathy, reflexivity, and justice as connections preserved beyond positivist paradigms
- **Emotional Reflexivity & Positionality**
  Thick description instead of generalization; as contextually layered inquiry
- **Triangulated Data Collection:**
  Interviews, learner artifacts: Emotional Presence Index, and reflexive journal
- **Validity & Reliability through Methodological Triangulation**
  Integration, convergence, transparency, and achievement—analytical credibility through triangulation
- **Constructivist Grounded Theory Analysis**
  Iterative coding cycles over emotional markers to elucidate relational dynamics
- **Theoretical Alignment with EAF@UDE Pillars**
  Acknowledges empathy as co-presence dialogically, cultural reflexivity, and ethical co-creation
- **Cultural & Contextual Sensitivity**
  Empathy acknowledges culture and context—sensitivity; resists essentialism or universalism
- **Narrative Understanding over Statistical Abstraction**
  Narratives are cherished in their complexity

6.2. Inquiry as Relational Praxis: Ethics, Reflexivity, and Empathic Research in Distance Education

This chapter establishes the methodological foundation of the study by positioning research not as a neutral, depersonalized pursuit, but as an ethically charged, relational act. Rooted in critical, humanistic, and affective educational paradigms, the inquiry is framed as both epistemological and moral—a stance that challenges the instrumental logic of positivist methodologies. The Empathic Anthropological Framework for Distance Education (EAF@UDE) is not only theorized as a pedagogical model but also enacted as a methodological stance. Research, much like teaching, becomes a site where empathy, reflexivity, and justice converge. This methodological orientation centers relationality as a principle of inquiry, treating knowledge not as discovered fact, but as co-constructed meaning emerging through emotionally and ethically situated interactions.

To support this relational stance, the research adopts a dual paradigmatic orientation that is both constructivist and emancipatory. In alignment with Charmaz's constructivist grounded theory, knowledge is treated as a product of interaction, narrative engagement, and reflexive dialogue. Simultaneously, the study is guided by an emancipatory epistemology that views research as a process of social critique and transformation. Drawing from Kincheloe and McLaren (2005), the research aims not only to understand educational phenomena but to intervene in them—to disrupt inequality, surface hidden emotional labor, and foster critical consciousness. Within this dual orientation, emotion is not bracketed out as bias; rather, it is recognized as the very force that animates ethical inquiry. Empathy, vulnerability, and narrative engagement are positioned not in opposition to rigor but as its foundation.

This ethical and philosophical positioning is operationalized through a qualitative methodology that integrates educational ethnography and case study. The case study provides the structural boundaries of the investigation—EPA524, a postgraduate distance learning course on critical education offered by the Open University of Cyprus in Fall 2023—while educational ethnography offers the interpretive tools to generate "thick description." This methodology seeks not generalizability but meaning, focusing on the emotional, cultural, and pedagogical textures of relational learning. The selected course exemplifies the values embedded in EAF@UDE, prioritizing reflective writing, narrative pedagogy, empathic feedback, and peer dialogue. Thus, the course is not only the object of analysis but a living embodiment of the framework under study.

The participants in this case were ten postgraduate students, aged 27 to 49, from varied cultural, professional, and educational backgrounds. Seven were women and three were men. All participants were enrolled in a distance learning program and engaged through a blend of synchronous and asynchronous modalities using platforms such as Moodle, Zoom, and Padlet. The course design emphasized emotional literacy, narrative reflection, and community-based interaction, making it particularly suited for studying the dynamics of empathy in digital education. Sampling followed a purposive strategy grounded in maximum variation, seeking not representativeness but richness of emotional expression, diversity in digital literacy, and engagement with reflexive learning.

Data collection embraced a triangulated design to capture the layered, intersubjective nature of empathy as it manifests across time, space, and modalities. Four primary sources informed the analysis: semi-structured interviews, student artifacts, a custom-designed questionnaire, and a reflexive journal maintained by the tutor-researcher. Interviews, lasting 45 to 60 minutes each, probed the students' emotional experiences, perceptions of empathy, feedback responses, and peer relationships. These were conducted via Zoom, transcribed verbatim, and coded for affective language and narrative inflection. Student artifacts, including discussion posts, Padlet contributions, reflective journals, and peer feedback, were analyzed not only for content but for tone, emotional cues, and relational gestures such as acknowledgment and mirroring.

The Emotional Presence Index (EPI), a custom questionnaire, provided another lens on affective engagement. It combined Likert-scale items with open-ended questions designed to surface students' felt sense of emotional recognition and safety within the course. While not intended for statistical generalization, the EPI served as a validation mechanism for qualitative themes, surfacing patterns in emotional resonance and student reflections on empathy. Finally, the reflexive journal functioned as both a methodological tool and a moral compass, documenting the tutor-researcher's emotional responses, classroom climate, and pedagogical dilemmas throughout the twelve-week course. This journal added a meta-analytic layer to the research, acknowledging the researcher's presence and emotional entanglement as integral to the inquiry.

Data analysis followed the iterative procedures of constructivist grounded theory, which honors co-construction, subjectivity, and emergent meaning. The analysis unfolded in three stages: initial coding, focused coding, and theoretical integration. In the initial coding phase, transcripts and artifacts were read line-by-line to identify relational and emotional indicators—statements such as "I felt heard," "She mirrored my thoughts," or "That stayed with me." These were treated as affective signposts pointing to moments of empathic connection. In the focused coding stage, these signposts were clustered into conceptual groupings such as narrative vulnerability, peer resonance, tutor responsiveness, and emotional mirroring. These groupings formed the analytical scaffolding for theoretical synthesis.

In the final stage, theoretical integration, the focused codes were aligned with the four central pillars of the EAF@UDE: emotional design, dialogic presence, cultural reflexivity, and ethical co-creation. Through memo-writing, constant comparison, and iterative coding validation, the analysis moved toward coherence without sacrificing complexity. NVivo 12 was used for organizing the codes and generating thematic maps, but the final interpretation remained grounded in dialogic reasoning, narrative flow, and rich description rather than abstract theorization or statistical abstraction.

6.3 Synopsis

This chapter synthesizes the key analytical findings related to the second research question of the study: how empathy functions as a transformative mechanism in distance education through pedagogical and technological processes. Drawing on a constructivist and emancipatory cognitive framework, and incorporating evidence from both UK-based and Greek-language ethnographic studies, the chapter reveals that empathy does not emerge spontaneously but is activated through deliberate educational design. Empathic engagement is scaffolded through dialogic rituals, multimodal storytelling, emotional feedback, authentic tasks, and sustained group collaboration.

Particular emphasis is placed on the role of the educator, reimagined as both a critical friend and emotional companion, whose vulnerability and presence foster an atmosphere of trust, reflexivity, and civic responsibility. Digital tools—when designed for humanizing rather than mechanizing learning—support emotional literacy, inclusivity, and affective co-presence. The Greek field study affirms that when authentic activities, collaborative learning, and technology are tightly integrated, empathy emerges as both a relational practice and a site of cognitive and ethical transformation.

Ultimately, the chapter argues that empathy must be structurally embedded into educational practice and policy—not as sentimentality, but as a rigorous, justice-oriented pedagogy that enables students not just to know the world, but to care for and act within it.

7. RESULTS AND DISCUSSION

7.1. Overview

This chapter presents the analytical core of the study by engaging with the two guiding research questions that underpin the Empathic Anthropological Framework for Distance Education (EAF@UDE): (1) To what extent can empathy be intentionally cultivated in university-level distance education? (2) Through what pedagogical and technological mechanisms can empathy function as a transformative force for inclusion, critical awareness, and humanization? Using constructivist grounded theory, a multi-modal set of qualitative data was collected and analyzed to surface emergent patterns and interpret how empathy was enacted, felt, and translated into deeper ethical and cognitive shifts. The study draws from interviews, student artifacts, reflective journals, and the Emotional Presence Index (EPI), ensuring a rich triangulation of voices and experiences. The findings are presented in two primary sections aligned with each research question, followed by an integrated discussion that situates these results within the theoretical and pedagogical commitments of the EAF@UDE.

7.2 Empathy as a Cultivable Practice in Digital Learning: A Multimodal and Cross-Cultural Perspective: Responding to Research Question 1

This section analyzes the findings emerging from the first research question, *(figure 5)* which asked to what extent empathy could be intentionally cultivated within asynchronous, university-level distance education. The question is more than pedagogical—it is fundamentally epistemological and ethical. It seeks to understand whether digital education, often critiqued as emotionally barren or depersonalized, can become a site where learners not only think and write critically but also feel, relate, and transform together.

Using constructivist grounded theory and triangulated data from both a UK-based study and a Greek-language field investigation, this discussion affirms that empathy is not merely possible in digital environments—it is cultivable, repeatable, and essential for relational and emancipatory learning. It emerges not as a spontaneous emotional event, but as a structurally supported practice: scaffolded through dialogic design, nurtured through emotional presence, and extended through ethical engagement.

7.2.1.Dialogic Scaffolding and the Conditions for Empathy

One of the strongest findings is that empathy flourishes within what we might call dialogic ecologies—carefully cultivated spaces of ongoing, reciprocal engagement. In both contexts, dialogic rituals such as weekly discussion threads, narrative journals, and asynchronous peer responses were not add-ons; they were the curriculum's emotional infrastructure. These activities helped convert the learning process from a transactional model of content delivery to a relational pedagogy of mutual recognition.

Freire's (1970) pedagogy of dialogue serves as a central anchor here. The students did not just post responses to prompts—they returned to earlier exchanges, quoted their peers, and evolved their arguments over time. Emotional resonance was not just permitted; it was legitimated through narrative continuity and the explicit valorization of personal voice.

In the Greek-language study, dialogic scaffolding was further intensified through a persistent social media thread on Viber. This space—semi-formal but educationally vital—offered students a site of immediacy, presence, and mutual care. Empathy was not confined to academic prose; it circulated through emojis, check-ins, shared worries, and collective coordination. The use of everyday digital tools in emotionally intentional ways transformed technology from a delivery system into what we might call a platform of belonging.

This aligns with Zembylas' (2018) work on emotional ecologies, which conceptualizes affect not as an internal, individual state but as a circulating, co-produced phenomenon shaped by space, power, and pedagogy. Emotional teaching and learning were clearly observed here, with students describing a shift from guarded detachment to expressive vulnerability over the course of the module.

7.2.2.. Authenticity and the Role of Real-World Relevance

The most effective empathy-building occurred within authentic, real-world learning activities—assignments that demanded not only intellectual analysis but also ethical and emotional engagement. These included case-based scenarios, social justice dilemmas, and reflective storytelling grounded in students' professional contexts.

In the UK study, activities such as reflective narratives on digital surveillance and peer interviews about educational exclusion helped learners "think through feeling" and "feel through thinking." In the Greek context, the professor's selection of themes—refugee rights, disability inclusion, and democratic schooling—connected directly to students' lived realities and professional ambitions.

Authentic activities offered what Omodan (2022) calls epistemic dissonance—cognitive and emotional discomfort that demands reflection and reorientation. Mezirow's (1991) concept of disorienting dilemmas becomes important here: the emotional disturbance caused by confronting unfamiliar perspectives prompted students to rethink ideological positions, engage in civic inquiry, and develop what we might term empathic worldviews.

These activities did more than teach content; they created empathetical environments where learners inhabited the moral complexity of real social issues. By practicing empathy as inquiry—not charity—students developed dispositions of ethical responsibility, critical humility, and civic imagination.

7.2.3. Collaboration as Co-Presence: The Value of Long-Term Grouping

One of the most compelling findings across both studies was the contrast between short-term groupings and sustained collaborative constellations. In courses where learners remained in the same peer groups over several weeks, there was significant evidence of co-regulated empathy. Students not only supported one another during emotionally charged assignments, but also offered logistical help during crises such as illness or connectivity issues.

These collaborative groups functioned as micro-communities of emotional accountability, reflecting the theory of "affective co-regulation" found in the work of Isaias et al. (2017). In moments of tension—conflict, misunderstanding, or workload imbalance—students in long-term groups demonstrated restorative dialogue practices, including naming emotion, seeking clarification, and offering repair.

In the Greek-language study, the continuity of group membership enabled the formation of deep interpersonal bonds. Students often spoke of "emotional security," "peer trust," and "learning with friends." The professor's role in designing the learning environment as a space of care and sustained interaction enabled these connections to flourish. Her presence as a "professor-as-friend" served as a cultural and affective bridge between students' emotional lives and the academic setting.

This model of the professor as a critical and emotional friend extends beyond traditional mentorship. It entails a reciprocal form of vulnerability—one where the educator's emotional honesty becomes both a model and an invitation. This ethos supports a more radical pedagogy: one where feelings are not distractions from learning but core tools of knowing.

7.2.4. Emotional Literacy Through Technology

ICT platforms such as Padlet, VoiceThread, and Moodle were used not simply as vessels for content delivery but as vehicles of emotional and multimodal expression. The integration of photos, symbols, music, and metaphors expanded the lexicon of empathy and allowed students to communicate affect in culturally and neurologically diverse ways.

Kalmar et al. (2022) argue that emotionally intelligent platform design must include affordances for voice, texture, and ambiguity. In the Greek study, Padlet boards became mosaics of collective identity and feeling—digital murals that blended personal and collective storylines. Students felt "seen" in these spaces, not just read.

This multimodal expression enabled the cultivation of emotional intelligence, particularly for those who found verbal academic discourse alienating or inaccessible. Rather than standardizing how empathy was shown, these tools supported plural literacies, making emotional teaching and learning more inclusive and participatory.

The emotional infrastructure of the course was also reflected in consistent tone and presence from the tutor—through both formal announcements and informal chats. In both studies, learners interpreted this presence not as surveillance but as care. Emotional visibility, rather than emotional neutrality, became a condition of ethical pedagogy.

7.2.5. Feedback as Emotional and Intellectual Dialogue

Feedback played a central role in cultivating empathy, both as a practice and as a pedagogical signal. Empathic feedback was dialogic, narrative-rich, and emotionally attuned. It acknowledged emotional labor, praised effort, and reflected students' affective language.

Rather than offering only evaluation, feedback became a moment of dialogue—a pedagogical site of what Freire (1970) might call "naming the world." Tutors echoed student voices back to them, reframed confusion as curiosity, and turned struggle into insight.

The Greek study added important nuance to this finding. Feedback was not merely unidirectional; it was also peer-generated. Students offered empathic affirmations—"thank you for sharing this," "this really made me think"—that reinforced affective belonging and intellectual solidarity. This horizontal exchange of empathy helped redistribute emotional labor across the learning community, decentering the authority of the tutor and creating shared responsibility for emotional safety.

Such practices align with Kalopana (2022) and Brookfield (2017), who suggest that feedback, when rooted in empathy, becomes a form of moral witnessing. It acknowledges the learner not just as a knower but as a feeler, situated in a world of uncertainty, history, and hope.

7.2.6. Ethical Dissonance and Civic Reflection

Perhaps the most transformative moments were those involving ethical dissonance—experiences where students encountered perspectives that disrupted their assumptions or challenged their comfort zones. Reading a classmate's story about racism, hearing about undocumented status, or encountering alternative epistemologies triggered discomfort, confusion, and eventually reflection.

These moments, often cited as turning points by participants, illustrate empathy's role as an emancipatory cognitive framework. It enables learners to not only feel with others but to think differently about themselves and the world. In both contexts, students responded to ethical dissonance with acts of civic reflection: writing opinion pieces, changing instructional practices, initiating social projects.

This proves that empathy, when scaffolded, does not dilute rigor—it deepens it. It moves learners beyond performative understanding into ethical inquiry and moral imagination. It converts discomfort into dialogue, emotion into ethics, and learning into lived action.

7.2.7. Empathy as a Systemic and Cultural Practice

The capacity to cultivate empathy at scale depends on institutional choices and cultural framing. It cannot rest solely on the charisma or values of individual educators. Instead, it must be embedded in pedagogical design, institutional culture, and platform architecture.

In both studies, empathy emerged where institutions permitted emotional experimentation, where tutors were trained (explicitly or implicitly) in emotional literacy, and where students were not penalized for affective expression. In contrast, environments with rigid deadlines, impersonal feedback, and fragmented peer interaction failed to support sustained empathic development.

Empathy must be systemic: embedded in curriculum, modeled by faculty, and supported by emotionally intelligent technology. It must be treated not as a soft supplement to critical pedagogy but as a core condition of it.

7.2.8. Conclusion: Empathy as Learning, Not Luxury

The answer to Research Question 1 is unequivocal: yes, empathy can be intentionally cultivated in distance education. But only when learning is structured as a relational, emotional, and cognitive ecology. When students are asked not just to analyze texts but to witness lives; not just to write essays but to narrate truths; not just to consume content but to co-create meaning—then empathy emerges, not as sentimentality, but as a rigorous, ethical, and transformative educational force.

This vision is not utopian. It is already happening—in the platforms students build together, in the courage they show in telling their stories, and in the quiet labor of professors who act as emotional architects, critical friends, and human companions in the learning journey.

In a world where distance education is growing rapidly, the challenge is not only how to teach remotely but how to feel, relate, and transform through digital space. Empathy is the pedagogy of that possibility.

Figure 5. Key Findings-Research Question 1

EMPATHY IN DIGITAL EDUCATION:

Key Findings from RQ1 (EAF@UDE)

• **Empathy is Cultivable**

Not spontaneous but structurally supported; embedded in design, presence, and ethics.

• **Dialogic Scaffolding as Emotional Infrastructure**

Peer feedback, narrative journaling, and persistent communication foster empathy.

• **Authenticity and Real-World Relevance**

Social justice themes and professional contexts promote emotional and civic engagement.

• **Collaborative Co-Presence through Long-Term Grouping**

Stable peer constellations create emotional trust and mutual care.

• **Emotional Literacy via Multimodal Technology**

Platforms like Padlet and Moodle support culturally diverse affective expression.

• **Feedback as Moral Witnessing**

Empathic, dialogic feedback—both tutors and peers—strengthening affective belonging

• **Ethical Dissonance Triggers Civic Reflection**

Exposure to discomforting perspectives prompt ethical growth and social action.

• **Empathy Requires Systemic Support**

Institutional culture, tutor training, and/or platform architecture must enable emotional expression

Conclusion: Empathy is Learning

7.3. Through What Pedagogical and Technological Mechanisms can Empathy Function as a Transformative Force for Inclusion, Critical Awareness, and Humanization in Distance Education? Responding to Research Question 2

The second research question *(figure 6)* directs attention not simply to whether empathy can exist in digital higher education but to the mechanisms by which it is enacted, sustained, and transformed into a pedagogical force capable of fostering inclusion, critical consciousness, and social transformation. Through a synthesis of empirical findings from the UK-based EAF@UDE study and a Greek-language ethnographic case study, this section critically examines the operationalization of empathy in online learning, offering a culturally grounded and pedagogically rich analysis. The goal is to uncover the structural, relational, and emotional architecture that supports empathy as a dynamic and emancipatory cognitive framework for distance learning.

7.3.1. From Disclosure to Transformation: The Empathic Cycle in Practice

At the core of this research lies the Empathic Cycle of Digital Learning—an iterative model comprising five recursive stages: vulnerable disclosure, peer resonance, empathic feedback, cognitive reframing, and ethical awakening. This model reflects the trajectory from interpersonal sharing to civic consciousness and resonates strongly with the findings from the Greek-language fieldwork, which emphasizes the social constructivist and emotionally driven logic of emancipatory learning communities.

In both case contexts, the journey begins with vulnerable disclosure. This is not a trivial starting point; it is a radical pedagogical act in an environment often characterized by depersonalized interaction. Learners tentatively share personal stories, emotions, or reflections, frequently couched in caveats like "I'm not sure if this fits." These disclosures act as catalysts for relational anchoring, inviting empathic responses and mutual recognition. The design of asynchronous forums, multimodal tools like Padlet, and emotionally attuned prompts all serve to scaffold these disclosures, making emotional openness a valid—and validated—epistemic move.

In the Greek study, vulnerable sharing was most evident in collaborative autobiographical assignments and reflection-based tasks that asked students to recount lived experiences and identity positions. This was not only encouraged by the thematic design of the course but amplified by the professor's willingness to be emotionally present—a dimension we return to below. It is this fusion of curricular design, tutor modeling, and student openness that creates what might be called authentic empathetical environments.

7.3.2. Dialogic Presence and Peer Resonance

The second mechanism, peer resonance, underscores the role of horizontal empathy. When students reciprocate disclosure with comments such as "I've felt this too," or reflect back a peer's emotional stance with respect and nuance, the dialogic space becomes charged with mutuality. In the Greek fieldwork, this emerged vividly in the Viber social media group created by the tutor. Far from being a peripheral communication channel, the group became a socio-emotional hub—offering immediate responses to queries, support during stress, and casual, non-academic interaction that fostered community bonding. Students described feeling that "someone is always there"—a crucial antidote to the emotional isolation that often characterizes online learning.

This finding aligns with research by Cartee (2021) and Bates (2019), who assert that sustained communication in digital learning enhances students' sense of belonging and academic confidence. What distinguishes this study is the way peer-to-peer empathy is framed not merely as relational balm but as a cognitive and emotional catalyst. Through dialogic rituals—repeated narrative engagement, emotional mirroring, and narrative remixing—students co-construct learning in ways that are affectively charged and ethically grounded.

7.3.3. The Tutor as Emotional and Critical Friend

Perhaps the most influential mechanism across both data sets is the presence of the educator—not just as a subject expert or facilitator, but as what might be termed a critical friend and emotional companion. This dual identity disrupts traditional academic hierarchies and recasts the professor as a co-learner in emotional and epistemic exploration.

In the EAF@UDE study, tutor feedback was cited as a key moment of emotional reflection, or what one student called "being mirrored back to myself." This feedback was less about correction and more about dialogue—connecting the student's intellectual struggle with emotional labor and growth. Similarly, in the Greek context, the professor deliberately avoided a stance of authority and instead modeled vulnerability, compassion, and intellectual humility. Her openness created a culture of permission—students felt free to disclose, to risk, and ultimately, to grow.

More than just offering feedback, the professor in the Greek study demonstrated a consistent empathic stance. She listened attentively, responded promptly, and treated students not as passive recipients but as collaborators in a shared educational journey. Importantly, she also cultivated friendly relationships without sacrificing academic rigor. This approach—what Fragkaki (2008) refers to as ethical authority—is an under-discussed but transformative dynamic in distance education. It reframes the professor not as a remote evaluator but as a human presence capable of holding emotional space, setting ethical norms, and nurturing relational safety.

This role is crucial within the emancipatory cognitive framework underpinning both studies. By dismantling positional power and replacing it with a posture of emotional attunement and critical solidarity, the professor becomes not only a guide but an affective catalyst. This has implications far beyond pedagogy: it models the very empathy that students are being asked to practice.

7.3.4. Multimodality and the Humanization of Technology

In both studies, digital platforms emerged not as neutral delivery tools but as emotional infrastructures—spaces that either enabled or inhibited the flow of empathy. Tools like Padlet, VoiceThread, and Viber were not valued merely for their technological sophistication, but for their capacity to humanize interaction. These platforms supported multimodal expression—image, voice, metaphor, gesture—which allowed students to articulate emotion beyond text, engaging emotional literacies that are often marginalized in academic discourse.

For neurodivergent or linguistically diverse learners, this proved transformative. The ability to use visual storytelling, music, or metaphor enabled deeper affective presence and greater epistemic inclusion. This echoes the work of Anderson and Dron (2011) and Bickmore et al. (2022), who argue that digital environments must be emotionally intelligent if they are to support authentic learning.

In the Greek study, the multimodal emphasis was further supported by the use of social media, which blurred the boundaries between formal and informal learning. The professor's willingness to communicate outside the LMS demonstrated a recognition that human connection does not follow institutional schedules. Such flexibility is part of what it means to humanize technology—to resist the reduction of education to algorithmic efficiency and instead foreground its emotional, unpredictable, and relational dimensions.

This insistence on emotional teaching and learning stands in stark contrast to the prevailing logic of datafication in education, where AI tools are increasingly seen as capable of managing teaching tasks. What these studies show, however, is that no algorithm can truly feel, nor respond to the nuanced emotional cues that make teaching ethical, affective, and human. The danger is not only in letting AI do everything for us, but in forgetting that learning is fundamentally about being seen, felt, and changed by others.

7.3.5.Empathy and Civic Awakening

The final mechanism, ethical or civic awakening, emerged when empathy moved beyond individual reflection and toward social consciousness. In both the UK and Greek contexts, students described moments when peer narratives triggered not only compassion but moral discomfort and a desire to act. These "disorienting dilemmas," to borrow from Mezirow (1991), reconfigured learners' understanding of power, privilege, and positionality.

In the Greek case study, civic engagement emerged through discussions of refugee rights, educational exclusion, and social justice. These were not hypothetical debates but emotionally anchored reflections sparked by peer stories and teacher modeling. Students did not merely understand the world differently—they committed to changing it, whether through advocacy, curricular reform, or community work.

This shift from emotional reaction to ethical action illustrates empathy's full pedagogical arc. It moves from relational presence to critical reflection, and ultimately to social transformation. In this light, empathy is not a soft skill but a radical educational force—a praxis of care, critique, and co-construction.

7.3.6. Integrated Framework: Empathy as a Systemic Imperative

The findings across both studies confirm that empathy is not a spontaneous byproduct of good teaching. It is a design imperative—a pedagogical, emotional, and technological architecture that must be built with intention, nuance, and care. Whether in Cyprus or the UK, the mechanisms are strikingly similar: dialogic rituals, emotional feedback, critical friendship, multimodal expression, and ethical scaffolding. Together, they form a model of empathic pedagogy that is both globally relevant and locally responsive.

This model aligns with and expands the theoretical scaffolding of the Empathic Anthropological Framework for Distance Education. It situates empathy within a broader ecology of humanizing pedagogy, critical digital literacy, and emancipatory cognitive practice. It challenges dominant paradigms of efficiency and objectivity and replaces them with a vision of education as relational, reflexive, and transformative.

From a policy and institutional perspective, these insights demand a fundamental rethinking of quality in distance education. Metrics must move beyond content mastery and include emotional presence, dialogic depth, and civic resonance. Platforms must be evaluated not only for usability but for their capacity to host vulnerability and support ethical engagement. And faculty development must include training not just in pedagogy, but in emotional literacy, trauma-informed practice, and intercultural sensitivity.

7.3.7 Conclusion: Designing for Empathy, Acting for Justice

Empathy is not a luxury in education—it is a necessity. As both studies show, empathy can be cultivated, scaffolded, and sustained in online environments, but only through the combined power of intentional design, authentic relationship, and humanizing technology. The role of the professor is central—not as an omniscient authority but as a critical and emotional friend who models care, challenges power and builds trust.

To foster truly authentic empathetical environments, institutions must resist the temptation to automate what is fundamentally human. AI can support learning, but it cannot feel, reflect, or care. If education is to be a humanizing force in an increasingly digitized world, we must commit to teaching not only through technology but beyond it—with heart, with reflexivity, and with justice in mind.

Empathy is not the opposite of rigor. It is the condition for it. And it is how we build learning spaces worth belonging to—and futures worth fighting for.

Figure 6. Key Findings-Research Question 2

EMPATHY AS A TRANSFORMATIVE FORCE IN DISTANCE EDUCATION (RQ2 - EAF@UDE)

- **The Empathic Cycle in Practice**
  Five stages: vulnerable disclosure → peer resonance; empathic feedback; cognitive reframing; ethical awakening.
- **Dialogic Presence & Peer Resonance**
  Sustained emotional exchange in semi-formal channels (e.g. Viber) fosters trust, belonging, and mutual reflection.
- **Tutor as Emotional & Critical Friend**
  The educator models vulnerability, care, and ethical authority—disrupting hierarchies and nurturing relational safety.
- **Multimodality as Emotional Infrastructure**
  Platforms like Padlet and VoiceThread enable diverse emotional literacies through visual, auditory, and symbolic expression.
- **Civic Awakening via Ethical Tension**
  Disorienting dilemmas trigger reflection, positional awareness, and civic engagement rooted in empathy.
- **Empathy as Systemic Design Imperative**
  Institutional commitment, platform design, and faculty training intentionally scaffold emotional presence.
- **Integrated Humanizing Pedagogy**
  Combines emotional literacy, cultural reflexivity, and ethical co-creation—beyond technocratic efficiency.
- **Conclusion: Empathy is Justice-Oriented Pedagogy**
  Empathy must be designed, modeled, and sustained—across the system. It enables inclusion, resistance, and ethical transformation.

7.4. Synopsis

Unit 7 presents the analytical findings of the study by addressing the two central research questions: whether empathy can be intentionally cultivated in university-level distance education, and through what pedagogical and technological mechanisms it fosters inclusion, critical awareness, and humanization. Drawing on a constructivist grounded theory approach and multiple qualitative data sources, the chapter reveals that empathy in digital learning is not accidental—it is co-constructed through structured, emotionally attuned practices. Empathy was fostered through dialogic rituals, affective feedback, tutor modeling of vulnerability, and the use of multimodal platforms as emotional infrastructures. These practices formed a recursive process, conceptualized as the Empathic Cycle of Digital Learning, wherein vulnerable disclosure leads to emotional resonance, cognitive reframing, and civic awakening. The findings underscore that empathy is not a soft supplement to academic rigor but a transformative pedagogical force. It redefines feedback, enriches peer interaction, and catalyzes ethical engagement. The study confirms that with intentional design and relational care, distance education can be reimagined as a space of critical reflection, emotional presence, and humanizing learning.

8. CONCLUSION AND FUTURE PROPOSALS

8.1. Overview

This chapter consolidates the key findings of the study and situates them within the broader context of digital pedagogy, answering two central research questions: (1) To what extent can empathy be cultivated in university-level distance education? and (2) Through what pedagogical and technological mechanisms can empathy function as a driver of inclusion, critical awareness, and humanization? Drawing on the Empathic Anthropological Framework for Distance Education (EAF@UDE), the chapter presents a synthesis of theoretical, empirical, and pedagogical insights, emphasizing empathy as both a cultivable capacity and a structural necessity within digital learning. It introduces a holistic model of empathic digital pedagogy encompassing pedagogical design, emotional infrastructure, cultural reflexivity, and civic intentionality. The chapter concludes with forward-looking proposals for research, platform design, and educator development, arguing for the integration of emotional teaching and learning within an emancipatory cognitive framework. It positions the educator as both a critical and emotional friend and calls for humanising technology in an era where AI risks detaching learning from relational meaning.

8.2. Conclusion

This study affirms that empathy, when situated within an emancipatory cognitive framework, is not an incidental emotion but a foundational pillar of ethical distance education. Grounded in the Empathic Anthropological Framework for Distance Education (EAF@UDE), the findings show that emotional teaching and learning can flourish in digital spaces—when designed with intentionality, cultural reflexivity, and ethical presence. Through dialogic rituals, multimodal expression, and narrative feedback, learners co-created authentic empathetical environments marked by trust, inclusion, and civic awareness. The professor's role was reimagined—not as a detached evaluator, but as both a critical friend and emotional ally—curating spaces where students could risk vulnerability, feel intellectually held, and engage with complexity. Crucially, this study warns against the trend of delegating relational pedagogy to artificial intelligence. Humanising technology is essential—not to mimic care, but to support emotional presence and prevent the pedagogical danger of not thinking and not feeling. Empathy must not be reduced to a metric or outsourced to automation. Instead, it must lead. As digital education continues to scale, only an education that reclaims empathy as a relational, reflective, and moral imperative can resist the flattening effects of datafication and cultivate truly humanizing learning experiences.

8.3. Future Research and Practice Proposals

Building on this study's emancipatory cognitive framework, future research should investigate how emotional teaching and learning can be sustained longitudinally across diverse cultural and technological contexts. Comparative ethnographic studies may deepen our understanding of how authentic empathetical environments adapt across linguistic and political differences. Further exploration is also needed into the ethical tensions of AI-mediated affective computing: Can emotion-aware systems support co-presence without displacing the human responsibility to feel, think, and relate? From a practical standpoint, institutions should prioritize the training of educators as both critical and emotional friends—mentors who model vulnerability, curate dialogic learning, and humanize digital space. Emotional literacy must become integral to faculty development, while Learning Management Systems should include features that support mood check-ins, multimodal storytelling, and peer recognition. Assessment frameworks must evolve to capture relational growth and civic resonance, not only content mastery. Educational technology must be redesigned not to replace empathy but to serve it. In doing so, we move from transactional learning to transformative co-existence. The future of distance education must not be governed by speed and scale alone, but by the courage to cultivate emotional depth and ethical imagination. In this, empathy is not a luxury—it is our pedagogical compass.

8.4. Synopsis

This chapter argues that empathy, when intentionally embedded into the design and facilitation of distance education, transforms digital classrooms into authentic empathetical environments. Grounded in an emancipatory cognitive framework, it demonstrates how emotional teaching and learning—through feedback, multimodal dialogue, and ethical presence—redefines the role of the professor as a critical friend and emotional ally. By analyzing five core pedagogical mechanisms, the chapter shows how empathy fosters inclusion, civic engagement, and transformative learning. It cautions against the dehumanizing drift of over-automated education, advocating instead for technologies that support—not substitute—relational presence. Empathy, ultimately, is presented not as an affective surplus but as the connective tissue of humanized, justice-oriented digital pedagogy.

REFERENCES

Anderson, T., & Dron, J. (2011). Three generations of distance education pedagogy. *International Review of Research in Open and Distance Learning, 12*(3), 80–97. DOI: 10.19173/irrodl.v12i3.890

Andreotti, V. (2014). Soft versus critical global citizenship education. In Abdi, A., Shultz, L., & Pillay, T. (Eds.), *Decolonizing global citizenship education* (pp. 21–31). Sense Publishers., DOI: 10.1007/978-94-6209-518-8_2

Apple, M. W. (2006). Understanding and interrupting neoliberalism and neoconservatism in education. *Pedagogies, 1*(1), 21–26. DOI: 10.1207/s15544818ped0101_4

Baran, E., & Correia, A. P. (2014). A professional development framework for online teaching. *TechTrends, 58*(5), 95–101. DOI: 10.1007/s11528-014-0791-0

Bates, A. W. (2019). *Teaching in a digital age* (2nd ed.). BCcampus. https://pressbooks.bccampus.ca/teachinginadigitalagev2/

Bickmore, T., Schulman, D., & Yin, L. (2022). Compassionate computing: Design and evaluation of relational agents. *Journal of Human–Computer Interaction, 38*(3), 201–227. DOI: 10.1080/07370024.2022.2035237

Biesta, G. (2021). *World-centred education: A view for the present.* Routledge., DOI: 10.4324/9781003098331

Borup, J., Graham, C. R., West, R. E., Archambault, L., & Spring, K. (2012). The adolescent community of engagement framework. *Journal of Technology and Teacher Education, 22*(1), 107–130.

Brookfield, S. D. (2017). *Becoming a critically reflective teacher* (2nd ed.). Jossey-Bass.

Bubandt, N., & Willerslev, R. (2015). The dark side of empathy. *Comparative Studies in Society and History, 57*(1), 5–34. DOI: 10.1017/S0010417514000589

Calloway-Thomas, C. (2018). Empathy in the global world: An intercultural perspective. *Sage (Atlanta, Ga.).*

Cartee, D. (2021). The pedagogy of empathy: Cultivating deeper learning. *Journal of Educational Research and Practice, 11*(1), 88–99. DOI: 10.5590/JERAP.2021.11.1.07

Charmaz, K. (2006). Constructing grounded theory: A practical guide through qualitative analysis. *Sage (Atlanta, Ga.).*

Freire, P. (1970). Pedagogy of the oppressed. *Continuum : an Interdisciplinary Journal on Continuity of Care.*

Fricker, M. (2007). *Epistemic injustice: Power and the ethics of knowing.* Oxford University Press., DOI: 10.1093/acprof:oso/9780198237907.001.0001

Garrison, D. R., Anderson, T., & Archer, W. (1999). Critical inquiry in a text-based environment: Computer conferencing in higher education. *The Internet and Higher Education, 2*(2–3), 87–105. DOI: 10.1016/S1096-7516(00)00016-6

Gay, G. (2018). *Culturally responsive teaching: Theory, research, and practice* (3rd ed.). Teachers College Press.

Habermas, J. (1987). *The theory of communicative action: Vol. 2. Lifeworld and system: A critique of functionalist reason.* Beacon Press.

Hodges, C., Moore, S., Lockee, B., Trust, T., & Bond, A. (2020). The difference between emergency remote teaching and online learning. *EDUCAUSE Review, 27,* •••. https://er.educause.edu/articles/2020/3/the-difference-between-emergency-remote-teaching-and-online-learning

hooks, b. (1994). *Teaching to transgress: Education as the practice of freedom.* Routledge.

Isaias, P., Miranda, P., & Pífano, S. (2017). Emotional presence in e-learning systems. *Information Systems Frontiers, 19*(4), 747–763. DOI: 10.1007/s10796-015-9614-x

Kincheloe, J. L. (2005). *Critical constructivism primer.* Peter Lang.

Kincheloe, J. L., & McLaren, P. (2005). Rethinking critical theory and qualitative research. In Denzin, N. K., & Lincoln, Y. S. (Eds.), *The SAGE handbook of qualitative research* (3rd ed., pp. 303–342). SAGE.

Mezirow, J. (1991). *Transformative dimensions of adult learning.* Jossey-Bass.

Murphy, C., Martin, M., & Parmar, N. (2022). Designing for emotion in online adult education. *Adult Education Quarterly, 72*(1), 34–56. DOI: 10.1177/07417136221080742

Ndlovu-Gatsheni, S. J. (2018). *Epistemic freedom in Africa: Deprovincialization and decolonization.* Routledge., DOI: 10.4324/9780429492204

Nhat Hanh, T. (1998). *Interbeing: Fourteen guidelines for engaged Buddhism.* Parallax Press.

Noddings, N. (2013). *Caring: A relational approach to ethics and moral education* (2nd ed.). University of California Press.

Owusu-Boampong, A., & Holmberg, C. (2015). *Distance education in European higher education – The state of play*. UNESCO Institute for Lifelong Learning.

Pillow, W. (2003). Confession, catharsis, or cure? *International Journal of Qualitative Studies in Education : QSE, 16*(2), 175–196. DOI: 10.1080/0951839032000060635

Rodriguez, M. (2022). Artificial affect: A critique of empathy in algorithmic education. *Learning, Media and Technology, 47*(2), 210–227. DOI: 10.1080/17439884.2022.2029280

Selwyn, N. (2016). *Education and technology: Key issues and debates* (2nd ed.). Bloomsbury.

Tsourli, E. B., Dalakoura, Z. D., & Fragkaki, M. (2022). Developing 21st-century skills in university distance education. *International Conference on Open & Distance Education, 11*(7A), 154–169. https://doi.org/DOI: 10.12681/icodl.3316

Williamson, B. (2017). Big data in education: The digital future of learning, policy and practice. *Sage (Atlanta, Ga.)*. Advance online publication. DOI: 10.4135/9781529714920

Young, H. (2018). *Digital humanism: Technology and the future of humanity*. Routledge.

Zembylas, M. (2013). The "emotional turn" in education. In Wong, P. T. P. (Ed.), *The human quest for meaning* (2nd ed., pp. 387–400). Routledge.

Zembylas, M. (2018). Reinventing critical pedagogy as decolonising pedagogy. *Review of Education, Pedagogy & Cultural Studies, 40*(5), 404–421. DOI: 10.1080/10714413.2019.1570794

Chapter 2
Anthropologic Perspective of ICT in Education:
How Society Shapes and Is Shaped by EdTech

Wasswa Shafik
http://orcid.org/0000-0002-9320-3186
Dig Connectivity Research Laboratory (DCRLab), Uganda

ABSTRACT

This chapter investigates the anthropologic perspective of Information and Communication Technology (ICT) in education by examining how cultural, social, and contextual factors influence the adoption, integration, and impact of digital technologies in learning environments. It explores the intersection of educational technology and human behavior, highlighting the role of local traditions, beliefs, and values in shaping technological practices. Through a multidisciplinary lens grounded in educational anthropology, this chapter critically analyzes how ICT transforms pedagogical methods, teacher-learner relationships, and knowledge dissemination in diverse cultural settings. Emphasis is placed on ethnographic case studies and theoretical frameworks that elucidate the socio-technical dynamics underpinning digital learning. The chapter aims to contribute to a more culturally responsive approach to ICT implementation in education, offering insights for policymakers, educators, and technologists seeking to bridge the gap between innovation and indigenous knowledge systems.

DOI: 10.4018/979-8-3373-1190-6.ch002

1. INTRODUCTION

Anthropology is a scientific field that, in terms of understanding, offers some possibly useful key concepts and insights. Anthropology traditionally studies and examines the interactions of different cultures. From the anthropological view, information and communication technology already appears as a culture of its own. As an anthropological culture, ICT is worth further examination concerning what happens when its elements are implemented in the educational culture (Tariq et al., 2024). Furthermore, from the anthropological viewpoint, ICT is fundamentally regarded as a tool for creating, sustaining, and, if necessary, changing the social bonds between individuals in society. Given these insights, it might prove necessary to investigate whether and how anthropological studies help to clearly observe and understand more or less hidden and/or unthought-of presuppositions and attitudes held by teachers and students that drive the implementation of and/or the attitudes towards ICT in educational processes in use today (Kasturi et al., 2024).

Whatever is considered the possible evidence of the decreasing interest in anthropological productions, the importance of being an able observer and participant in the use of educational technology should not, all the same, be undervalued. After all, if theories help to lead to understanding, the use of educational technology has to bridge the association of anthropological knowledge and practical usage in the field of educational technology (Chen & Zhang, 2024). The beginning of educational technology that realizes a mesh between anthropological theory and practice is what these chapter hopes for. Thus, arguments about anthropological concepts and why they are so relevant for computer-mediated learning should not merely be read, reflected upon, and forgotten. They should remain in the minds of educators and educational management staff as accountants of the continuously evolving and sensitive use of computer technology in higher education (Wojewska et al., 2024).

It is appropriate and necessary to analyze educational systems in their particular cultural context. In order to gain insight into the role of ICT in the learning and teaching process and values and knowledge construction, the text makes use of anthropology. Anthropology is a generalizing field of study that tries to find the specific qualities of humans within our cultural framework. The focus on cultural aspects enables anthropology to provide us with an important explanatory tool that helps to understand the complexity of the phenomena related to education (Liu & Liang, 2024). Educational anthropology brings anthropology to the field of education and seeks to understand how people use educational culture, teaching, or text to maintain and interpret information about their cultural differences and similarities. Not only educational practices but also the relationships that shape and maintain them must be described. Most educational anthropological research is based on school experiences and practices, but learning takes place in many different places and environments

(Kaewsaeng-on & Mehmood, 2024). For this reason, we will make use of research on processing ICT work situated in the field of technical and vocational education, which is not present to a great extent in educational anthropology and which will provide a new perspective to both the professional public and policymakers in the area of vocational and technical education.

1.1. Definition and Scope of Anthropology in Education

Anthropology is the study of the complex systems of meaning that are contained within and managed by people in today's world. These cultural systems contain the key to understanding the thinking patterns, actions, and relationships between people and represent important knowledge to draw on in understanding the function of educational settings. From an anthropological perspective, a range of interconnected critiques of both theory and practice are at work. For example, anthropology offers an insight into one particular part of what constitutes contemporary educational diversity (Ma, 2024). The in-depth anthropological research and description of educational practices in some parts of societies have demonstrated that there is not now a generalized best way to do education. There is, instead, a range of alternative 'best' practices that have validity to those who work within them. In educational settings, anthropology is a theory of and method for designing learning processes that starts from the needs, interests, and desires of the people who are the grassroots of educational systems (Lalwani, 2024).

It is also a type of approach to research that seeks to articulate possible directions for a commonly agreed future in shared engagement between educators and their students. Anthropologists today can be researchers and educators in academic institutions, specialists in a range of global agencies, and have specific roles as policy advisors and members of government, identifying the need for social and cultural informants within ministries and departments (Wang et al., 2024). In many locations and spaces, a much more complex range of opportunities presents itself modeled on the exceptional and differential nature of the global city in which they may find themselves. Social anthropologists often work in the policy world, activated by what the discipline knows about people, and managing various institutional and infrastructural elements to influence policy aimed at improving the lives of populations across the world (Nohong et al., 2024). This capacity to extend and apply social insights is part of the tenacity of social anthropology in the face of today's intellectual and practical crises, whether in education or science.

1.2. Theoretical Frameworks in Anthropology and Education

University has become a dynamic and global model of people's educational needs. This chapter covers various possibilities for bridging anthropological and educational theoretical frameworks to address key questions on how ICT could be understood from an anthropological perspective. Some of the theoretical foundations that support both anthropology and education are explained first through some of the influential theories, such as cultural relativism and social constructivism (Guo et al., 2024). This is followed by an explanation of the anthropological comparative perspective as a framework for understanding cultures in educational practice norms. Then, we explain how both anthropological theories and comparative perspectives help us comprehend the complex interaction between culture, learning, and technology (Joshipura et al., 2024).

In the last two sections of the chapter, we delve more deeply into these theoretical frameworks to enhance our understanding of anthropology and education and the implications for how research on people's ICT practices has evolved in education. Theory conveys the basic elements, frameworks, or philosophical points of view underlying research or practice in education. Through analysis and interpretation, theory identifies and interprets the central concepts, structures, and assumptions in research, proposes a series of examinations to help guide a research study, and conceptualizes a framework for observations (Horn, 2024). Understanding the theory relieves us from simply reacting to the present and opens us to interpret the present from the task, the desires, and the prowess constructed in an entirely different time. In effect, entering an education program is to enter the conversation as a participant. Argument theory works as a guide to uncertainty; when a situation is vague, and the competence to act is unknowable, the theory will empower analysis in order to decide (Hamzah Nasution et al., 2024). In a mythical world where building-based education delivers the content we need, ICT represents the content at our fingertips.

1.3. Historical Evolution of ICT in Education

The use of technologies in education has experienced several stages of development over time; thus, it is feasible to talk about the educational use of these technologies – separated from the social, economic, or political implications – in different historical periods. In general terms, one can assert that the use of Information and Communication Technologies in education has been present since it was possible: first, it was radio, which allowed a massive delivery of education even at a distance; then television, the star medium in educational expansion, created national and international networks, which began to be operational in the 1960s; followed by high-tech in the form of cassettes and videotapes; then, computing, and binary and hexadecimal code for personal training on minicomputer systems, which appeared at the end of the 1970s, began to operate in the 1980s as a microcomputer and linked to more central supercomputer complexes (Xiao et al., 2024).

Computers represented an authentic technological revolution in the 1990s. In addition, they gave rise to a new paradigm: from Closed Distributed Education and training to massive Open training through the Internet. More recently, computers have enabled non-linear or hypermedia training, artificial intelligence, audiovisual touch-up training, or simulations on CD-ROMs and DVDs. Indeed, at this time, in many places, Information and Communication Technologies have become available to a greater or lesser extent in schools and educational institutions, apart from the power of therapy for the different types of pupils and other uses and applications (C. Zhang et al., 2024). Information and technologies, online interactivity, and the Internet could be added, representing a highly relevant educational phenomenon. This boom in the application of Information and Communication Technologies to education has caused reforms and counter-reforms of educational policies that have accompanied educational reforms and counter-reforms brought about by other factors and criteria of social evolution. These educational criteria have had economic, labor, vocational, democratic, and cultural policies behind them. Some of them have been linked to the "information society" or the "knowledge society." In this research project, we intend to explain an anthropological point of view (LUTSIV et al., 2024).

2. CULTURAL DIMENSIONS OF ICT INTEGRATION IN EDUCATION

In educational literature, a few theories suggest cultural factors are influential in the integration of ICT. One side of technological innovation lies in the existing cultural beliefs and practices, deeply rooted in culture and strongly determined by events and experiences. A wider perspective on the cultural phenomena finds ICT as fundamental in resignifying the concept of "local" and the idea of local culture (Bondarchuk et al., 2024). This integration also enables education and its stakeholders to integrate activities along local dimensions further. Used in education, ICT is not, and cannot be, neutral as it is based and built on cultural rules and values. Context is defined as the physical, social, and psychological background that frames any situation, idea, object, or event. Moreover, educational commemoration and its future may reveal a potential risk: the friction between fundamental principles of modern education and traditional educational paradigms that will not decrease as fast as the spread of ICT tools (Cancio-Méndez et al., 2024).

Important for the successful use of ICT in school is the linkage between cultural identity and individual identity. A number of authors analyze ICT initiatives addressing issues of social and cultural inclusion and exclusion in education. Communities of learning with particular cultural characteristics are included, but several pedagogical approaches are used to communicate from national levels towards regional and local learning communities. An extensive body of literature describes how personal identity is socially and culturally shaped and co-constructed, focusing on the importance of diversity in teacher-student relationships and educational settings, involving students and parents with different socio-economic, ethnic, and linguistic backgrounds (Cabellos et al., 2024). The importance of a culturally responsive pedagogic approach is exemplified as well. Educators are urged to adapt ICT practices and tools to the values, needs, and skills of diverse students, creating an "intercultural education" in which cultural and intercultural dimensions are integrated. A few authors outline the need to embed ICT initiatives within the cultural and social context in which they occur and to ensure that they "remain culturally manageable (Maryuni et al., 2024)." Thus, it is suggested that the impact of cultural aspects should be examined in the future development of effective, ICT-mediated learning approaches.

2.1. Key Theories in Anthropology and Education

Cultural Theory Different cultural constructs have been delineated worth consulting from the field of anthropology when discussing technology and education, proffering different lenses through which to view the respective analyses. Cultural theory holds that culture consists of meaning systems: the symbols, texts, and norms that underpin society. When applied to education, this theory allows us to understand how culture conceptualizes knowledge, how meaning is made in educational encounters, and the connection between pedagogical practice and cultural and social orders (Taxiarhis Vouglanis & Anna Maria Driga, 2024). Noting that individuals' meaning-making of schools will be mediated by their own experiences of school and their cultural expectations of these institutions, irregularity in educational outcomes and experiences at schools that are seemingly homogeneous can be attributed to these cultural variations. Social Constructivism Conversely, social constructivist theories also claim a cultural base to knowledge and meaning systems; however, these systems are constructed in and through social relationships. When utilized in educational spheres, social constructivism allows for an understanding of educational practices, including knowledge and learning being co-created within social environments (Budiarto et al., 2024). Social constructivism is particularly applicable to how ICTs can create environments that enhance learning, aligning with the potential to transcend the borders of traditional social interaction in educational contexts to create a quality learning experience.

Shifts in power should perpetuate a dynamic steer from the conventional approach to teaching, where the competent are taught about the common in standard settings, towards the novice further engaged in authentic activities typically available for the expert. This is at the heart of the pedagogical shift induced by the apprenticeship model or socio-cultural theory, both of which mesh neatly with social-cultural theories. In extrapolating new opportunities in education systems, the non-linearity of social development can now be catered for, since individuals are able to construct their understanding and learning (Ibrahim, 2024). Central to ideologies in anthropology is the idea that schools are 'cultural constructs' as there is deemed to be a sharp contrast of educational experience within a park of social and cultural boundaries. It is the larger cultural, economic, and social frame that shapes the experience and the outcome for why self-worth and racial/ethnic notions are distinct as teachers find reactions considerably diverse than expressed in that case on conversations among relevant fellow peers. In an applied sense, understanding the way in which cultural beliefs are operative in educational practices brings into focus the need for teachers to be reawakened, or to reawaken themselves, to the cultural contexts of our teaching in order to reposition teaching and student learning away from traditional ethnocentric notions and activities (Gómez-Devís & Herranz-Llácer, 2024). By aligning these cultural constructs with the interpretivist critical stance, a subtle way in which positioning students metaphysically can purport to skew to the end and to determine these students to be situated as ontological objects and seen as 'other' is also unveiled.

2.2. Societal Impact of ICT in Education

Both the utilization of ICT in enhancing educational practice and the outcomes of using ICT in the education system may be analyzed from sociocultural and societal angles. The provision of ICT infrastructure, digital resources, specialized personnel, and ICT didactic material in schools is usually regarded as a positive means of improving a society's educational system. For all children, but especially for those who are socially and educationally disadvantaged, the school's promotion of ICT and media competencies ensures access to and greater equity in and for the education system (Asqui et al., 2024). Investments in ICT promise not only individuals but entire societies competitive advantages in the local and global labor market. ICT is mainly regarded as an innovation capable of transforming the labor market by calling for high-performing, creative, knowledgeable workers. On a local level, schools receiving superior financial support may provide their students not only with cutting-edge digital facilities but also with personnel who have acquired specialized pedagogical and didactical knowledge, know-how, and competencies for effectively integrating ICT into teaching and learning activities (Erik Pezo, 2024). For these students, education may occur at a time, in a place, and at a pace that increases their interest and learning outcomes, gives them a sense of ownership of their learning materials, and increases their creativity. Along with these benefits, ICT may contribute to the intermittent pains of childhood, such as boredom, homesickness, rejection, bullying, or maltreatment. In a digital educational space, students may feel more grown-up by communicating or chatting with friends from all over the world. Digital messages, avatars, open-ended chats, forums, or digital postings facilitate self-disclosure while promoting new behaviors and patterns of thinking and feeling that can lead to new insights and self-understanding, a delicate prerequisite for effective learning (Reddy & Babu, 2024).

2.3. Ethical Considerations in the Use of ICT in Education

What are some of the ethical concerns we should take into consideration when discussing and implementing ICT in education? Who owns data about students and teachers, when and why? What consent should be requested, from whom, and what are the security risks? Furthermore, what are the educational outcomes of these new students' and teachers' digital footprints? In a society shaped by surveillance, ICTs in education have the potential to normalize surveillance, short-circuiting the abilities of individuals to assert their rights to freedom from interference (Rantyuk & Vakaliuk, 2024). The efficiency and convenience of new technologies can shortchange the capacity of educators to alter their practice for individuals and to pursue moral and political ends; technologies of surveillance can diminish an end's moral value. Will all pupils be natively positioned and, concomitantly, positioned to be more powerful surveillants (Alrikabi et al., 2024)? These power dynamics are certainly threats.

What differentiates surveillance in wider society from surveillance in educational technologies? Or perhaps it does? Any networked technology can be co-opted to surveil; the advantage of the educational context is a potentially normatively rich and transparent environment. Professional associations and organizations should publish guidelines and policies about digital non-refoulement and the data's right to be forgotten. Teachers have a responsibility to prevent and redress the impact of digital surveillance, information, or attention practices that deny, epistemically injure, profile, sort, or stigmatize learners. Because of the relationship between surveillance and normalization, the outcomes of digital gaze or 'thumbprint' should be used purely formatively. Our professional organizations must begin seriously scrutinizing why we have disregarded the diminution of privacy and openness and what to do about it once we open our eyes (Shafik, 2024d). Lastly, to help these students navigate systems of power and to gain some leverage or recourse when these systems malfunction or deviate from the common good, e-lit and digital citizenship courseware should be incorporated into the curriculum.

3. ANTHROPOLOGICAL STUDIES ON THE DIGITAL DIVIDE IN EDUCATION

Introduction. The fact of using a computer and the Internet yields privileges and power in many dimensions of life, including education. Hence, many emphasize the necessity of adequate research on the conditions of PCs and Internet use. Much of anthropological research on ICT focuses on the availability of access to technology. The main scientific problem in such studies is to understand which social tensions and relations are crucial for determining the so-called "digital divide." The digital divide is primarily about unequal opportunities for access to information, education, the job market, or health services. The root causes of the digital divide are not only technological problems but also social and cultural ones. Consequently, access to technology depends on social structure (Abubakari et al., 2024). The main social determinants that bar access to technology are the following: personal education, household income, and personal characteristics.

Socio-economic status is also critically important in another study on educational ICT, a research project conducted in rural and urban areas in India. In one of the case studies, access and opportunities to use computers and the Internet in schools were measured, specifically in Delhi and Bangalore. The schools were ranked according to their costs for parents, the number of computers in them, the number of computer teachers, how many days a week the computer lessons were given, and the number of hours each week. Daily travel distances and time needed to reach the schools, as well as the absence rates, were also measured (Shafik, 2024c). These revealed that in the most expensive schools in Delhi, there was an absence rate of about 30%, which dropped to 15% in the more expensive schools in Bangalore. Only in the outlying areas of Bangalore, where the schools were cheaper, did the percentage of children not attending regularly again rise to around 30%. This research is important for the European Union, as stated in action 47 of their E-learning Communication, which states that "Member States and the Commission have to finance targeted measures to bridge the digital divide in education and training" to "improve the opportunities for all for access and participation (Shafik, 2024a)."

3.1. Gender Perspectives on ICT in Education

The gender perspective builds on the multiple observations and analyses of the role of women and men in relation to information, communication technology, science and technology, and technical professions. Gender starts already in early childhood through learning and socialization. Therefore, there are some key questions to be asked: How do stereotypes and prejudices towards female and male thinking and acting form, and how can they change? How do girls and boys, women and men act in educational contexts (Shafik, 2024b)? What about educational materials, tests, and performance in school? How do women and men use ICT? What about educators, and what about public discourses and political considerations? Are 'girls' rooms phenomena of necessity, or are they symptoms of discrimination? What about the 'digital divide'? Generally speaking, women are profoundly underrepresented in the technical field, and the rate of women in tertiary ICT-related education has continuously decreased (Cancio-Méndez et al., 2024).

When women's voices are less often heard than men's, the reasons lie in the scarcity of female representation, which is constructed by gender dynamics. For the pursuits of ICT in education, the expectations of female students should play an important role, and ideas could be applied. So far, a focus has been placed on successful projects and innovations, illustrating the feminization of new technologies and the empowerment of women and girls. Prejudices and gender roles stereotype female and male learners and educators. These expectations, however, may give an advantage to the educator in facilitating the learning process and the professional teaching situation (Cabellos et al., 2024). Therefore, stereotypes may correlate with and perpetuate inequality. Consequently, we need to deal with how the various educational cultures linked to gender in different countries are inextricably related to the school system, the design of curricula and educational reforms, the teaching profession and its working conditions, as well as prejudices and gender roles in society and the design of ICT. Being visible does not constitute superiority or high numbers of women in leadership positions. At school, in vocational and mathematics education, girls and boys can be encouraged to make the best use of their skills, also in the field of technology (Maryuni et al., 2024). Success in mathematics, natural sciences, engineering, and design can be seen as a stated aim of all who conduct or support teaching based on an anthropological perspective. In effects of sexism and racism in educational software can be addressed rather than waiting for the government or other central instances to draw up a policy guideline. Programmatic approaches can be carried out at the theoretical level, in concept development, by those who are active in educational projects, educational administration, government policy, public interest group advocacy, and activity reporting in the media, with planners of justice and human rights education, and with others who are engaged in shaping multicultural educational scholarship (Pérez-Jorge et al., 2024; Umar et al., 2024). A social ethos definitely intervenes, however, in the potential success of these diversified approaches. When prioritizing educational design by educators working with target groups directly and determining their funding, they favor the potential staying power of girls and women in the system. A vital issue is how many social connections an educational practice can have. The pivotal locus is a disagreement in the sociology of cultural appraisals: To what extent, and how, do different cultures and social strata generate a variety of social scores about an issue, and in what way do either the sameness or the difference of these cultural scores shape the interrelated cultures, though it is likely that they also shape the social spheres that interconnect through them, as presented in Table 1 (Nasrullah Rahimi et al., 2024).

Table 1. Key Aspects of ICT in Education from an Anthropological Perspective

Aspect	Definition	Cultural Impact	Challenges	Example Initiatives	Reference
Digital Literacy	Ability to use ICT tools effectively	Expands knowledge access	Digital divide	National digital literacy programs	(Lubuva et al., 2024)
Technology and Social Learning	Learning through online collaboration	Encourages cross cultural learning	Loss of traditional interaction	MOOCs, virtual study groups	(Martínez-Pérez et al., 2024)
Digital Divide	The gap in access to ICT resources	Exacerbates inequality	Limited infrastructure	Low-cost mobile learning devices	(Kumar & Priyanka, 2024)
ICT and Indigenous Knowledge	Use of digital tools to preserve cultural heritage	Supports cultural sustainability	Risk of cultural homogenization	Online Indigenous archives	(Adekoya et al., 2024)
Human-Computer Interaction (HCI)	Relationship between humans and technology	Enhances learning efficiency	Usability issues for diverse users	Adaptive learning systems	(Eghrefuvwoma et al., 2024)
Online Learning Communities	Virtual spaces for education & knowledge sharing	Promotes global learning	Potential loss of local context	Social media in education	(Bondoc, 2024)
Socio-technical Systems	Interaction between ICT and social structures	Shapes educational institutions	Resistance to tech adoption	Smart classrooms	(Y. Zhang et al., 2024)
ICT & Traditional Education	Integration of digital tools in learning	Modernizes pedagogy	Tension with traditional methods	Blended learning programs	(Uleanya & Prinsloo, 2024)
AI & Personalized Learning	Adaptive learning based on AI algorithms	Supports individualized learning	Data privacy concerns	AI-driven tutoring platforms	(Bondarchuk et al., 2024)
Equity in Digital Education	Ensuring ICT access for all students	Reduces education gaps	Economic and policy constraints	Government ICT subsidies	(Taxiarhis Vouglanis & Anna Maria Driga, 2024)

continued on following page

Table 1. Continued

Aspect	Definition	Cultural Impact	Challenges	Example Initiatives	Reference
Cross-cultural E-learning	Online education adapted to different cultures	Encourages diversity	Language and content barriers	Multilingual online courses	(Othman et al., 2024; Uleanya & Prinsloo, 2024)
Virtual Reality in Education	Using VR for immersive learning experiences	Enhances experiential learning	High costs	VR-based historical reconstructions	(Ibrahim, 2024)

3.2. Inclusive Education and ICT: An Anthropological Approach

Interest in inclusive education has spread worldwide. Unprecedented developments in information and communication technologies within diverse social groups have raised expectations about their potential contribution to the international agenda and framework for action. There is a moral, technical, educational, and social convergence in developing education systems, where policies and practices build the capacity for an inclusive educational response to the diversity of individuals (Kurti et al., 2024). It is clear that including signals for the implementation of technology to diversify, individualize, modify, or adapt educational situations in such a way that disabilities do not prevent access to the curriculum is inherent in this evolution of ideas. Inclusive educational thinking is fundamentally geared towards addressing disability. Although inclusion and proposals for emerging technologies have adapted to educational diversity, especially in specific cultures and contexts, disability remains the most significant cultural addition to the world stage today (Othman et al., 2024).

It is no longer about choosing special education or inclusive education. The affinity between these movements lies in the identification of the needs and contexts of each group of children who access the education system. A defining priority of the identity of each pedagogy is also to ensure that each learner, within an evolutionary process, can gradually access the contextualization of knowledge in usable knowledge. By choosing technology as a means of enabling more efficient and comprehensive access to the curriculum, an anthropological perspective can be used to draw attention to certain technological resources as assistive devices for students with disabilities (Uleanya & Prinsloo, 2024). In this perspective, cultural and technological influence help readers understand how each culture imagines complementing human diversity through technology. The development of inclusive educational perspectives is one of the responses to teach the largest number of students who are intended to learn within the education system. This implies that the diverse developing nations hope for education systems that must move beyond the pressures of exclusion towards inclusive policies and best practices (Y. Zhang et al., 2024).

3.3. Methodological Approaches in the Anthropology of Education

Qualitative and quantitative approaches constitute the principal methodological distinction, each embracing a variety of sub-approaches. Qualitative research is often, but not exclusively, interpretative and inductive. It generates thick descriptions and is primarily interested in understanding and describing the experiences, meanings, and processes of social and individual life. Different approaches have developed to carry out such research, such as qualitative description, grounded theory, phenomenology, and ethnography. Of these, ethnography might be considered the most closely connected with anthropological research (Adekoya et al., 2024). Ethnography and participant observation have become increasingly widespread in educational research. The very complexity of educational environments suggests the need to use a variety of methods to capture the various aspects of what goes on, and even then, it may be difficult to discern the intricate connections among them. Indeed, some suggest that educational research should be rooted in multiple research traditions, including such methods as surveys and ethnography. Mixing methods in this way increases the possibility of triangulation, validation, and corroborating findings. This may be the key to enhancing our understanding of learning, especially for research on teaching and learning processes. The context, meaning, and value ascribed to particular educational practices often vary from one culture to another and may even be contested within a given social group. Culture, then, is not only relevant to what we research but also to how we research and to what we find (Nasrullah Rahimi et al., 2024). It seems that culture can have an effect on our own practice, acting as both a constraint and an opportunity for educational research.

3.3.1. Qualitative and Quantitative Research Methods

Anthropological research in education uses a range of different research methods, both qualitative and quantitative, to answer the questions 'What is education?'; 'How can it be understood from an analytical and critical perspective?'; and 'How can the findings be used to improve and reform it?' It uses qualitative research methods to explore individual cases (which can be institutions, groups, or individuals) in depth. Many of the methods normally associated with qualitative research (like participant observation, in-depth interviews, and unstructured questionnaires) are found in education studies, which are primarily situated in the ethnographic tradition (Alrikabi et al., 2024). The findings from this research are unique and allow readers to gain a rich and comprehensive insight into the experiences of those within the culture being studied. The fact that researchers use the individual case as their main unit of analysis means that they can draw out the nuanced and subtle characteristics that differentiate one context from another. This method can give valuable insights into the way education and learning are influenced by beliefs, attitudes, and values within particular frameworks of communities and institutions. Anthropological education research in qualitative methods is concerned with these unique contexts and the human phenomena that are tied to them (Asqui et al., 2024). Quantitative research methods provide another possibility for the systematic investigation of educational systems. These methods are designed to examine large numbers of subjects and offer statistical analyses that can demonstrate correlations and relationships between key variables that can be measured and are often tested to see if relationships are statistically significant. Some qualitative researchers are wary of the quantitative approach because of what they see as its 'reductionist tendencies and the unidimensional nature of the data.' Often, researchers use questionnaires to collect data using a combination of scales and closed questions that provide the numerical and categorical data used for the statistical analyses (Ibrahim, 2024). In order to provide a richer decoding of experiences and views, this is often triangulated with the use of open-ended questions.

4. TEACHER PROFESSIONAL DEVELOPMENT IN ICT: AN ANTHROPOLOGICAL LENS

There are two things that an anthropological perspective can bring to the study of teachers and technology. First, training in and the conduct of teaching are cultural practices. Marking who is and who is not a teacher, what counts as appropriate or inappropriate training for teachers, what makes for a legitimate or illegitimate student, and what gives one the right to evaluate the performance of a teacher — each of these questions is answered in different ways in different societies. Second, as a professional group, teachers are often caught between the values and expectations of two different communities — those of academic specialists and those of the practical specialists (Budiarto et al., 2024). Framed in this way, the development of a university program that studies language teachers and technology can provide useful insights into the sociocultural context in which the tools for professional development are used.

Given the need for ongoing professional development and the change process that teachers have to undergo in order to incorporate it into their practice, it becomes crucial to understand the roles of the cultural conceptions of technologies and the cultural work of tools. This becomes central for the design of ongoing professional development to support this change. With some teachers expressing anxieties over starting a teacher professional development program, universities need to produce available programs that reflect a specific student body's cultural relations to technology in terms of technologies and their educational uses (Gómez-Devís & Herranz-Llácer, 2024).

The stand-apart character of teacher education programs means a significant departure from the often copious coursework in the areas of technology integration and/or second language learning, but also other sociocultural influences that have a direct or indirect bearing on teacher practice. It affords preservice and in-service teachers an opportunity to immerse themselves in a personal exploratory journey to better inform their professional practice. Designing culturally responsive teacher professional development programs that go beyond the engineering domain of ICT tools and build an understanding of the cultural territory is essential. This is anchored in an anthropological frame of reference that recognizes that information technology will not impact the diverse educational and business digital terrains in the same way (Erik Pezo, 2024). Such a perspective consists of examining the tensions, contradictions, and absorptive states that educational, business, and home digital terrains and the relationships within constitute.

Lack of quality training around the use of ICT in conjunction with pedagogical applications produces academia informers who assist in the practical re-imaginings of professional development. For deep subject matter knowledge, produce grounding and context for the professoriat by inviting them into the cultural domain of struggle outlined in an anthropological study into the relationship between ICT and the societies they inhabit, including an avenue into another terrain. Work with respective funding institutions to increase the number and size of grants administered. It is time for systemic change that recognizes fundamental differences in values and attributes that have as their substrata different conceptions of what it genuinely means to be a teacher and student. From this orientation, think anew about the incongruent values within the digital terrain in systemic teacher learning and work towards a consonance between third-level professional education that affords existing teachers the same social, cultural, and informational capital as preservice teachers are trained with (Reddy & Babu, 2024). This means valuing research and professional identity, which are the two pillars of true teacher knowledge. Reclaiming these means moving towards education as research-based and research-informed.

4.1. Student-Centered Learning and ICT in Anthropological Context

In recent years, the integration of ICT into student-centered learning models has offered the promise to turn learning into a personalized experience for each student.

Such approaches rest on a belief that teachers can maximize learning outcomes for each student if they are aware of and tailor education to students' diverse needs and backgrounds. To integrate such approaches into practice, teachers need to learn from and about their students. ICT provides a range of tools and resources to mine, manage, and act on student data (Alrikabi et al., 2024). At the outset, it is suggested that such personalized approaches to education are context-sensitive. Educators must acknowledge that there may be a number of contradictory factors affecting students' school experiences and real-world interactions; some students may, for instance, not wish to connect their identity out of school with their school identity. With its focus on the local and contextual knowledge of students, such Culturally Relevant Pedagogy attends to the cultural, social, and language backgrounds of students (Pérez-Jorge et al., 2024).

In progressive education and today's global knowledge economy, we can identify a number of practical tools and paradigms that create a culturally appropriate and context-sensitive model for pedagogy. It is advocated that a 'learning by doing' or experiential education approach is beneficial. Additionally, it is suggested that students' experiences create significant shifts in how students navigate, organize, and remediate information. In their collaboration, students need to collect, organize, and share information, which demands them to read and write complex, multifaceted texts on a daily basis. Additionally, as a result of the web's ubiquitous and rapidly changing information, these students are engaged in hypothesis-free experiments that deepen their emergent knowledge of information tools and the metaphors that shape them (Lubuva et al., 2024). Following this line of thought, it is suggested that we have in the tools an opportunity to help classrooms build skills of digital-age literacy in the 3 Cs: collaboration, critical thinking, and creativity.

4.2. Cultural Diversity and Inclusion in Educational Technology

Cultural diversity and inclusion ensure that rich learning experiences are based on the learner's unique personality and the knowledge that exists in each culture. Educational technology cannot continue to reflect the homogeneity of the dominant elite society of the nineteenth and twentieth centuries and expect us to understand and meet the needs of all learners. To do so does not recognize the full diversity of individuals and learners who are unique from one another and express learning in a wide variety of ways. There is enormous individual diversity in learning characteristics that we must attend to by appealing to as many diverse learning styles as possible (Martínez-Pérez et al., 2024). Statistics indicate that about 50% of adults have a multimodal learning preference.

Therefore, any technology used in education must accommodate and reflect the diversity of our users, or else we block their ability to learn effectively through that technology. One of the greatest challenges we have today is the digital divide caused between teachers, students, and the community when the technology does not reflect or support diverse learning styles and experiences (Kumar & Priyanka, 2024). The modes and ways that people process and categorize the world are different, and those tests and measurements that are delivered only in written form must recognize that oral communication has often been the mode used in many cultures, which are culturally specific presentation formats. The global reach of technology challenges our abilities to be a globalized and multicultural society that promotes equity through the opportunities and services we can provide to our learners (Eghrefuvwoma et al., 2024). Hence, creating opportunities and strategies that incorporate lessons learned from the cognitive perspectives and pedagogy of diverse cultural communities would only strengthen the learning experience offered.

4.3. Case Studies and Best Practices in ICT Integration in Education

Introduction and Purpose this section aims to showcase various successful case studies and best practices of information and communication technology (ICT) integration in different educational contexts. The case studies will showcase successful initiatives that have been implemented by stakeholders in various parts of the world with the objective of enhancing students' skills and providing them with the tools to adapt to the society in which they live. Importantly, the case studies should address the local challenges, educational contexts, and unique institutional, cultural, and socio-economic influences that have led to the success of the initiatives (Y. Zhang et al., 2024). Each case study presents a situation in which ICT is used to meet the specific needs of a local community, to aid in the creation of a local economy, or to improve living conditions in general. Each best practice case study focuses on the efforts to improve the education of students within the framework of a specific university by incorporating the use of ICT. Methods for Case Studies The case studies in this work are presented consistently. Consistent with their local focus, each case study is written from an external perspective, describing a project that was developed and implemented by different actors (Budiarto et al., 2024). The case study presents the motivation for the project, local factors, challenges faced, attempts to overcome the challenges, and lessons learned. Each case is presented in a way that the reader will have a clear picture of the background that drove the project, the process, software, and technology used in the project, possible partnerships, resources available to support the project, potential barriers, and pragmatic considerations, and finally, an analysis of the successes of the projects and lessons learned, with recommendations for future ICT-related developments in similar areas (Nasrullah Rahimi et al., 2024).

4.4. Research Methodologies in Anthropological Studies of ICT in Education

Both qualitative and quantitative methodologies have been used to assess the influence of technology on learning. Mixed-method approaches that combine qualitative and quantitative research are used in order to provide a more complete evaluation of a subject. This methodological chapter frames the discussion from an anthropological perspective and shall therefore pay close attention to the who, what, where, when, and why of conducting studies on Information and Communication Technologies and education. In curriculum studies in particular, an increased awareness has emerged in recent years of the importance of background contexts within which technology is implemented (Lubuva et al., 2024). As determined by society and shaped by human negotiations, the education-society intersection is immensely complex; ideas, values, and attitudes mixed in myriad patterns are supplemented by sociological, economic, pedagogical, and, of course, technological aspects.

Research into technology and education generally falls into two kinds: first, studies on how and whether technology "works" in education, or studies on how education "works" in technology. This methodological chapter helps the reader gain a clearer glimpse of the research we report elsewhere. Setting out broader agendas and establishing the agendas of ordinary, non-academic persons, topics, discussions, and issues that were not at the focal point of the research are already surfacing. In ordinary people's or practitioners' views, topics and issues reflect aspects of educational technology that academic researchers neglect (Kumar & Priyanka, 2024). Research on education and technology has been conducted in several educational environments, most of them characterized by cultural and language diversity. Posing questions to young trainee teachers in a British university with no foreign students yields useful data but does not provide insight into the complex issues of researchers focused on wider questions of ICT in education across many cultures and educational systems. Only by combining case studies and other approaches could tackling the repertoire of these different aspects and their related sub-questions be possible. Additionally, using top-end research in non-Western countries may reveal aspects that would otherwise be difficult to discern through low-budget research such as small-scale studies in schools, publishers, or Internet-based networks (Bondoc, 2024; Othman et al., 2024). As far as possible, findings were verified in an everyday context outside the research and other impacts, as presented in Table 2.

Table 2. Impact of ICT on Education from an Anthropological Lens

Educational Factor	Traditional Learning Model	ICT-Integrated Learning Model	Benefits	Challenges	Reference
Knowledge Transmission	Oral, book-based learning	Digital, multimedia learning	Diverse learning resources	Digital dependency	(Kurti et al., 2024)
Cultural Knowledge Preservation	Passed down through generations	Digital documentation	Wider cultural accessibility	Risk of cultural dilution	(Alrikabi et al., 2024)
Teacher-Student Interaction	Face-to-face classroom setting	Online and hybrid learning	Greater flexibility	Reduced personal interaction	(Budiarto et al., 2024)
Educational Accessibility	Limited by location and resources	Global access via ICT	Expands education outreach	Digital divide	(Cancio-Méndez et al., 2024)
Learning Adaptability	Fixed curriculum	Personalized, AI-driven learning	Individualized education	Data privacy concerns	(Taxiarhis Vouglanis & Anna Maria Driga, 2024)
Role of Community	Local community-based knowledge sharing	Online learning communities	Peer learning opportunities	Loss of local educational influence	(Ibrahim, 2024)
Skill Development	Emphasis on memorization	Interactive, skill-based learning	Hands-on tech training	Dependence on technology	(Reddy & Babu, 2024)
Pedagogical Approaches	Lecture-based methods	Gamified, interactive learning	Increased engagement	Implementation barriers	(Rantyuk & Vakaliuk, 2024)
Multilingual Education	Language barriers in traditional texts	AI-driven translation tools	Improves accessibility	Accuracy of translations	(Pérez-Jorge et al., 2024)
Assessment Methods	Paper-based exams	Digital assessments	Real-time feedback	Cheating and academic integrity	(Nasrullah Rahimi et al., 2024)
Economic Disparities	Costly physical resources	Cost-effective e-learning	Reduces material costs	Initial tech infrastructure investment	(Martínez-Pérez et al., 2024)
Institutional Adoption	Slow adaptation to new teaching methods	ICT-driven educational reforms	Accelerates innovation	Faculty training require	(Kumar & Priyanka, 2024)

5. FUTURE TRENDS AND INNOVATIONS IN ICT IN EDUCATION FROM AN ANTHROPOLOGICAL PERSPECTIVE

Technology continues to develop rapidly, and emerging technologies may have the potential to significantly influence formal and informal educational practices. Educational practices are measured from the perspective of teaching and learning in the formal and informal sense of acquiring knowledge (Y. Zhang et al., 2024). In teaching practices, the emergence of artificial intelligence, machine learning, communications, and mobile learning is increasingly resonating. These developments will continue to grow with the user base ever expanding in society in the future. From a practical learning perspective, this trend raises a question: where will students gain knowledge about education or technologies in communities with underdeveloped technologies (Uleanya & Prinsloo, 2024)?

From the perspective of the anthropology of education, the emergence of technology needs to be observed in relation to education within a certain culture. Education can still be considered to rarely involve formal sectors. The use of technology in technological communities still appears limited. This occurs because technological literacy in society is still low. However, in the context of formal education, access to ICT with a variety of options is already available. The role of teachers in developing knowledge and skills related to technology for their students is quite important at this time (Maryuni et al., 2024). Not only that, but discussions on the past have explored how future research could be conducted from the perspective of the anthropology of education. Technological innovation brings about the phenomenon of indifference and rejection from teachers in implementing them into the learning process. The role of teachers in educational practices reflects the important factor of mediators in educational technology implementation. Technological innovations are related to the knowledge and skills of teachers to use them to achieve learning objectives, supporting their effective performance and interpersonal communication in lively learning situations. It should be emphasized that this socio-cultural phenomenon aims to bring together the interests of researchers, developers, and users (including teachers and students) to use ICT as a functional, effective, constructive, and efficient tool (Ibrahim, 2024). How to anticipate the negative impact, minimize costs, and hold meetings and technical workshops on functional education in the mainstream of effective use of ICT in education is also a focus of research in the field where the anthropology of education should orient their skills.

5.1. Intertwining of Anthropology

New research venues might include, as mentioned above, an intertwining of anthropology, computer science, and educational technology, for a more in-depth look at the circulations of learning theories, technologies, and capitalism. Through investigating questions related to digital inequalities, diversity, power, and change, anthropology is likely to play an important and parallel role with global divides and global relations of power. More powerful innovations and large-scale research programs have been suggested, specifically a trilingual and triage age panel study based on a scheme for comparative analysis proposed to look at the acquisition of an ICT-related working culture and ethic that are age- and sex-related (Erik Pezo, 2024). Ongoing concerns that are likely to persist for most, if not all, of the next five years relate to the availability and accessibility of technologies, the ability of all to adapt to rapid changes in these technologies, and the ability of people in and out of the formal educational structure to use technology for their own benefit. Only 14 percent of people introducing new information and communication and learning technologies into the classrooms were professional educators, showing there is still much resistance from professional bricoleurs who want to do it their way (Alrikabi et al., 2024). Identify and work with other disciplinary positions within anthropology and outside it, set within its context in the older and more critically developed literature. The references associated with a domain of de-materialization indicate the point from which at least some of this domain begins to unravel. These will include forms of resistance and prohibition emerging from a normalized state of subjection. Given prior positivisms and the logical consequences of a particular epistemology, which sees ICT as almost unproblematically productive of market-oriented unities, I predict some of the most significant barriers to equity, and threats to distinctive ontologies and epistemologies, will indeed come from within the new generation of so-called liberatory tools and resources. Hence society, prior to culture, the pedagogical also poses as a focus for our inquiry. Ongoing critical reflection suggests areas for further research. Consider cultural strategies that might emerge as ICT becomes more commonplace across the social fabric (Lubuva et al., 2024). By fairly universalist, I don't mean in any way to imply a largely shared set of understandings or techno-rhetoric about them, rather that many users across the UK have been dramatically politicized in ways that demand a deeper understanding, determination, and political commitment.

5.2. Privacy and Data Security Issues

Educational technology allows the acquisition of detailed data about the cognitive and affective variables of the students. Some of the data collected and processed by ICT includes the times at which the user is connected, the IP address, the queries they make on the Internet, and the content they access. Moreover, it can also provide data related to cognitive and affective issues, as in the case of e-learning services or applications that require the typing pattern for positive identification. Federating all this data, only part of which is directly identifying, can allow the building of a personality or user profile from digital behavior, especially when information is stored ad hoc for this case (Bondarchuk et al., 2024). In addition, the increased data to be managed in ICT makes the security management more complex in order to minimize the risk of unintentional leaks and protection against potential threats via cyberattacks. A data breach may carry serious consequences for the students affected by it. For example, data breaches and privacy violations involving educational technology affected at least 192 K-12 public and private schools across 30 states from 2016 to 2017, with about 718 school-record breaches. Moreover, educational institutions could face financial penalties, lawsuits, and public damage to trust. Last but not least, security breaches may have a high cultural cost (Cabellos et al., 2024). People need to believe educational institutions form a new generation of citizens and professionals capable of solving emerging problems today. A lack of trust in university student management may lead to a waste of human and public resources in the short term. The right to the protection of personal data is enshrined in the EU Charter of Fundamental Rights and is regulated by the General Data Protection Regulation. Some areas merit particular attention, like person-oriented game-based environments, because special category data can be accidentally inferred. Some games include malware or security vulnerabilities that may lead to a data breach. Participants are generally informed about the identity of the researchers and the aim of the research. Research involving identifiable category data is always subject to additional privacy considerations (Maryuni et al., 2024). At least the following informal rule must be applied: the more privacy-confidential the data, the stronger the protection and the higher the consent requirement. A culture of privacy can be established by addressing privacy and security issues in a systemic and collaborative manner. For instance, collaboration and discussion of privacy and security issues are important. At the very least, in addition to educating about good privacy practices and finding the right balance in the cost-benefit analysis of protections, user and institutional awareness of surveillance culture, the risks involved, and the motives behind potential attackers are important weapons to combat privacy attacks (Asqui et al., 2024).

5.3. Cross-Cultural Studies

The approach of comparative education stresses the need to study various educational systems in order to reveal universal principles, to develop theoretical approaches explaining the mechanisms of the interaction between culture and education, and to understand culture-specific mechanisms that deepen theoretical knowledge about the principles that are useful in developing a practical approach to the use of technologies in education in each specific case. In this light, the section presents cross-cultural studies that have been conducted across various countries and regions: Australia, Bangladesh, India, the US, Kazakhstan, Russia, Canada, Ethiopia, and Kyrgyzstan, as well as our study that includes sites in the three western regions and two eastern regions of each country respectively (Rantyuk & Vakaliuk, 2024). Cross-cultural research often observes symbolic transfer and the impact of ICT characteristics on instruction and educational outcomes. A study of integration and teaching practices in Kyrgyzstan found that local educational values and the cultural context of teaching often persevered, despite the policy and certain trends towards ICT integration. A study conducted in Ethiopia on the integration of radio technology into instructional practices was proposed as a way to overcome language barriers. Radio-based instruction aligned with authentic Ethiopian pedagogy and learning values, making information culturally compatible and also offering a bridge to technological advancements. The study demonstrated that technology can align with, and draw on, cultural models, values, and norms, and produce meaningful learning experiences (Umar et al., 2024). In each of these studies, the specifics of the cultural environment and local educational systems led to specific outcomes.

5.4. Emerging Trends

The past half century has seen a rapidly evolving educational landscape. As with other domains, it has been forcefully shaped by educational technologies. There are so many possible trends to consider that we cannot possibly discuss them here. We highlight here just a few with respect to their common impact on our understanding of how we approach conducting anthropological research. We observe the ongoing and emerging integration of artificial intelligence and machine learning in educational technologies. Online learning platforms are now firmly in place and growing; it is the rare university, library, or museum that is not actively involved with efforts to digitize and disseminate some (if not all) components of their collections and pedagogy online (Nasrullah Rahimi et al., 2024). We are starting to appreciate more fully the impact of social media; while this is today focused on the enormous wealth of shared knowledge and the democratization of access to scholarly and other institutions – we are also becoming increasingly aware of the differences digital affordances make in terms of digital literacy and the vast distances we did not even recognize ten years ago. We need to make sure that our future anthropologists are capable of engaging with these spaces and become fluent in the methods and ethical considerations that go along with such work. These new educational channels are reshaping our understandings of how members of existing and emergent communities actively engage and share always-on, big data gathered from wearable and smart technologies (Lubuva et al., 2024). It is important for us educational anthropologists to consider these new forms of engagement, as they represent powerful new sites of learning online. We are also seeing shifts in remote learning and the restraints of traditional face-to-face education and the concomitant demographic changes in who can (and does) have access to what forms, levels, and scope of formal and informal education. These shifts are anticipated to yield accessibility and inclusivity gains in the educational system although, as with every system, there remain potential barriers to entry. The impacts from the global pandemic across educational systems worldwide is a massive natural experiment that can only further inform these transformations (Eghrefuvwoma et al., 2024). Therefore, interacting with all these aspects is an emerging interest in participatory research methods and research that partners directly with the communities.

6. CONCLUSION

In an attempt to respond to the discourse of technology determinism prevalent in the field of ICT in education, this paper has attempted to provide anthropology of education perspectives to examine both the understanding of what it is to 'know' and to 'learn', in addition to suggesting the characteristics of successful technology integration. Throughout, we have argued that for any discussion of the role of ICT in education, it is essential to recognize that education is shaped by and is instrumental in the expression and formation of wider cultural and societal values. As such, the paper has included discussions of how to know and deal with ICT in education, the impact different lenses have on understanding the nature of education, and the problems of 'successful' technology-facilitated change. Throughout, it is clear that learning and education need to look at more than just issues of technological innovation and advantage but need to take into account the rich mosaic of academic, cultural, social, and ethical formations that intersect this central concern. It is clear, therefore, that for the majority of the world's students, successful learning in general, and advancements within existing power relations and disparities in investment, pedagogy, sociocultural capital, time, and resource access are crucial. Consequently, the focus of the e-strategies of the majority of countries in the 'peripheries' and the 'semi-peripheries' of the world should cease to be based on issues of integration per se, and be more squarely faced toward the investment in, for example, the development and supply of local language resources as part of efforts to increase the spread of meaningful, culturally responsive education technologies. Access to a successful educational model based on issues of cultural and equitable educational distribution and use can only be met from a perspective that sees culture as more than just 'differences in learning and knowing processes.' Rather, such a guarantee resides in the acknowledgment of education as cultural, and must then be seen and dealt with both in terms of the cultural values that it reinforces and the frameworks that dictate social activity within learning.

REFERENCES

Abubakari, M. S., Shafik, W., & Hidayatullah, A. F. (2024). Evaluating the Potential of Artificial Intelligence in Islamic Religious Education: A SWOT Analysis Overview. In *AI-Enhanced Teaching Methods* (pp. 216–239). IGI Global. DOI: 10.4018/979-8-3693-2728-9.ch010

Adekoya, C. O., Fasae, J. K., & Alade, A. V. (2024). Academic libraries, ICT use and sustainable higher education development. *Information Discovery and Delivery*, *52*(1), 1–10. Advance online publication. DOI: 10.1108/IDD-01-2022-0002

Alrikabi, H. T. H. S., Al-Dujaili, M. J., Majeed, B. H., & Alrubeei, I. R. N.Mohammed Jawad Al-DujailiBan Hassan MajeedIbtihal Razaq Niama ALRubeei. (2024). Information and Communication Technology and its Impact on Improving the Quality of Engineering Education Systems. *International Journal of Engineering Pedagogy*, *14*(1), 4–19. Advance online publication. DOI: 10.3991/ijep.v14i1.46943

Asqui, J. D. C., Quichimbo, D. M. D., Ajila, L. E. V., Molero, D. A. L., & Campos, M. E. V. (2024). Inclusive Education from Cultural Diversity and ICT. *Migration Letters : An International Journal of Migration Studies*, *21*(S2), 400–412. Advance online publication. DOI: 10.59670/ml.v20i8.6627

Bondarchuk, O. I., Balakhtar, V. V., Pavlenok, K. S., Kyrichuk, V. O., Yakubovska, N. O., & Grubi, T. V. (2024). Monitoring the psychological component of higher education quality using ICT tools. *CTE Workshop Proceedings,* 11. DOI: 10.55056/cte.685

Bondoc, R. S.Jr. (2024). ICT-Driven instructional and assessment strategies for physical education in the new normal. *Environment and Social Psychology*, *9*(4). Advance online publication. DOI: 10.54517/esp.v9i4.2155

Budiarto, M. K., Gunarhadi, G., & Rahman, A. (2024). Technology in education through mobile learning application (MLA) and its impact on learning outcomes: Literature review. *Journal of Education and Learning*, *18*(2), 413–420. Advance online publication. DOI: 10.11591/edulearn.v18i2.20976

Cabellos, B., Siddiq, F., & Scherer, R. (2024). The moderating role of school facilitating conditions and attitudes towards ICT on teachers' ICT use and emphasis on developing students' digital skills. *Computers in Human Behavior*, *150*, 107994. Advance online publication. DOI: 10.1016/j.chb.2023.107994

Cancio-Méndez, M., Rodríguez-Fernández, J. E., López-García, S., & Rico-Díaz, J. (2024). The influence of information and communication technologies on physical activity in primary education: A systematic review. *Educar*, *60*(1). Advance online publication. DOI: 10.5565/rev/educar.1854

Chen, J., & Zhang, J. (2024). The nexus between green finance and energy consumption in regional comprehensive economic partnership countries. *Environmental Science and Pollution Research International*, *31*(9), 14071–14087. Advance online publication. DOI: 10.1007/s11356-024-32003-y PMID: 38270766

Eghrefuvwoma, O. G., Agada, S. A., Ighodaro, I. H., & Olasupo, S. F. (2024). ICT in Higher Education: A Tool for Sustainable Development in West Africa. *International Journal of Research and Innovation in Applied Science*, *IX*(I), 237–244. Advance online publication. DOI: 10.51584/IJRIAS.2024.90122

Gómez-Devís, M. B., & Herranz-Llácer, C. V. (2024). New Technologies: ICT and Education in future Early Childhood Education teachers. Characterization from lexical availability. *Ogigia*, *35*. Advance online publication. DOI: 10.24197/ogigia.35.2024.7-30

Guo, M., Pang, D., & Dong, J. (2024). Promoting green growth: Resource management and reosurce efficiency in East Asian and Pacific ecnomies. *Resources Policy*, *90*, 104705. Advance online publication. DOI: 10.1016/j.resourpol.2024.104705

Hamzah Nasution, B., Siregar, M., Agustina, R., & Sitompu, U. (2024). *Green Banking Concept Implementation in Banking Credit Governance in Indonesia: Comparison Between Indonesia and China*. KnE Social Sciences., DOI: 10.18502/kss.v8i21.14761

Horn, M. (2024). The European green deal, retail investors and sustainable investments: A perspective article covering economic, behavioral, and regulatory insights. *Current Research in Environmental Sustainability*, *7*, 100241. Advance online publication. DOI: 10.1016/j.crsust.2024.100241

Ibrahim, U. (2024). Assessing the Impact of Mobile Applications on Student Engagement in ICT and Computer Science Education. [IJAER]. *International Journal of Applied Educational Research*, *2*(1), 51–62. Advance online publication. DOI: 10.59890/ijaer.v2i1.1457

Joshipura, M., Mathur, S., & Kedia, N. (2024). Sustainable investing and financing for sustainable development: A hybrid review. In *Sustainable Development*. DOI: 10.1002/sd.2912

Kaewsaeng-on, R., & Mehmood, S. (2024). Quantile modeling for environmental risk: SAARC's journey with green finance, policies, and regulations. *Journal of Cleaner Production*, *434*, 140234. Advance online publication. DOI: 10.1016/j.jclepro.2023.140234

Kasturi, D., Lakshmi, D. V. G., Naidu, D. M. P., & Sheela, P. (2024). Evolution of the Green Investment: Prospectus to the Companies. *Migration Letters : An International Journal of Migration Studies*, *21*(4), 581–590. Advance online publication. DOI: 10.59670/ml.v21i4.7534

Kumar, S., & Priyanka. (2024). The Effects of Information and Communication Technology (ICT) on Pedagogy and Student Learning Outcome in Higher Education. *EAI Endorsed Transactions on Scalable Information Systems, 11*(2). DOI: 10.4108/eetsis.4629

Kurti, E., Ferati, M., & Kalonaityte, V. (2024). Closing the gender gap in ICT higher education: Exploring women's motivations in pursuing ICT education. *Frontiers in Education*, *9*, 1352029. Advance online publication. DOI: 10.3389/feduc.2024.1352029

Lalwani, V. (2024). Incorporating green assets in equity portfolios. *Finance Research Letters*, *59*, 104815. Advance online publication. DOI: 10.1016/j.frl.2023.104815

Liu, G., & Liang, K. (2024). The role of technological innovation in enhancing resource sustainability to achieve green recovery. *Resources Policy*, *89*, 104659. Advance online publication. DOI: 10.1016/j.resourpol.2024.104659

Lubuva, E. E., Ndibalema, P., & Mbwambo, E. (2024). Designing ICT integrated lesson activities: An intervention to bolster tutors' pedagogical use of ICT competences in Tanzania teacher education. *Arab Gulf Journal of Scientific Research*, *42*(4), 1917–1936. Advance online publication. DOI: 10.1108/AGJSR-05-2023-0191

LUTSIV, B., MAYOROVA, T., & LUTSIV, P. (2024). "GREEN FINANCE" IN THE PARADIGM OF SUSTAINABLE BANKING DEVELOPMENT OF THE ECONOMY OF UKRAINE. *WORLD OF FINANCE, 3(76)*. DOI: 10.35774/sf2023.03.064

Ma, H. (2024). Role of green innovation and green finance on green recovery: Analysis of natural resources rents. *Resources Policy*, *89*, 104571. Advance online publication. DOI: 10.1016/j.resourpol.2023.104571

Martínez-Pérez, A., Lezcano-Barbero, F., Casado-Muñoz, R., & Zabaleta-González, R. (2024). ICT training in Spanish non-formal education: A revolution in the making. *European Journal of Social Work*, *27*(1), 177–191. Advance online publication. DOI: 10.1080/13691457.2022.2162006

Maryuni, Y., Ribawati, E., Prasadana, M. A. F., Nurhasanah, A., Wibowo, T. U. S. H., Nashar, N., Fauzan, R., Putra, A. P., & Sari, I. J. (2024). The Effect of STEM-based History Podcast Media on Literacy in Digital Era through ICT Courses for History Education Students. *International Journal of STEM Education for Sustainability*, *4*(1), 70–78. Advance online publication. DOI: 10.53889/ijses.v4i1.310

Nohong, M., Sobarsyah, M., Sohilauw, M. I., & Herman, B. (2024). Nexus Between Strategic Green Finance and Green Competitive Advantage: Study at Indonesia Corporate. *Journal of Environmental Assessment Policy and Management*, *26*(2), 2450003. Advance online publication. DOI: 10.1142/S1464333224500030

Othman, W., Makrakis, V., Kostoulas-Makrakis, N., Hamidon, Z., Keat, O. C., Abdullah, M. L., Shafie, N., & Mat, H. (2024). Predictors of Motivation and Barriers to ICT-Enabling Education for Sustainability. *Sustainability (Basel)*, *16*(2), 749. Advance online publication. DOI: 10.3390/su16020749

Pérez-Jorge, D., González-Herrera, A. I., Alonso-Rodríguez, I., & Rodríguez-Jiménez, M. del C. (2024). Challenges and Opportunities in Inclusive Education with ICT: Teachers' Perspectives in the Canary Islands during the COVID-19 Pandemic. *Education Sciences*, *14*(3), 283. Advance online publication. DOI: 10.3390/educsci14030283

Pezo, E. (2024). Adoption Of ICT By Basic Education Teachers: A Systematic Review. *Journal of Namibian Studies : History Politics Culture*, *40*, 314–331. Advance online publication. DOI: 10.59670/3tn2an17

Rahimi, N., Rashidi, A., Hakimi, M., & Shirpoor, M. (2024). Navigating E-Governance and ICT Dynamics in Afghanistan Higher Education. *International Journal of Scientific Multidisciplinary Research*, *3*(2), 137–152. Advance online publication. DOI: 10.55927/ijsmr.v3i2.8011

Rantyuk, I. I., & Vakaliuk, T. A. (2024). Selecting project management ICT tools for informal education of IT professionals. *Educational Dimension*, *10*, 167–197. Advance online publication. DOI: 10.55056/ed.630

Reddy, K. S., & Babu, N. S. (2024). Navigating the digital divide: Assessing faculty ICT skills and challenges in shaping the future of higher education in India. *Brazilian Journal of Development*, *10*(1), 355–378. Advance online publication. DOI: 10.34117/bjdv10n1-023

Shafik, W. (2024a). Enhancing AI-Enabled Education Through Mobile Learning and Bring Your Own Device (BYOD) Integration. In *Exploring Youth Studies in the Age of AI* (pp. 140–173). IGI Global.

Shafik, W. (2024b). Generative AI for Social Good and Sustainable Development. In *Generative AI: Current Trends and Applications* (pp. 185–217). Springer. DOI: 10.1007/978-981-97-8460-8_10

Shafik, W. (2024c). Mobile Learning and Bring Your Own Device (BYOD): Enhancing Education in the Digital Age. In *Integrating Cutting-Edge Technology Into the Classroom* (pp. 240–267). IGI Global.

Shafik, W. (2024d). Towards Trustworthy and Explainable AI Educational Systems. In *Explainable AI for Education: Recent Trends and Challenges* (pp. 17–41). Springer. DOI: 10.1007/978-3-031-72410-7_2

Tariq, G., Sun, H., & Ali, S. (2024). Environmental footprint impacts of green energies, green energy finance and green governance in G7 countries. *Carbon Footprints*, *3*(1). Advance online publication. DOI: 10.20517/cf.2023.48

Uleanya, C., & Prinsloo, P. (2024). Mapping the imaginary of Information Communication Technologies (ICTs) in education: The case of Botswana. *Learning, Media and Technology*. Advance online publication. DOI: 10.1080/17439884.2024.2306553

Umar, I., U, I., J. A, I., B, S., & I. M, M. (. (2024). Advancing Teachers' Professional Development: Exploring Models and Impact Assessment on Ict Integration in Colleges of Education. *International Journal of Integrative Research*, *2*(1). Advance online publication. DOI: 10.59890/ijir.v2i1.792

Vouglanis, T., & Driga, A. M. (2024). The use of ICT in the education of students with alexia/acquired dyslexia. *International Journal of Science and Research Archive*, *11*(1), 116–123. Advance online publication. DOI: 10.30574/ijsra.2024.11.1.0004

Wang, R., Destek, M. A., Weimei, C., Albahooth, B., & Khan, Z. (2024). Drivers of Sustainable Green Finance: Country's Level Risk and Trade Perspective for OECD Countries. *Journal of Environment & Development*, *33*(1), 125–148. Advance online publication. DOI: 10.1177/10704965231217046

Wojewska, A. N., Staritz, C., Tröster, B., & Leisenheimer, L. (2024). The criticality of lithium and the finance-sustainability nexus: Supply-demand perceptions, state policies, production networks, and financial actors. *The Extractive Industries and Society*, *17*, 101393. Advance online publication. DOI: 10.1016/j.exis.2023.101393

Xiao, Z., Duritan, M. J. M., & Jia, R. (2024). Resourceful futures: Integrating responsible mining and green education for sustainable development in developing and emerging economies. *Resources Policy*, *88*, 104377. Advance online publication. DOI: 10.1016/j.resourpol.2023.104377

Zhang, C., Zhang, L., Liu, L., & Du, C. (2024). The study of the relationship between green finance and resource efficiency in east asian economies. *Resources Policy*, *89*, 104658. Advance online publication. DOI: 10.1016/j.resourpol.2024.104658

Zhang, Y., Adams, D., & Lee, K. C. S. (2024). Assessing academic staff's ICT integration in higher education: Instrument development and validation. *Journal of Computers in Education*, *11*(1), 75–94. Advance online publication. DOI: 10.1007/s40692-022-00248-6

Chapter 3
Embracing Digital Transformation:
Impacts on Teaching and Learning Practices

Sam Ramaila
http://orcid.org/0000-0002-7351-477X
University of Johannesburg, South Africa

ABSTRACT

The rapid advancement of digital technologies has transformed teaching and learning. This chapter explores the impacts of digital transformation, focusing on technologies like AI, AR/VR, big data, and cloud computing, which reshape pedagogy, personalize learning, and enhance engagement. It also emphasizes the need for educators' professional development, digital literacy, and adaptive strategies. The chapter addresses digital equity, ensuring accessible education for all, and advocates for a strategic approach to technology, driving meaningful educational change.

1. INTRODUCTION

The rapid advancement of digital technologies has reshaped the educational landscape, fundamentally altering how teaching and learning take place. Digital transformation in education refers to the comprehensive integration of technology into pedagogical practices, institutional operations, and student engagement strategies to enhance learning experiences and outcomes (McCarthy et al., 2023). Unlike simple digitization, which involves converting traditional teaching materials into digital formats, digital transformation represents a fundamental shift in educational approaches, leveraging artificial intelligence (AI), big data, cloud computing,

DOI: 10.4018/979-8-3373-1190-6.ch003

virtual and augmented reality (VR/AR), and adaptive learning systems to create more personalized, interactive, and accessible learning environments (Mukul & Büyüközkan, 2023).

The significance of digital transformation in education extends beyond technological adoption; it is a response to the evolving demands of the 21st century, where digital literacy, critical thinking, and problem-solving skills are essential for success. The transition from traditional to technology-enhanced learning models has been accelerated by global events such as the COVID-19 pandemic, which underscored the need for resilient and flexible educational systems (Lemay, Bazelais & Doleck, 2021). As institutions worldwide embrace digital tools, educators must adapt their teaching methodologies to effectively integrate technology, ensuring that learning remains engaging, inclusive, and relevant.

This chapter explores the profound impacts of digital transformation on teaching and learning practices. It examines how emerging technologies are reshaping instructional strategies, assessment methods, and student engagement while addressing challenges such as digital equity, data privacy, and the need for continuous professional development. Additionally, the chapter provides insights into best practices for harnessing technology to enhance learning outcomes, emphasizing the importance of strategic implementation and policy development. By critically analyzing the opportunities and challenges of digital transformation, this chapter aims to equip educators, researchers, and policymakers with the knowledge needed to navigate and maximize the benefits of a rapidly evolving digital educational landscape.

2. BACKGROUND

In an era defined by rapid technological advancement, the education sector stands at a pivotal crossroads. The traditional classroom, once characterized by blackboards, printed textbooks, and face-to-face instruction, is undergoing a profound transformation. Digital technologies are no longer supplementary tools but essential drivers of teaching and learning, reshaping how knowledge is accessed, delivered, and assessed (Haleem et al., 2022). This shift is not merely about integrating technology; it is about reimagining education to align with the demands of a digital society. The digital revolution has already disrupted numerous industries, from healthcare to finance, and education is no exception. The emergence of artificial intelligence (AI), cloud computing, big data analytics, virtual and augmented reality (VR/AR), and adaptive learning platforms has unlocked new possibilities for personalized learning experiences (Ciarli et al., 2021). These innovations have redefined the roles of educators and students, fostering environments where learning is not confined to physical spaces but extends into virtual, interactive, and globally connected landscapes. With digital tools, students can now engage in immersive learning experiences, collaborate across borders, and access a wealth of knowledge at their fingertips.

The urgency to embrace digital transformation in education was further highlighted by the COVID-19 pandemic, which forced institutions worldwide to transition rapidly to online learning (Mulaudzi, 2024). While this shift exposed existing digital divides and infrastructure challenges, it also demonstrated the resilience and adaptability of the education sector. It underscored the need for robust digital strategies, scalable infrastructure, and the development of digital competencies among educators and learners. However, digital transformation is not without its challenges. Issues such as digital equity, data privacy, cybersecurity, and the ethical implications of AI in education must be carefully navigated (Raab, 2020). Additionally, the role of educators is evolving, requiring continuous professional development to effectively leverage technology while maintaining the human-centred aspects of teaching (Rahimi & Oh, 2024). The risk of technology replacing meaningful teacher-student interactions must be addressed through well-balanced digital integration strategies.

This chapter explores the far-reaching impacts of digital transformation on teaching and learning practices. It delves into how digital tools are reshaping pedagogy, assessment, and student engagement while critically examining the challenges that accompany this shift. By analyzing case studies, emerging trends, and best practices, the chapter provides insights into harnessing technology for meaningful educational reform. The goal is to highlight the transformative potential of digital technologies while ensuring that innovation in education remains equitable, inclusive, and pedagogically sound.

3. ISSUES, CONTROVERSIES, AND PROBLEMS

The integration of digital transformation into education has sparked a range of issues, controversies, and challenges that impact its effectiveness and equitable implementation. While digital tools and emerging technologies offer immense potential for enhancing teaching and learning, their adoption is not without complications. These challenges stem from technological, pedagogical, social, and ethical considerations that require critical reflection and strategic intervention.

a. Digital Divide and Equity Concerns

One of the most pressing issues in digital education is the digital divide, which refers to disparities in access to digital resources, infrastructure, and connectivity. While some schools and institutions have seamlessly incorporated advanced digital tools into their teaching and learning environments, others—especially in underprivileged and rural areas—struggle with limited internet access, outdated devices, and inadequate funding for technological upgrades (Mhlongo et al., 2023). This unequal access exacerbates educational disparities, preventing students from fully engaging in digital learning experiences. The controversy lies in whether digital transformation truly fosters inclusivity or further entrenches existing educational inequalities.

b. Resistance to Technological Change

Despite the push for digital transformation, resistance from educators, students, and institutions remains a significant challenge. Some educators feel unprepared to integrate digital tools effectively due to a lack of training or concerns about the pedagogical value of certain technologies (Munje & Jita, 2020). Students, particularly those accustomed to traditional learning methods, may struggle to adapt to self-directed and technology-driven learning models (Dahal & Bhat, 2024). Institutions may also resist change due to the high costs of digital adoption and uncertainty about long-term sustainability (Dibrov, 2015). The debate centres on whether digital education enhances learning or disrupts established educational practices without adequate support structures.

c. Pedagogical Effectiveness and Learning Outcomes

The effectiveness of digital transformation in improving learning outcomes is a topic of ongoing debate. While some studies suggest that blended learning and adaptive technologies enhance engagement and comprehension, others argue that over-reliance on digital tools can lead to superficial learning, decreased critical thinking skills, and reduced face-to-face interaction (Balalle, 2024; De Bruijn-Smolders & Prinsen, 2024). The challenge lies in ensuring that digital technologies are integrated meaningfully, rather than being used as mere replacements for traditional instructional methods.

d. Privacy, Security, and Ethical Concerns

The widespread adoption of digital tools raises critical questions about data privacy, cybersecurity, and ethical considerations. Learning management systems, online assessments, and educational applications collect vast amounts of student data, making them potential targets for cyberattacks and unauthorized data breaches (Akpen et al., 2024). Additionally, concerns about surveillance, data tracking, and the ethical use of artificial intelligence in education create controversies about student autonomy and digital rights (Power, Heavin & O'Connor, 2021). Institutions must navigate these ethical dilemmas while ensuring compliance with data protection laws and fostering responsible digital citizenship.

e. Technological Dependence and Digital Fatigue

As education becomes increasingly digitized, concerns about over-reliance on technology and the negative effects of prolonged screen time have emerged. Digital fatigue, characterized by cognitive overload, decreased attention spans, and mental exhaustion, poses challenges for both educators and students (Huang, 2023). While technology facilitates convenience and flexibility in learning, excessive dependence on digital tools may reduce hands-on, experiential learning and limit opportunities for social interaction (Haleem et al., 2022). Striking a balance between digital and traditional learning experiences remains a contentious issue.

f. Cost and Sustainability of Digital Transformation

Implementing digital transformation requires significant financial investment in infrastructure, software, teacher training, and ongoing maintenance. The high cost of technology adoption raises concerns about long-term sustainability, particularly in low-resource settings. Educational institutions face difficult decisions about prioritizing budget allocations while ensuring that digital transformation initiatives remain effective and adaptable to future technological advancements (Joseph, Onwuzulike & Shitu, 2024).

g. The Role of Artificial Intelligence and Automation in Education

The increasing use of artificial intelligence (AI) in education has sparked debates about its impact on teacher roles, student autonomy, and ethical decision-making. While AI-driven learning platforms can personalize education and automate administrative tasks, concerns arise regarding the potential dehumanization of learning, bias in AI algorithms, and the risk of replacing human educators with technology (Sayeed Salih et al., 2025). The challenge is to leverage AI ethically and effectively without compromising the fundamental human aspects of teaching and mentorship.

The issues, controversies, and problems surrounding digital transformation in education underscore the complexities of integrating technology into teaching and learning. While digital tools offer numerous benefits, they also introduce challenges related to accessibility, resistance to change, pedagogical effectiveness, privacy, and sustainability. Addressing these concerns requires thoughtful policy interventions, capacity-building efforts, and a commitment to equitable and ethical digital practices. As education continues to evolve in the digital age, stakeholders must engage in ongoing dialogue to ensure that digital transformation enhances, rather than hinders, meaningful learning experiences.

4. CHAPTER OBJECTIVES

The rapid evolution of digital technologies has significantly influenced the educational landscape, reshaping how teaching and learning occur in contemporary classrooms. This chapter seeks to provide a comprehensive exploration of digital transformation in education by examining its drivers, assessing its impact on pedagogical approaches, and identifying both opportunities and challenges associated with its implementation.

First, the chapter aims to examine the key drivers of digital transformation in education, including technological advancements, changing learner expectations, and institutional and policy frameworks. Understanding these factors provides a foundation for evaluating how digital integration is shaping modern education and the extent to which these drivers influence decision-making at institutional, national, and global levels.

Second, the chapter will analyze the impact of digital transformation on teaching methodologies and instructional design. Innovative approaches such as blended learning, flipped classrooms, and the integration of digital platforms have revolutionized the way educators facilitate learning. By assessing these shifts, this chapter will highlight how educators are adapting to new technologies to enhance student engagement and learning outcomes.

In addition, the chapter will explore how digital transformation enhances student learning experiences. The incorporation of personalized learning pathways, adaptive learning technologies, and virtual collaborative environments has provided students with more tailored and interactive educational opportunities. This objective will focus on the ways in which digital tools promote active learning, foster critical thinking, and improve accessibility for diverse learners.

Despite its benefits, digital transformation is not without its obstacles. Therefore, this chapter will identify challenges and barriers to the effective implementation of digital tools. Issues such as the digital divide, resistance to technological change, and concerns over privacy and security present significant hurdles to the widespread adoption of digital education. By highlighting these challenges, the chapter will provide insights into how educators and institutions can navigate and mitigate potential roadblocks.

Finally, this chapter seeks to propose strategies for leveraging digital transformation to create inclusive and equitable educational opportunities. Ensuring that all students, regardless of socioeconomic background or geographical location, have access to quality digital learning resources is paramount. The discussion will centre on practical solutions, including digital literacy initiatives, policy recommendations, and the use of emerging technologies to bridge gaps in access and participation.

By addressing these objectives, this chapter will contribute to a deeper understanding of the evolving digital landscape in education and provide a roadmap for stakeholders seeking to harness digital transformation for meaningful and sustainable educational advancements.

5. DRIVERS OF DIGITAL TRANSFORMATION

Digital transformation in education does not occur in isolation. It is driven by a confluence of factors that extend beyond mere technological advancements, encompassing shifts in learner needs, institutional priorities, and broader policy frameworks. Understanding these key drivers is crucial for harnessing the full potential of digital technologies in educational settings. This section explores the core drivers of digital transformation: technological advancements, changing learner needs and expectations, and policy frameworks and institutional priorities.

5.1 Technological Advancements

The rapid evolution of digital technologies has been a primary catalyst for the transformation of educational practices. Innovations such as cloud computing, artificial intelligence (AI), big data analytics, mobile learning applications, and immersive technologies like virtual and augmented reality (VR/AR) have created new avenues for teaching and learning (Bin Rashid & Karim Kausik, 2024). These technologies are not merely enhancing traditional methods; they are revolutionizing the way knowledge is disseminated and acquired. Cloud computing, for instance, allows students and educators to access learning resources and tools from anywhere, breaking down geographical barriers and fostering a more flexible and inclusive learning environment (Samyan & St Flour, 2021). AI-driven learning platforms are increasingly being used to personalize education, adapting to individual learners' needs and providing real-time feedback that can improve learning outcomes (Das, Malaviya & Singh, 2023). Big data analytics enables the collection and analysis of vast amounts of student performance data, helping educators identify trends and make informed decisions to improve curriculum design and instructional strategies (Bose, Dey & Bhattacharjee, 2022). Moreover, VR and AR are transforming the learning experience by creating immersive environments where students can explore complex concepts in ways that were previously unimaginable (Al-Ansi et al., 2023).

For example, learners can virtually dissect a human body, explore historical sites, or conduct scientific experiments, all from the comfort of their classrooms. These advancements not only enhance engagement but also promote deeper learning and higher-order thinking by allowing learners to interact with content in dynamic, hands-on ways.

5.2 Changing Learner Needs and Expectations

The evolving needs and expectations of today's learners are another significant driver of digital transformation. The current generation of students, often referred to as "digital natives," has grown up with technology and expects its seamless integration into all aspects of their lives, including their education. They are accustomed to on-demand access to information, social interactions through digital platforms, and personalized experiences in every area of life. As a result, the traditional, one-size-fits-all approach to teaching is no longer sufficient to meet the diverse learning needs of modern students (Yang et al., 2019). Learners today demand more flexible, accessible, and engaging educational experiences that reflect the realities of the digital world. Online learning, blended learning, and mobile learning offer students greater autonomy in how, when, and where they learn (De Bruijn-Smolders & Prinsen, 2024). These modalities allow learners to engage with course materials at their own pace, revisit content as needed, and collaborate with peers and instructors in new, dynamic ways. Additionally, the use of gamification, interactive simulations, and digital assessments has been shown to enhance motivation and engagement, which is vital in a time when attention spans are shorter, and distractions are abundant (Balalle, 2024). Moreover, learners are increasingly focused on developing skills that are aligned with the demands of the digital economy. This includes not only technical competencies, such as coding, data literacy, and digital collaboration, but also critical thinking, problem-solving, and adaptability. As a result, education systems are under pressure to provide learning opportunities that equip students with the competencies needed to thrive in an increasingly complex and interconnected world.

5.3 Policy Frameworks and Institutional Priorities

The role of policy frameworks and institutional priorities in driving digital transformation cannot be overstated. Government policies, national education strategies, and institutional goals shape the direction of educational reform and the adoption of digital technologies (Gabriel et al., 2022). Policymakers are increasingly recognizing the importance of digital literacy and the need to integrate technology into curricula to ensure that students are adequately prepared for the demands of the future workforce. In many countries, educational policy frameworks now emphasize the need to bridge the digital divide, ensuring that all learners, regardless of socio-economic background, have equal access to technology (Vindigni, 2024). Initiatives aimed at expanding broadband access, providing devices for underserved populations, and developing digital literacy programs are critical in making digital transformation an inclusive and equitable process. Additionally, policies that support teacher professional development in digital technologies are essential for equipping educators with the skills and knowledge needed to integrate technology into their teaching effectively.

At the institutional level, universities, colleges, and schools are increasingly prioritizing the adoption of digital tools and technologies as part of their strategic vision. Many institutions are aligning their goals with national and international trends in digital transformation, seeking to enhance their reputation, improve student outcomes, and remain competitive in a globalized education market (McCarthy et al., 2023). This has led to a greater focus on the development of digital campuses, the integration of e-learning platforms, and the incorporation of technology into administrative processes to streamline operations and improve efficiency. Furthermore, institutional priorities are evolving to place greater emphasis on learner-centred, flexible, and personalized learning. Educational institutions are beginning to recognize that digital transformation is not just about adopting new tools but about rethinking the entire approach to teaching and learning (Mukul & Büyüközkan, 2023). By embedding digital technologies into institutional strategies, schools and universities can create more responsive, agile educational environments that are better equipped to meet the demands of a rapidly changing world.

Technological advancements, shifting learner needs and expectations, and evolving policy frameworks and institutional priorities are all driving the digital transformation of education. These factors are interconnected, with each influencing the others in profound ways. The integration of digital technologies into education offers significant opportunities for enhancing teaching and learning practices, improving accessibility, and preparing students for success in the digital age (Haleem et al., 2022). However, for these advancements to be fully realized, educators, institutions, and policymakers must work together to overcome challenges related to equity, infrastructure, and professional development. By doing so, they can ensure that digital transformation is not just a technological shift but a comprehensive reimagining of education that benefits all learners.

6. IMPACT ON TEACHING PRACTICES

The integration of digital technologies into education has had a profound impact on teaching practices, leading to the development of innovative pedagogical approaches, enhanced teacher-student interactions, and a greater emphasis on professional development for educators. As technology continues to reshape the educational landscape, these changes are influencing how teachers design, deliver, and assess learning experiences. This section explores three critical areas where digital transformation is impacting teaching practices: innovative pedagogical approaches, enhanced teacher-student interactions through digital platforms, and the need for professional development and digital competencies for educators.

6.1 Innovative Pedagogical Approaches

Digital transformation has given rise to a range of innovative pedagogical approaches that challenge traditional models of teaching and learning. Among the most prominent are flipped classrooms and blended learning, both of which leverage technology to create more dynamic, flexible, and student-centred learning environments. In a flipped classroom, traditional teaching methods are inverted. Rather than delivering content during class time, instructors assign pre-recorded lectures, readings, or videos for students to engage with outside of class. This allows class time to be used for active learning activities such as group discussions, problem-solving exercises, and collaborative projects. The flipped classroom model capitalizes on digital tools to enable students to learn at their own pace, revisit content as needed and engage more deeply with the material during in-class activities (Liu, 2019). This approach fosters critical thinking, enhances student engagement, and encourages greater student responsibility for their own learning.

Blended learning combines the benefits of both online and face-to-face learning. It offers a flexible and adaptive learning environment that can be tailored to meet the diverse needs of students. In blended learning, traditional classroom instruction is complemented with online courses, digital resources, and virtual collaboration tools. This hybrid approach allows students to engage with learning materials in a variety of formats, whether through video lectures, interactive simulations, or digital assessments (Singh, Steele & Singh, 2021). Blended learning also enables personalized learning experiences by providing students with more control over their learning pace and the ability to access content on demand (Bhadri & Patil, 2022). This pedagogical approach not only enhances student learning outcomes but also allows educators to use data-driven insights to adjust their teaching methods and provide targeted support to students.

Both flipped classrooms and blended learning are examples of how technology can transform teaching by shifting the focus from teacher-centred instruction to student-centred learning. These approaches emphasize active participation, collaboration, and the development of higher-order cognitive skills, such as critical thinking, problem-solving, and creativity. By incorporating digital tools into the learning process, educators can create more engaging and meaningful learning experiences that cater to the diverse needs and preferences of today's students.

6.2 Enhanced Teacher-Student Interactions Through Digital Platforms

Digital platforms have significantly enhanced the ways in which teachers and students interact, fostering more dynamic, personalized, and continuous communication. In traditional classroom settings, interactions are often limited to face-to-face exchanges during scheduled class periods. However, digital platforms such as Learning Management Systems (LMS), discussion boards, video conferencing tools, and messaging apps have expanded the opportunities for interaction beyond the physical classroom. One of the most significant advantages of digital platforms is the ability to facilitate asynchronous communication, allowing students to reach out to instructors and peers at any time (Alshammari & Alhalafawy, 2023). This fosters a more flexible and accessible learning environment, where students can ask questions, clarify doubts, and participate in discussions at their own convenience. Platforms such as Moodle, Blackboard, and Google Classroom enable students to submit assignments, engage with course materials, and receive feedback from their teachers without the constraints of time and location (Josué et al., 2023). This flexibility not only supports different learning styles and schedules but also empowers students to take more ownership of their learning process.

Synchronous communication tools, such as video conferencing platforms like Zoom and Microsoft Teams, also play a critical role in enhancing teacher-student interactions. These tools enable real-time interactions, offering opportunities for live lectures, group discussions, and virtual office hours. Through these platforms, teachers can provide immediate feedback, facilitate collaborative projects, and create interactive learning experiences. The face-to-face element of video conferencing helps to preserve the human connection that is often lost in purely asynchronous online learning environments, fostering a sense of community and engagement among students (Gruber, Hargittai & Nguyen, 2022). Additionally, digital platforms enable more personalized and differentiated learning experiences (Josué et al., 2023). Teachers can use learning analytics and data from digital platforms to monitor student progress, identify areas where students may need additional support, and tailor their teaching to meet individual needs (Hase & Kuhl, 2024). For example, if a student is struggling with a particular concept, a teacher can provide additional resources, suggest extra practice exercises, or offer one-on-one support through virtual meetings. This data-driven approach not only enhances student learning but also strengthens the teacher-student relationship by providing timely, targeted interventions.

6.3 Professional Development and Digital Competencies for Educators

As digital technologies continue to evolve, there is an increasing demand for educators to develop digital competencies and integrate technology into their teaching practices effectively. The rapid pace of technological change means that professional development is no longer a one-time event but an ongoing process of learning and adaptation. Educators must continuously update their skills to stay abreast of new tools, platforms, and pedagogical strategies that can enhance teaching and learning. Professional development programs focused on digital literacy and pedagogical technology are essential for supporting educators in this transformation (Anis, 2024). These programs offer teachers the opportunity to learn how to use digital tools effectively, create engaging online content, design interactive learning activities, and incorporate new pedagogical approaches such as flipped classrooms and blended learning. In addition to enhancing technical skills, professional development initiatives also provide teachers with the pedagogical knowledge needed to make informed decisions about when and how to use technology in the classroom (Sancar, Atal & Deryakulu, 2021). Effective professional development helps educators understand the theoretical underpinnings of digital learning and equips them with practical strategies for applying technology in ways that align with best teaching practices (Postholm, 2012).

Moreover, fostering a culture of digital collaboration among educators is essential for sharing knowledge and experiences. Online communities, virtual workshops, and professional learning networks provide platforms for teachers to exchange ideas, discuss challenges, and collaborate on the design of digital learning experiences (Trust, Krutka & Carpenter, 2016). These collaborative environments help educators stay motivated and engaged in their professional development while also ensuring that digital transformation is approached in a collective, community-driven way. The development of digital competencies among educators is critical for the success of digital transformation in education. Teachers who are comfortable with technology and understand its pedagogical potential are more likely to create engaging, innovative learning experiences that foster student success (Sun & Yan, 2025). In contrast, teachers who lack digital skills or are resistant to technological change may struggle to adapt to new teaching methods, hindering the overall effectiveness of digital transformation efforts (Wagner et al., 2024). Therefore, investing in professional development for educators is a fundamental step in ensuring that digital transformation leads to meaningful improvements in teaching and learning practices.

Digital transformation is reshaping teaching practices by introducing innovative pedagogical approaches, enhancing teacher-student interactions, and emphasizing the need for ongoing professional development. Pedagogical models such as flipped classrooms and blended learning, supported by digital tools, are creating more dynamic and personalized learning environments (Schmid et al., 2023). Digital platforms are also facilitating more frequent and flexible communication between teachers and students, enabling real-time feedback, collaboration, and personalized learning experiences (Josué et al., 2023). However, for educators to successfully embrace these changes, they must continually develop their digital competencies through targeted professional development programs. Ultimately, the impact of digital transformation on teaching practices is far-reaching, fostering a more engaging, flexible, and responsive educational experience for both teachers and students.

7. IMPACT ON LEARNING EXPERIENCES

The digital transformation in education has dramatically reshaped the learning experiences of students, making education more personalized, collaborative, and focused on developing higher-order cognitive skills. By leveraging advanced technologies, educators now have the tools to create dynamic, flexible, and engaging learning environments that cater to the individual needs of students. This section explores three significant ways digital transformation has impacted learning experiences: personalized and adaptive learning pathways, collaborative learning through virtual environments, and the enhancement of critical thinking and problem-solving skills.

7.1 Personalized and Adaptive Learning Pathways

One of the most significant contributions of digital transformation to learning experiences is the ability to personalize education. Personalized learning pathways allow students to learn at their own pace, on their own terms, and according to their individual strengths, weaknesses, and interests (Bernacki, Greene & Lobczowski, 2021). With digital tools, educators can tailor learning experiences to meet the diverse needs of learners, fostering greater engagement and improving educational outcomes (Josué et al., 2023). Adaptive learning technologies play a central role in personalized learning by using data and algorithms to adjust the pace, content, and difficulty of learning materials based on the learner's performance (Kabudi, Pappas & Olsen, 2021). Platforms such as Khan Academy, Duolingo, and various Learning Management Systems (LMS) offer real-time feedback and adjust content dynamically, ensuring that students are constantly challenged without being overwhelmed (Turnbull, Chugh & Luck, 2020). These platforms track students' progress, identify areas where they may need additional support, and provide tailored resources to help them overcome challenges.

For example, if a student is struggling with a particular concept, an adaptive learning system may offer supplementary lessons, quizzes, or practice exercises that focus on that specific area of difficulty. Conversely, if a student demonstrates mastery over a topic, the system can provide more advanced materials, ensuring that the student is continually challenged and engaged. This personalized approach not only helps students progress at their own pace but also promotes greater autonomy in their learning journey. Moreover, personalized learning pathways foster a sense of ownership and responsibility for learning (Bayly-Castaneda, Ramirez-Montoya & Morita-Alexander, 2024). Students are empowered to make choices about the content, pace, and format of their learning, which increases motivation and engagement. The flexibility offered by digital tools also accommodates a wide range of learning styles and preferences, allowing students to learn through videos, interactive simulations, readings, and other multimedia resources. This individualized approach ensures that each student's unique needs are met, promoting deeper learning and greater success.

7.2 Collaborative Learning Through Virtual Environments

Digital transformation has also significantly enhanced the opportunities for collaborative learning. Traditional classroom settings, while fostering some degree of collaboration, often limit students' ability to work together outside of class time or in groups with diverse perspectives (Le, Janssen & Wubbels, 2017). Digital platforms, however, have created virtual environments that enable collaboration across time, space, and geographical boundaries (Lane et al., 2024). Learning management systems (LMS) such as Moodle, Canvas, and Google Classroom offer tools for group discussions, file sharing, and collaborative project work, allowing students to interact with one another in real-time or asynchronously (Al-Sharhan et al., 2020). These platforms encourage students to engage in group projects, participate in online discussions, and collaborate on problem-solving tasks, all of which are integral to developing teamwork and communication skills.

In addition to traditional tools, more sophisticated digital platforms such as Microsoft Teams, Zoom, and Slack have introduced virtual classrooms, video meetings, and instant messaging capabilities that facilitate real-time collaboration (Lane et al., 2024). These platforms enable students to work together on group assignments, participate in live discussions, and receive immediate feedback from peers and instructors. Video conferencing tools allow for face-to-face interactions in virtual environments, creating a more personal and dynamic space for collaboration (Morrison-Smith & Ruiz, 2020). This real-time engagement fosters a sense of community among students, even in fully online or hybrid learning environments.

Collaborative learning through digital platforms also extends beyond the classroom, enabling students to connect with peers from different cultural and geographical backgrounds (Lin & Wang, 2024). This global aspect of virtual collaboration exposes students to diverse perspectives, enriching their learning experiences and preparing them for the interconnected, multicultural world they will encounter in the workforce. Furthermore, students can engage with a wider range of resources and expertise, accessing digital libraries, research databases, and online communities of practice to enhance their learning and problem-solving abilities.

7.3 Enhancing Critical Thinking and Problem-Solving Skills

Critical thinking and problem-solving skills are essential for success in the 21st century, and digital transformation in education plays a pivotal role in fostering these skills. Technology provides students with opportunities to engage in complex, real-world problems and simulations that challenge their thinking, encourage exploration, and foster the development of higher-order cognitive skills (Lu & Xie, 2024). Interactive tools, simulations, and virtual labs have become integral in teaching critical thinking and problem-solving. For example, in science education, digital platforms like PhET Interactive Simulations allow students to experiment with scientific concepts in a virtual environment, testing hypotheses, observing outcomes, and refining their understanding of the subject matter (Diab et al., 2024). These tools provide immediate feedback, encouraging students to think critically about the problems they encounter and the strategies they use to solve them.

Similarly, online problem-solving platforms such as coding exercises, design challenges, and math puzzles engage students in complex tasks that require them to apply knowledge, think strategically, and explore multiple solutions (Xu, Wang & Wang, 2023). These tasks encourage students to break down problems into manageable parts, evaluate different approaches, and work through obstacles to find the best solution. This process helps students develop the skills necessary for tackling challenges in various fields, including science, technology, engineering, and mathematics (STEM). Furthermore, digital tools enable the use of inquiry-based learning strategies, where students pose questions, conduct research, and explore issues on their own or in collaboration with others (Hinostroza, Armstrong-Gallegos & Villafaena, 2024). By engaging in inquiry-based activities, students learn how to evaluate evidence, analyze information, and draw conclusions based on sound reasoning. These activities are designed to develop critical thinking by prompting students to question assumptions, explore alternative perspectives, and justify their conclusions with evidence.

The integration of artificial intelligence (AI) and machine learning tools in education is also enhancing critical thinking and problem-solving abilities. AI-powered platforms can simulate real-world scenarios, providing students with interactive experiences that require them to think critically and make decisions in complex situations (Song, Wu & Ding, 2024). These platforms offer a safe space for students to experiment, fail, and learn from their mistakes, which is crucial for developing problem-solving skills. Moreover, digital platforms foster a more active and participatory approach to learning, encouraging students to engage with the content in meaningful ways (Balalle, 2024). By working on digital projects, solving problems collaboratively, and using technology to conduct research and analysis, students develop the cognitive skills necessary for success in an increasingly digital and knowledge-based society.

Digital transformation is reshaping the learning experiences of students by creating personalized, adaptive learning pathways, fostering collaborative learning environments, and enhancing critical thinking and problem-solving skills. Technologies such as adaptive learning platforms, virtual collaboration tools, and interactive simulations allow students to engage with content in more meaningful, flexible, and student-centred ways (Josué et al., 2023). These changes not only improve the learning experience but also prepare students for the demands of the modern workforce, where critical thinking, collaboration, and technological proficiency are essential. As digital tools continue to evolve, they will further revolutionize the learning process, providing students with even greater opportunities for personalized and transformative educational experiences.

8. CHALLENGES AND BARRIERS

While the digital transformation in education offers immense potential to enhance teaching and learning practices, it also introduces a range of challenges and barriers that can hinder the effective implementation of technology in the classroom. These challenges often arise from systemic issues such as disparities in access to digital resources, resistance to adopting new technologies, and concerns surrounding privacy, security, and ethical implications. This section examines these key obstacles in more detail and considers how they affect the integration of digital tools in education.

8.1 Digital Divide and Access Disparities

One of the most significant barriers to digital transformation in education is the digital divide, which refers to the gap between individuals or communities with access to technology and those without (Afzal et al., 2023). This divide is not only geographical but also socioeconomic, as students from lower-income families or rural areas often lack access to the necessary hardware, internet connectivity, and digital resources required for online learning. The COVID-19 pandemic exacerbated this issue, with many students worldwide facing difficulties in accessing online education due to insufficient technological infrastructure at home (Tan et al., 2025). The digital divide can result in unequal educational opportunities, where students who lack access to technology are unable to fully participate in digital learning experiences (van de Werfhorst, Kessenich & Geven, 2022). This creates a disparity in learning outcomes, as students with limited access may miss out on important resources such as interactive learning platforms, video lectures, and online assessments. Moreover, even when students have access to devices, they may still face challenges related to the quality of internet connections, which can affect their ability to engage in real-time lessons or access cloud-based resources experiences (van de Werfhorst, Kessenich & Geven, 2022). In response to this challenge, institutions and governments are working to bridge the digital divide by providing low-cost devices, subsidized internet access, and funding for infrastructure development (Mulaudzi, 2024). However, the solution to this issue requires significant investment in technology, as well as policies that prioritize equity in educational access. Without addressing these access disparities, the potential benefits of digital transformation cannot be fully realized, and the gap between privileged and disadvantaged students will continue to widen.

8.2 Resistance to Change and Technological Adoption

Another major challenge in the digital transformation of education is resistance to change, particularly from educators and institutions. While many teachers are enthusiastic about incorporating digital tools into their classrooms, others may resist due to a variety of reasons (Boonmoh, Jumpakate & Karpklon, 2021). These can include lack of familiarity with technology, concerns about the effectiveness of digital tools, and fear of replacing traditional teaching methods that have been effective for years. For some educators, the transition to digital learning may seem overwhelming, particularly if they have limited experience with technology (Sarva & Purina-Bieza, 2023). There is a learning curve associated with mastering new tools and platforms, and many teachers may feel anxious about using technology in a way that could disrupt their established teaching practices (Josué et al., 2023). Additionally, teachers who have not received adequate training or professional development on how to use digital tools effectively may struggle to integrate them into their pedagogy (Ajani & Govender, 2023).

This resistance can also extend to students, particularly those who are less comfortable with technology or who prefer traditional face-to-face learning experiences. Some students may feel disconnected from digital learning environments, especially if they have limited experience with online tools or if they feel that technology detracts from their ability to engage in meaningful learning (Novak, McDaniel & Li, 2023). To overcome this resistance, it is essential to provide educators with the necessary professional development and training to enhance their digital competencies. Schools and universities must create supportive environments that encourage experimentation with new technologies, while also offering guidance and assistance. By fostering a culture of continuous learning and collaboration, institutions can help educators embrace the potential of digital tools to enrich teaching and learning. Additionally, students should be equipped with the skills and confidence needed to navigate digital learning environments, ensuring that they are active participants in their educational experiences.

8.3 Privacy, Security, and Ethical Considerations

As education becomes increasingly digital, concerns about privacy, security, and ethical issues have become more pronounced. The collection, storage, and sharing of student data raise significant privacy concerns, particularly as educational platforms and digital tools often require the submission of personal information, academic records, and behavioural data. With the increasing reliance on cloud-based platforms and third-party services, there is a heightened risk of data breaches, cyberattacks, and unauthorized access to sensitive information. Moreover, the ethical implications of using technology in education extend beyond data privacy and security. There are concerns about the potential for algorithmic bias in adaptive learning systems and the fairness of automated assessments (Min, 2023). For example, AI-powered learning platforms may inadvertently reinforce existing biases if the data they rely on is not diverse or representative. Similarly, automated grading systems may overlook the nuances of student work, leading to inaccurate assessments that fail to reflect students' true capabilities.

In addition to data privacy and algorithmic bias, the use of digital tools raises questions about the broader ethical implications of technology in education. For instance, the increasing reliance on digital platforms and surveillance technologies may lead to concerns about the erosion of students' autonomy and the potential for over-monitoring of their activities (Nishnianidze, 2024). Educators and institutions must carefully consider the ethical consequences of digital transformation and strive to ensure that technology is used in ways that respect students' rights, dignity, and individuality. To address these concerns, educational institutions must prioritize the implementation of robust data protection policies, adhere to relevant privacy laws (such as the General Data Protection Regulation, or GDPR), and ensure that students' personal information is handled securely. Additionally, educators and policymakers must be mindful of the ethical implications of using AI and other technologies in the classroom, ensuring that digital tools are designed and implemented in a way that promotes fairness, inclusivity, and equity.

The digital transformation of education holds great promise, but it is not without its challenges and barriers. The digital divide, resistance to change, and privacy, security, and ethical considerations are significant obstacles that must be addressed to fully realize the benefits of digital tools in education. By investing in infrastructure, providing professional development for educators, and ensuring that privacy and security concerns are adequately addressed, institutions can create an environment where digital transformation can thrive. Overcoming these barriers will require collaboration among governments, educational institutions, and technology providers, as well as a commitment to ensuring that all students have access to the tools and resources they need to succeed in the digital age.

9. FUTURE DIRECTIONS AND STRATEGIES

As digital transformation continues to reshape education, it is imperative to develop forward-thinking strategies that ensure sustainable and equitable integration of technology in teaching and learning. The future of digital education will be shaped by efforts to build digital literacy among educators and students, promote equitable access to digital resources, and harness emerging technologies to foster sustainable and innovative educational practices. This section explores these key strategies, offering insights into how institutions and policymakers can effectively navigate the evolving digital landscape in education.

9.1 Building Capacity for Digital Literacy Among Educators and Students

Digital literacy is the foundation for effective engagement with technology in education. Both educators and students must develop the necessary skills to navigate, critically assess, and effectively use digital tools for teaching and learning. For educators, digital literacy extends beyond basic technological competence—it encompasses the ability to integrate digital tools into pedagogical practices, assess the credibility of online resources, and utilize data-driven insights to personalize learning experiences. To build capacity for digital literacy among educators, institutions should prioritize professional development programs that focus on integrating technology into teaching. These programs should not only provide technical training but also emphasize pedagogical strategies for leveraging digital tools effectively. Schools and universities can implement continuous learning opportunities, such as workshops, online courses, and peer mentoring programs, to help educators stay updated with emerging technologies. Additionally, educational institutions should foster a culture of digital innovation by encouraging experimentation with new teaching approaches, such as gamification, AI-driven learning analytics, and immersive virtual environments.

For students, digital literacy should be embedded into the curriculum from an early stage, equipping them with skills such as information literacy, cybersecurity awareness, and digital collaboration. Beyond technical proficiency, students should be taught how to critically evaluate digital content, identify misinformation, and navigate ethical challenges associated with digital learning. By integrating digital literacy across disciplines, institutions can prepare students not only for academic success but also for the demands of a technology-driven workforce.

9.2 Promoting Equity in Access to Digital Resources

Ensuring equitable access to digital resources remains one of the most pressing challenges in the digital transformation of education. Socioeconomic disparities, geographic barriers, and infrastructural limitations continue to create gaps in students' access to digital tools, limiting their ability to fully engage in online and blended learning environments. Without targeted efforts to bridge this digital divide, the benefits of digital transformation will remain unevenly distributed. To promote equity, policymakers and institutions must implement initiatives that provide affordable and accessible digital resources for all students, particularly those from underprivileged backgrounds. Governments can play a crucial role by investing in national broadband infrastructure, subsidizing internet access for low-income families, and partnering with private sector stakeholders to provide cost-effective digital devices. Schools and universities should explore innovative models such as device loan programs, mobile learning solutions, and community-based digital learning hubs that allow students to access technology outside of the classroom.

In addition to physical access, digital inclusion requires culturally and linguistically relevant content that reflects the diverse needs of learners. Educational technology should be designed to accommodate different learning styles, languages, and abilities, ensuring that all students—regardless of their backgrounds—can benefit from digital learning experiences. Institutions should also adopt open educational resources (OER), which provide freely accessible and adaptable learning materials that reduce financial barriers for students. Moreover, partnerships between governments, non-profit organizations, and technology companies can play a critical role in addressing digital inequities. Collaborative initiatives that provide digital literacy training, infrastructure development, and resource-sharing programs can significantly enhance access to quality digital education, particularly in underserved communities.

9.3 Harnessing Emerging Technologies for Sustainable Education

The rapid evolution of emerging technologies presents new opportunities for re-imagining education in ways that are more engaging, personalized, and sustainable. Technologies such as artificial intelligence (AI), virtual reality (VR), augmented reality (AR), blockchain, and cloud computing are transforming the educational landscape, offering innovative solutions for both teaching and learning. However, the effective and ethical implementation of these technologies requires a strategic approach that prioritizes sustainability, accessibility, and pedagogical relevance. AI-powered learning platforms, for example, can offer personalized and adaptive learning experiences, helping students progress at their own pace while receiving targeted support based on their individual needs. AI can also assist educators in automating administrative tasks, analyzing student performance data, and providing real-time feedback, allowing them to focus on more meaningful instructional interactions. However, institutions must ensure that AI-driven systems are transparent, fair, and free from algorithmic biases that could disadvantage certain student groups.

VR and AR have the potential to create immersive learning environments that enhance student engagement and understanding of complex concepts. These technologies can be particularly beneficial in STEM education, allowing students to conduct virtual experiments, explore 3D simulations, and interact with historical or scientific models in ways that traditional classroom settings cannot replicate. To make these technologies widely accessible, institutions must explore cost-effective implementation models, such as shared VR labs and open-source AR content. Blockchain technology offers promising solutions for secure and verifiable digital credentialing. By using blockchain for academic records and certification, institutions can enhance transparency, reduce fraudulent credentials, and streamline cross-institutional recognition of qualifications. Additionally, cloud-based platforms can facilitate seamless collaboration and resource-sharing among educators and students across different locations, fostering a more interconnected and globally inclusive learning ecosystem.

As emerging technologies continue to shape the future of education, sustainability should remain a key consideration. The environmental impact of digital transformation, including energy consumption and electronic waste, must be addressed through responsible technology policies. Institutions should adopt energy-efficient digital solutions, promote the responsible disposal and recycling of electronic devices, and encourage the development of low-power computing systems that reduce environmental footprints. The future of education is undeniably digital, and the successful transformation of teaching and learning practices depends on strategic approaches that prioritize digital literacy, equity, and sustainability. By investing in professional development for educators, integrating digital literacy into curricula, and addressing disparities in access to digital resources, institutions can ensure that all learners can benefit from technology-enhanced education. Furthermore, harnessing emerging technologies in a responsible and pedagogically meaningful manner can create more engaging, inclusive, and future-ready learning environments.

As we move forward, collaboration among governments, educational institutions, technology developers, and global organizations will be essential in shaping policies and frameworks that support sustainable digital education. By embracing these strategies, the education sector can not only keep pace with technological advancements but also ensure that digital transformation fosters meaningful learning experiences for all.

10. CONCLUSION

Digital transformation has become an integral force shaping the future of education, profoundly impacting teaching methodologies, learning experiences, and institutional frameworks. As educators, learners, and policymakers navigate this evolving landscape, it is evident that the effective integration of digital technologies is not merely an enhancement of traditional educational practices but a fundamental redefinition of how knowledge is accessed, shared, and constructed.

Throughout this chapter, I have explored the various dimensions of digital transformation in education. The drivers of this shift—including rapid technological advancements, changing learner expectations, and policy imperatives—underscore the necessity for institutions to adopt forward-thinking strategies that align with the digital era. The impact on teaching practices has been particularly significant, with educators leveraging innovative pedagogical models such as flipped classrooms, blended learning, and interactive digital platforms to foster more dynamic and engaging instructional experiences. Additionally, professional development and digital competency training for educators have emerged as essential factors in ensuring the successful adoption of technology-enhanced learning.

For learners, digital transformation has unlocked new possibilities for personalized and adaptive learning, enabling students to engage with content in ways that align with their unique needs and learning styles. Collaborative learning environments, facilitated by virtual platforms, have expanded opportunities for global knowledge exchange, peer interaction, and interdisciplinary problem-solving. Moreover, digital education has fostered the development of critical 21st-century skills, including analytical thinking, creativity, and technological fluency, preparing students for success in an increasingly digitalized world.

Despite these advancements, challenges remain. Issues such as the digital divide, resistance to technological change, and concerns related to privacy, security, and ethical considerations highlight the complexities of implementing digital transformation equitably and sustainably. Addressing these barriers requires a multi-faceted approach that prioritizes inclusivity, ensures robust digital infrastructure, and fosters a culture of adaptability within educational institutions.

Looking ahead, the future of digital education will be shaped by efforts to build digital literacy among both educators and students, promote equitable access to digital resources, and harness emerging technologies in meaningful and pedagogically sound ways. By fostering a culture of continuous learning, collaboration, and responsible technological innovation, educational institutions can create a learning ecosystem that is not only technologically advanced but also inclusive, accessible, and responsive to the needs of diverse learners.

In conclusion, embracing digital transformation in education is no longer an option—it is a necessity for institutions seeking to remain relevant in an era of rapid technological change. By leveraging the potential of digital tools while addressing challenges through thoughtful policy and practice, education systems can ensure that digital transformation serves as a catalyst for enhanced learning experiences, empowered educators, and a future-ready generation of learners.

REFERENCES

Afzal, A., Khan, S., Daud, S., Ahmed, Z., & Butt, A. (2023). Addressing the digital divide: Access and use of technology in education. *Journal of Social Sciences Review, 3*(2), 883–895. DOI: 10.54183/jssr.v3i2.326

Ajani, O. A., & Govender, S. (2023). Impact of ICT-driven teacher professional development for the enhancement of classroom practices in South Africa: A systematic review of literature. *Journal of Educational and Social Research, 13*(5), 116. DOI: 10.36941/jesr-2023-0125

Akpen, C. N., Asaolu, S., Atobatele, S., Okagbue, H., & Sampson, S. (2024). Impact of online learning on students' performance and engagement: A systematic review. *Discover Education, 3*(1), 205. DOI: 10.1007/s44217-024-00253-0

Al-Ansi, B. M., Jaboob, M., Garad, A., & Al-Ansi, A. (2023). Analyzing augmented reality (AR) and virtual reality (VR) recent development in education. *Social Sciences & Humanities Open, 8*(1), 100532. DOI: 10.1016/j.ssaho.2023.100532

Al-Sharhan, S., Al-Hunaiyyan, A., Alhajri, R., & Al-Huwail, N. (2020). Utilization of learning management system (LMS) among instructors and students. In Advances in electronics engineering (pp. 2). Lecture Notes in Electrical Engineering. DOI: 10.1007/978-981-15-1289-6_2

Alshammari, F., & Alhalafawy, W. (2023). Digital platforms and the improvement of learning outcomes: Evidence extracted from meta-analysis. *Sustainability (Basel), 15*(2), 1305. DOI: 10.3390/su15021305

Anis, M. (2024). Teacher professional development in the digital age: Addressing the evolving needs post-COVID. *International Journal for Multidisciplinary Research, 6*(1), 1–14. DOI: 10.36948/ijfmr.2024.v06i01.12386

Balalle, H. (2024). Exploring student engagement in technology-based education in relation to gamification, online/distance learning, and other factors: A systematic literature review. *Social Sciences & Humanities Open, 9*, 100870. DOI: 10.1016/j.ssaho.2024.100870

Bayly-Castaneda, K., Ramirez-Montoya, M. S., & Morita-Alexander, A. (2024). Crafting personalized learning paths with AI for lifelong learning: A systematic literature review. *Frontiers in Education, 9*, 1424386. DOI: 10.3389/feduc.2024.1424386

Bernacki, M. L., Greene, M. J., & Lobczowski, N. G. (2021). A systematic review of research on personalized learning: Personalized by whom, to what, how, and for what purpose(s)? *Educational Psychology Review*, *33*(4), 1675–1715. DOI: 10.1007/s10648-021-09615-8

Bhadri, G., & Patil, L. (2022). Blended learning: An effective approach for online teaching and learning. *Journal of Engineering Education Transformations*, *35*(is1, S1), 53–60. DOI: 10.16920/jeet/2022/v35is1/22008

Bin Rashid, A., & Karim Kausik, M. A. (2024). AI revolutionizing industries worldwide: A comprehensive overview of its diverse applications. *Hybrid Advances*, *7*, 100277. DOI: 10.1016/j.hybadv.2024.100277

Boonmoh, A., Jumpakate, T., & Karpklon, S. (2021). Teachers' perceptions and experience in using technology for the classroom. *CALL-EJ*, *22*(1), 1–24.

Bose, S., Dey, S. K., & Bhattacharjee, S. (2022). Big data, data analytics and artificial intelligence in accounting: An overview. In Akter, S., & Wamba, S. F. (Eds.), *Handbook of big data methods* (pp. 1–34). Edward Elgar Publishing.

Ciarli, T., Kenney, M., Massini, S., & Piscitello, L. (2021). Digital technologies, innovation, and skills: Emerging trajectories and challenges. *Research Policy*, *50*(7), 104289. DOI: 10.1016/j.respol.2021.104289

Dahal, A., & Bhat, N. (2024). Self-directed learning, its implementation, and challenges: A review. *Nepal Journal of Health Sciences*, *3*(1), 102–115. DOI: 10.3126/njhs.v3i1.63277

Das, A., Malaviya, S., & Singh, M. (2023). The impact of AI-driven personalization on learners' performance. *International Journal on Computer Science and Engineering*, *11*(8), 15–22. DOI: 10.26438/ijcse/v11i8.1522

De Bruijn-Smolders, M., & Prinsen, F. R. (2024). Effective student engagement with blended learning: A systematic review. *Heliyon*, *10*(23), e39439. DOI: 10.1016/j.heliyon.2024.e39439 PMID: 39687123

Diab, H., Daher, W., Rayan, B., Issa, N., & Rayan, A. (2024). Transforming science education in elementary schools: The power of PhET simulations in enhancing student learning. *Multimodal Technologies and Interaction*, *8*(11), 105. DOI: 10.3390/mti8110105

Dibrov, A. (2015). Innovation resistance: The main factors and ways to overcome them. *Procedia: Social and Behavioral Sciences*, *166*, 92–96. DOI: 10.1016/j.sbspro.2014.12.489

Gabriel, F., Marrone, R., Van Sebille, Y., Kovanovic, V., & de Laat, M. (2022). Digital education strategies around the world: Practices and policies. *Irish Educational Studies*, *41*(1), 85–106. DOI: 10.1080/03323315.2021.2022513

Gruber, J., Hargittai, E., & Nguyen, M. H. (2022). The value of face-to-face communication in the digital world: What people miss about in-person interactions when those are limited. *Studies in Communication Sciences*, *22*(3), 1–19. DOI: 10.24434/j.scoms.2022.03.3340

Haleem, A., Javaid, M., Qadri, M. A., & Suman, R. (2022). Understanding the role of digital technologies in education: A review. *Sustainable Operations and Computers*, *3*, 275–285. DOI: 10.1016/j.susoc.2022.05.004

Hase, A., & Kuhl, P. (2024). Teachers' use of data from digital learning platforms for instructional design: A systematic review. *Educational Technology Research and Development*, *72*(4), 1925–1945. DOI: 10.1007/s11423-024-10356-y

Hinostroza, J. E., Armstrong-Gallegos, S., & Villafaena, M. (2024). Roles of digital technologies in the implementation of inquiry-based learning (IBL): A systematic literature review. *Social Sciences & Humanities Open*, *9*, 100874. DOI: 10.1016/j.ssaho.2024.100874

Huang, L. (2023). Ethics of artificial intelligence in education: Student privacy and data protection. *Science Insights Education Frontiers*, *16*(2), 2577–2587. DOI: 10.15354/sief.23.re202

Joseph, O. B., Onwuzulike, O. C., & Shitu, K. (2024). Digital transformation in education: Strategies for effective implementation. *World Journal of Advanced Research and Reviews*, *23*(2), 2785–2799. DOI: 10.30574/wjarr.2024.23.2.2668

Josué, A., Bedoya-Flores, M. C., Mosquera-Quiñonez, E. F., Bautista-Sánchez, J. V., & Bautista-Sánchez, J. V. (2023). Educational platforms: Digital tools for the teaching-learning process in education. *Ibero-American Journal of Education & Society Research*, *3*(1), 259–263. DOI: 10.56183/iberoeds.v3i1.626

Kabudi, T., Pappas, I., & Olsen, D. H. (2021). AI-enabled adaptive learning systems: A systematic mapping of the literature. *Computers and Education: Artificial Intelligence*, *2*, 100017. DOI: 10.1016/j.caeai.2021.100017

Lane, J. N., Leonardi, P. M., Contractor, N. S., & DeChurch, L. A. (2024). Teams in the digital workplace: Technology's role for communication, collaboration, and performance. *Small Group Research*, *55*(1), 139–183. DOI: 10.1177/10464964231200015

Le, H., Janssen, J., & Wubbels, T. (2017). Collaborative learning practices: Teacher and student perceived obstacles to effective student collaboration. *Cambridge Journal of Education, 48*(1), 103–122. DOI: 10.1080/0305764X.2016.1259389

Lemay, D. J., Bazelais, P., & Doleck, T. (2021). Transition to online learning during the COVID-19 pandemic. *Computers in Human Behavior Reports, 4*, 100130. DOI: 10.1016/j.chbr.2021.100130 PMID: 34568640

Lin, Y. L., & Wang, W. T. (2024). Enhancing students' online collaborative PBL learning performance in the context of co-authoring-based technologies: A case of wiki technologies. *Education and Information Technologies, 29*(2), 2303–2328. DOI: 10.1007/s10639-023-11907-1 PMID: 37361760

Liu, C. K. (2019). A holistic approach to flipped classroom: A conceptual framework using e-platform. *International Journal of Engineering Business Management, 11*, 1847979019855205. DOI: 10.1177/1847979019855205

Lu, D., & Xie, Y. N. (2024). The application of educational technology to develop problem-solving skills: A systematic review. *Thinking Skills and Creativity, 51*, 101454. DOI: 10.1016/j.tsc.2023.101454

McCarthy, A. M., Maor, D., McConney, A., & Cavanaugh, C. (2023). Digital transformation in education: Critical components for leaders of system change. *Social Sciences & Humanities Open, 8*(1), 100479. DOI: 10.1016/j.ssaho.2023.100479

Mhlongo, S., Mbatha, K., Ramatsetse, B., & Dlamini, R. (2023). Challenges, opportunities, and prospects of adopting and using smart digital technologies in learning environments: An iterative review. *Heliyon, 9*(6), e16348. DOI: 10.1016/j.heliyon.2023.e16348 PMID: 37274691

Min, A. (2023). Artificial intelligence and bias: Challenges, implications, and remedies. *Journal of Social Research, 2*(11), 3808–3817. DOI: 10.55324/josr.v2i11.1477

Morrison-Smith, S., & Ruiz, J. (2020). Challenges and barriers in virtual teams: A literature review. *SN Applied Sciences, 2*(6), 1096. DOI: 10.1007/s42452-020-2801-5

Mukul, E., & Büyüközkan, G. (2023). Digital transformation in education: A systematic review of Education 4.0. *Technological Forecasting and Social Change, 194*, 122664. DOI: 10.1016/j.techfore.2023.122664

Mulaudzi, L. (2024). Evaluating impact: South Africa's e-education white paper on digital equity and post-COVID education leadership. *Cogent Education, 11*(1), 2415749. Advance online publication. DOI: 10.1080/2331186X.2024.2415749

Munje, P., & Jita, T. (2020). The impact of the lack of ICT resources on teaching and learning in selected South African primary schools. International Journal of Learning. *Teaching and Educational Research, 19*(7), 263–279. DOI: 10.26803/ijlter.19.7.15

Nishnianidze, A. (2024). Surveillance in the digital age. *European Scientific Journal, 20*(37), 1–25. DOI: 10.19044/esj.2024.v20n37p1

Novak, E., McDaniel, K., & Li, J. (2023). Factors that impact student frustration in digital learning environments. *Computers and Education Open, 5*, 100153. DOI: 10.1016/j.caeo.2023.100153

Postholm, M. B. (2012). Teachers' professional development: A theoretical review. *Educational Research, 54*(4), 405–429. DOI: 10.1080/00131881.2012.734725

Power, D. J., Heavin, C., & O'Connor, Y. (2021). Balancing privacy rights and surveillance analytics: A decision process guide. *Journal of Business Analytics, 4*(2), 155–170. DOI: 10.1080/2573234X.2021.1920856

Raab, C. D. (2020). Information privacy, impact assessment, and the place of ethics. *Computer Law & Security Review, 37*, 105404. DOI: 10.1016/j.clsr.2020.105404

Rahimi, R. A., & Oh, G. S. (2024). Rethinking the role of educators in the 21st century: Navigating globalization, technology, and pandemics. *Journal of Marketing Analytics, 12*(2), 182–197. DOI: 10.1057/s41270-024-00303-4

Samyan, N., & St Flour, P. O. (2021). The impact of cloud computing on e-Learning during COVID-19 pandemic. [IJSES]. *International Journal of Studies in Education and Science, 2*(2), 146–172.

Sancar, R., Atal, D., & Deryakulu, D. (2021). A new framework for teachers' professional development. *Teaching and Teacher Education, 101*, 103305. DOI: 10.1016/j.tate.2021.103305

Sarva, E., & Purina-Bieza, K. E. (2023). Educators' perspectives on the main challenges and opportunities for implementing digital solutions in learning and teaching. International Journal of Emerging Technologies in Learning (iJET, 18(24), 62-80. https://doi.org/DOI: 10.3991/ijet.v18i24.44745

Sayeed Salih, O., Husain, M., Hamdan, M., Abdelsalam, S., Elshafie, H., & Motwakel, A. (2025). Transforming education with AI: A systematic review of ChatGPT's role in learning, academic practices, and institutional adoption. *Results in Engineering, 25*, 103837. DOI: 10.1016/j.rineng.2024.103837

Schmid, R. F., Borokhovski, E., Bernard, R. M., Pickup, D. I., & Abrami, P. C. (2023). A meta-analysis of online learning, blended learning, the flipped classroom, and classroom instruction for pre-service and in-service teachers. *Computers and Education Open*, *5*, 100142. DOI: 10.1016/j.caeo.2023.100142

Singh, J., Steele, K., & Singh, L. (2021). Combining the best of online and face-to-face learning: Hybrid and blended learning approach for COVID-19, post vaccine, & post-pandemic world. *Journal of Educational Technology Systems*, *50*(2), 140–171. DOI: 10.1177/00472395211047865

Song, A., Wu, K., & Ding, J. (2024). Developing an immersive game-based learning platform with generative artificial intelligence and virtual reality technologies – "LearningverseVR." Computers & Education: X Reality, 4, 100069. DOI: 10.1016/j.cexr.2024.100069

Sun, Y., & Yan, Z. (2025). Factors influencing teachers' technology adoption in technology-rich classrooms: Model development and test. *Educational Technology Research and Development*, *73*(3), 1481–1496. Advance online publication. DOI: 10.1007/s11423-025-10454-5

Tan, C. Y., Jang, S. T., Lam, S. M., An, A. Q., & Lo, U. K. V. (2025). Teaching and learning challenges due to the COVID-19 pandemic: A systematic review. *Educational Research Review*, *100*, 100667. DOI: 10.1016/j.edurev.2025.100667

Trust, T., Krutka, D. G., & Carpenter, J. P. (2016). "Together we are better": Professional learning networks for teachers. *Computers & Education*, *102*, 15–34. DOI: 10.1016/j.compedu.2016.06.007

Turnbull, D., Chugh, R., & Luck, J. (2020). Learning management systems, an overview. In *Encyclopedia of education and information technologies* (p. 248). Springer Nature., DOI: 10.1007/978-3-030-10576-1_248

van de Werfhorst, H. G., Kessenich, E., & Geven, S. (2022). The digital divide in online education: Inequality in digital readiness of students and schools. *Computers and Education Open*, *3*, 100100. DOI: 10.1016/j.caeo.2022.100100 PMID: 40478103

Vindigni, G. (2024). Strategies for overcoming barriers to inclusive and equitable education: A global analysis towards achieving sustainable development goal 4. European Journal of Arts. *Humanities and Social Sciences*, *1*(5), 3–47. DOI: 10.59324/ejahss.2024.1(5).01

Wagner, M., Ley, T., Kammerer, L., & Helm, C. (2024). Exploring teacher educators' challenges in the context of digital transformation and their self-reported TPACK: A mixed methods study. *European Journal of Teacher Education*, •••, 1–19. DOI: 10.1080/02619768.2024.2340689

Xu, E., Wang, W., & Wang, Q. (2023). The effectiveness of collaborative problem solving in promoting students' critical thinking: A meta-analysis based on empirical literature. *Humanities & Social Sciences Communications*, *10*(1), 16. DOI: 10.1057/s41599-023-01508-1

Yang, S., Tian, H., Sun, L., & Yu, X. (2019). From one-size-fits-all teaching to adaptive learning: The crisis and solution of education in the era of AI. *Journal of Physics: Conference Series*, *1237*(4), 042039. DOI: 10.1088/1742-6596/1237/4/042039

Chapter 4
The Future of Education in the Post-Democratic Era:
Technology, Governance, and Learning

Mustafa Kayyali
http://orcid.org/0000-0003-3300-262X
Maaref University of Applied Sciences, Syria

ABSTRACT

As democratic values increasingly come under strain globally, education systems are quietly entering a new era—one marked not by open civic participation, but by algorithmic oversight, technocratic governance, and controlled narratives. This chapter explores the contours of education in a post-democratic context, where the promises of digital transformation coexist with subtle forms of surveillance and behavioral standardization. It interrogates the evolving relationship between learners and power, showing how technologies once celebrated for democratizing knowledge may now reinforce hierarchical structures and limit dissent. At the same time, it examines how educational institutions can either reproduce systems of compliance or become sites of critical resistance. Through an interdisciplinary lens that weaves together political theory, technology ethics, and educational praxis, the chapter advocates for a reimagining of ICT-driven education—one that safeguards human agency, nurtures ethical awareness, and resists reductionist narratives.

DOI: 10.4018/979-8-3373-1190-6.ch004

INTRODUCTION

The 21st century has brought us face-to-face with paradoxes we are only beginning to understand. We are more connected than ever, yet more fragmented. We have unprecedented access to information, yet less consensus on truth. Nowhere are these contradictions more starkly visible than in the domain of education—an arena historically tied to the ideals of democracy, civic engagement, and human development. But as democracies weaken and governance becomes increasingly algorithmic, education finds itself in the crosshairs of a profound transformation. We are witnessing the subtle but significant transition from educational systems built to empower autonomous, critically thinking citizens, to frameworks designed to produce compliant, surveilled, and datafied subjects—individuals increasingly molded by platforms, policies, and protocols rather than guided by open discourse and participatory ideals. This chapter begins with a provocation: What does education look like when democracy is no longer the assumed framework? In many parts of the world, democratic backsliding is no longer a marginal phenomenon—it is becoming a central feature of governance. Institutions that once upheld accountability are increasingly controlled or weakened; public debate is either suppressed or algorithmically manipulated; and notions of collective decision-making are replaced by managerialism, centralization, or platform rule (Haggard & Kaufman, 2021). In this shifting landscape, education is not merely affected—it becomes instrumentalized. Curricula are reframed to reflect nationalist or utilitarian goals, teacher autonomy is eroded, and student engagement becomes something to be monitored and optimized, rather than nurtured and expanded.

Simultaneously, the rapid integration of Information and Communication Technologies (ICTs) into education systems—while often framed as progressive—has brought with it new forms of power that escape traditional checks and balances. Learning analytics, facial recognition in classrooms, predictive behavioral tools, and AI-driven content delivery systems are no longer speculative—they are increasingly normalized (Singh, 2024). While these technologies can offer efficiency and scale, they also risk reducing learning to quantifiable metrics and behavior to data points. Under the guise of personalization and performance, what often emerges is a subtle shift: from learning as a deeply human, dialogical experience to one that is monitored, curated, and pre-scripted by algorithmic logic (Elliott, 2022). But the issue is not simply technological—it is political and philosophical. If we are to understand the trajectory of education in the post-democratic era, we must go beyond the tools and platforms to interrogate the underlying values and structures that now shape educational priorities. Who decides what knowledge is valuable? What is the purpose of education when civic agency is de-emphasized? Can education still serve as a space of emancipation, or has it become a space of social conditioning?

These are not merely abstract questions—they have direct implications for learners, educators, and communities worldwide. This chapter seeks to grapple with these complex dynamics, offering a critical lens through which to view the entanglement of governance, technology, and learning. It situates current developments within a broader historical and political context, recognizing that the very idea of "education for democracy" has always been contested and unevenly realized. However, the current moment marks a shift in both speed and scope—what was once gradual now feels abrupt, and what was once localized is now globalized through digital infrastructures.

Throughout the chapter, we will examine how education is being reshaped by forces that are simultaneously technological and ideological. We will analyze how algorithmic governance redefines roles and responsibilities within education systems, and how these changes interact with broader trends in civic disengagement, policy centralization, and the erosion of public space. Importantly, we will also explore paths of resistance—how educators, students, and communities are reimagining education as a space of ethical reflection, human dignity, and political renewal, even in adverse conditions. Ultimately, this chapter does not aim to offer easy answers. Instead, it invites readers to dwell in the complexity of the present, to question the narratives of inevitability surrounding technology, and to reassert the role of human judgment and ethical responsibility in the design of educational futures. If education is to matter in the post-democratic era, it must be redefined not simply as a tool of governance or economic utility, but as a terrain of struggle, imagination, and hope.

RETHINKING CITIZENSHIP AND LEARNING IN POST-DEMOCRATIC SYSTEMS

To speak of citizenship today is to enter a conversation in flux. What once carried the promise of participation, voice, and rights is now laced with tension, ambiguity, and quiet disillusionment. Across the globe, democratic ideals—long held as the moral compass of modern societies—are being reconfigured, reframed, and in many instances, quietly eroded (Millefiorini, 2025). It is in this unsettling backdrop that we must begin to rethink not only what it means to be a citizen, but what it means to learn as a citizen—especially in systems where democratic engagement is no longer presumed, and where the contours of power are increasingly shaped by algorithms, closed-door policymaking, and strategic misinformation. In post-democratic systems—societies where democratic rituals may persist, but democratic substance is diluted—citizenship takes on a different shape. It becomes more procedural than participatory, more about compliance than contribution. Citizens are encouraged to engage, but only within tightly defined parameters; to speak, but only in ways that are sanctioned; to learn, but only what aligns with dominant narratives (Wilson & Brezicha, 2023). The result is a form of civic life that appears democratic on the surface but lacks the depth, complexity, and open-endedness that true democratic practice requires. Education, in this landscape, becomes both a battleground and a mirror. On one hand, it is tasked with reproducing the values of the society it serves. On the other, it holds the latent potential to disrupt, question, and reimagine those very values. The challenge, then, is immense: how can educational systems foster meaningful citizenship in contexts where political agency is curtailed, dissent is managed, and governance is increasingly opaque? How can learning remain liberatory in spaces that are designed to be instrumental?

The answer is not straightforward. For much of the modern era, education and citizenship were tightly linked through a particular narrative: that schools would prepare individuals to participate in the civic life of democratic nations, equipping them with the knowledge, skills, and dispositions necessary to think critically, vote responsibly, and contribute to public life (Wagner & Kössler, 2022). But that story, while powerful, was always partial. It often excluded marginalized voices, assumed a neutral public sphere, and failed to interrogate the deeper structures of inequality that shaped access to both learning and participation. What we face now is not simply the decay of that ideal, but its reconfiguration. In many contemporary societies, education is being repositioned as a tool of economic efficiency and national competitiveness, rather than as a means of nurturing civic agency or ethical reflection (BAMIDELE, 2024). Citizenship, in turn, is being reduced to acts of consumption, data-sharing, and behavior management. The citizen is no longer the engaged participant but the optimized user, the compliant subject, the data point in a larger algorithmic apparatus (Bruun, 2024). This shift has profound implications for what and how we teach. In a post-democratic context, curricula are often sanitized—stripped of political complexity, rendered ideologically safe, and repackaged as neutral. Historical narratives are revised or suppressed; controversial issues are avoided; and standardized testing becomes the dominant measure of "success." This depoliticization of education, however, is not neutral—it is itself a political act (Mendonça dos Santos, 2024). It shapes what students are allowed to know, what questions they are permitted to ask, and what futures they are able to imagine. And yet, even in constrained environments, spaces for critical citizenship can be carved out—but doing so requires intention, courage, and often subversive pedagogy. It means creating classrooms where complexity is not feared but welcomed, where disagreement is not silenced but explored, and where students are encouraged to see themselves not just as recipients of knowledge but as co-constructors of meaning. It also means equipping learners with the tools to decode power, to recognize how structures—both visible and invisible—shape their lives, and to ask not just how things work, but *why* they are the way they are, and *who* benefits from them remaining that way.

In post-democratic systems, such acts of learning can be deeply political. They disrupt the logic of control and conformity that underpins much of contemporary governance (Bickerton & Accetti, 2021). They remind us that education is not just about preparing for jobs or performing well on exams—it is about cultivating the capacity to live thoughtfully, ethically, and collectively. In other words, it is about reclaiming a deeper, more expansive notion of citizenship. But what might that citizenship look like in this new context? If traditional forms of civic engagement—voting, protesting, organizing—are limited or surveilled, what alternatives remain? Here, we must begin to broaden our imagination. Citizenship does not have to be tethered only to the state or formal institutions. It can also be relational, cultural, and affective. It can be enacted in the ways people care for one another, resist injustice in their everyday lives, create alternative spaces of belonging, and build solidarities that transcend borders.

Education, in turn, can support these practices—not by prescribing them, but by opening space for their emergence (Fawns et al, 2021). This might involve teaching students how to critically analyze media narratives, how to identify and challenge systemic injustice, how to collaborate across differences, and how to engage in ethical dialogue. It might also mean rethinking what counts as "knowledge" in the first place, and who gets to produce it. Too often, educational systems privilege certain forms of expertise while dismissing others—especially the lived knowledge of marginalized communities. A more inclusive vision of citizenship must challenge these hierarchies and recognize that wisdom, insight, and agency are not confined to official texts or elite institutions (Abdou, 2023). Of course, pursuing such an education is not without risk—especially in post-democratic regimes where critical inquiry is viewed with suspicion, and where educators themselves may face repercussions for stepping outside prescribed lines. But history shows us that even under repressive conditions, learning can be an act of resistance. From underground schools to digital forums, from street pedagogy to diasporic networks, communities have always found ways to reclaim education as a tool for liberation. The key is to hold on to that legacy, to nurture it, and to adapt it to the present moment. Technology, paradoxically, plays a dual role in this story. On one hand, it is a central instrument of control. In many post-democratic contexts, surveillance is no longer a secretive affair—it is embedded into educational platforms, learning management systems, and even classroom architecture. Students' movements are tracked, their behaviors analyzed, their emotions monitored (Sharma et al, 2022). Algorithms determine what content they see, what pace they learn, and sometimes even what feedback they receive. The line between education and surveillance is becoming dangerously thin. But on the other hand, technology also opens new possibilities for connection, creativity, and resistance. Decentralized learning networks, open-source knowledge platforms, and digital communities offer ways to circumvent traditional gatekeepers and foster alternative forms of civic engagement (Kayyali, 2025 a). The challenge is to use these tools in ways that empower rather than pacify, that invite exploration rather than enforce conformity.

In rethinking citizenship and learning, we must also grapple with our own complicity (Miles, 2024). Too often, educators and policymakers adopt the language of innovation and efficiency without questioning its assumptions. We celebrate "21st-century skills" while ignoring 20th-century injustices. We invest in digital tools without investing in the human relationships that make learning meaningful. We talk about preparing students for the future without asking what kind of future we are actually creating. To move forward, we need a different ethic—one that centers dignity, dialogue, and democratic possibility, even in the absence of formal democracy. We need to ask not just what students need to know, but *who* they are becoming through their education. Are they becoming more attuned to injustice, more capable of empathy, more willing to challenge power, more prepared to collaborate in building a more just world? If not, then we must question whether we are truly educating them at all.

Citizenship, after all, is not just a legal status or a political identity—it is a practice. It is something we do, not something we have. It is lived through our relationships, our choices, our commitments. Education should prepare us for that practice—not by telling us what to think, but by inviting us to think deeply, courageously, and ethically (Facer, 2021). In post-democratic systems, such an education is both rare and urgently needed. It will not come from top-down reforms or flashy technologies. It will come from the quiet, persistent efforts of educators who believe that learning can still be a space of freedom. It will come from students who refuse to be reduced to metrics. It will come from communities who understand that education is not just preparation for life—it *is* life, in all its complexity, uncertainty, and beauty. To rethink citizenship and learning, then, is not to abandon hope—it is to reclaim it. Not naively, but knowingly. Not as a promise, but as a struggle. It is to recognize that while the structures of democracy may be weakening, the spirit of democratic learning—the urge to question, to connect, to create—still lives in classrooms, in conversations, in acts of courage both large and small. Our task is to nurture that spirit, to protect it, and to let it guide us toward futures that are more just, more humane, and more worthy of our shared humanity.

FROM PARTICIPATION TO COMPLIANCE: THE SHIFTING ROLE OF EDUCATION

There was a time—idealized perhaps, but nevertheless formative—when the purpose of education was envisioned as the cultivation of free-thinking, active, and engaged individuals. The school was not just a place to acquire skills or prepare for employment; it was a civic institution, rooted in the values of participation, dialogue, and responsibility. Learners were considered future citizens, entrusted with the task of sustaining and deepening democracy. Education was meant to be transformative—not merely of individuals, but of societies (Hunter, 2023). Yet today, that vision is quietly being displaced. Across many parts of the world, we are witnessing a subtle yet profound redefinition of education's role: from a space of participatory development to one of behavioral conditioning; from fostering agency to ensuring conformity; from encouraging questions to promoting quiet, data-driven obedience. This transformation is not always visible on the surface. The buildings still stand, the curricula still refer to critical thinking, and the slogans in school mission statements still invoke values like empowerment and creativity. But underneath these familiar veneers, a different logic is taking hold—one shaped not by civic ideals, but by performance metrics, behavioral algorithms, institutional risk management, and state or corporate imperatives for control. Education is being retooled to serve new masters: efficiency, predictability, stability, and above all, compliance.

Compliance is not, in itself, a new concern in education. Schools have long been places where rules are taught, authority is exercised, and norms are instilled (Tuhuteru, 2023). But what distinguishes the current moment is the systematic and technological intensification of this trend, coupled with a retreat from democratic ideals. In the past, discipline and conformity existed alongside spaces for dissent, creativity, and collective meaning-making. Today, those spaces are shrinking. The balance has tipped—and participation, once the soul of democratic education, is becoming a managed illusion. Much of this shift can be traced to the growing dominance of audit cultures in education. The language of accountability, performance, outcomes, and benchmarking—borrowed largely from corporate management—has become central to how educational institutions define success. Schools, universities, and even early childhood centers are increasingly governed by metrics (Ozalp et al, 2022). Teachers are evaluated through standardized test scores; students are sorted and ranked according to quantifiable indicators; and institutions are compared in global league tables that reduce complex educational realities to numerical abstractions. In this environment, the richness of educational experience—the unmeasurable moments of curiosity, doubt, discovery, and transformation—is pushed to the margins. What matters is what can be counted. And what can be counted is often what can be controlled. This is where compliance enters as a new form of pedagogy. Students are trained not just in content, but in correct behaviors: punctuality, attention, submission to surveillance, adaptability to systems (Trabelsi et al, 2023). From biometric attendance systems to classroom cameras, from AI-based emotion detection software to learning analytics platforms that track every keystroke, students are becoming objects of continuous observation and adjustment. The classroom is no longer a sanctuary from external power—it is increasingly its extension.

Even digital learning platforms—often celebrated for their potential to democratize access—can serve as instruments of compliance. The algorithms behind them are not neutral; they encode particular pedagogical assumptions, favor certain learning paths over others, and reward specific types of behavior. When education is delivered through platforms designed primarily to monitor, optimize, and assess, the learner becomes less a participant and more a respondent to a script. Their journey is choreographed; their freedom, conditional. Ironically, this scripted learning is often cloaked in the rhetoric of personalization. Students are told that the system adapts to their needs, that their pathways are unique. But in practice, personalization often means the opposite: the narrowing of possibilities based on past behavior, the reduction of human complexity to data profiles, the quiet elimination of options deemed statistically unviable (Tariq, 2024). Rather than opening space for new forms of participation, technology in this mode functions as a sorting mechanism—subtle, opaque, and relentlessly efficient. Teachers, too, are drawn into the compliance machinery. Their professional autonomy is curtailed by scripted curricula, rigid assessment frameworks, and managerial oversight. They are expected to deliver content efficiently, ensure behavioral discipline, and meet ever-expanding bureaucratic demands—all while maintaining the appearance of student-centered learning. The teacher as intellectual, as critical interlocutor, as shaper of public discourse is increasingly replaced by the teacher as data manager, behavior enforcer, and facilitator of pre-approved content (Giroux, 2024).

This is not simply the result of technological change; it reflects deeper shifts in governance. In many post-democratic settings, education is being reimagined as a tool of national security, economic productivity, or social cohesion—as defined by central authorities or powerful private actors (Casey, 2024). Participation, in this context, is redefined as alignment: with state narratives, institutional protocols, or market expectations. To participate is to perform well, to comply, to not disrupt. One might ask: what is lost when education moves from participation to compliance?

The answer is not just philosophical—it is existential. What is lost is the ability to think otherwise. The courage to question power. The space to imagine alternatives. The sense of collective belonging that comes from engaging with difference, with ambiguity, with the open-endedness of democratic life (Lacelle-Webster, 2023). When education ceases to be a site of shared inquiry and becomes a mechanism for behavioral programming, it ceases to be education in the fullest sense. This loss has generational consequences. Young people raised in environments of managed learning may become highly skilled and highly adaptable—but at what cost? Without exposure to contestation, to struggle, to the messy work of co-creation, they may struggle to see themselves as political beings. They may become efficient workers, compliant citizens, loyal consumers—but their capacity to act collectively, to dissent ethically, to build something different may wither. And yet, it is precisely in such a moment that the promise of education must be reasserted—not nostalgically, but critically. Participation cannot be reduced to occasional feedback surveys or controlled classroom discussions (Ahmad, 2021). It must be reconceived as the foundation of education itself. This means fostering environments where students are not just taught content, but are invited to help shape it. Where questioning is not penalized but encouraged. Where difference is not managed away but made meaningful through dialogue.

Participation also means trusting students—and teachers—as capable moral and political agents. This is not about romanticizing youth or downplaying the challenges of facilitation. It is about recognizing that democracy is not something students will encounter *after* they graduate; it is something they must experience within their education. If we want a world where people can think critically, care deeply, and act collectively, we must create educational spaces where those practices are possible—and valued. To reclaim participation in education also requires resisting the seductions of control. It means asking hard questions about why we are so invested in monitoring, in standardization, in performance-based incentives. It means examining the fears and anxieties that underlie our embrace of compliant learners: fear of disruption, fear of difference, fear of failure. And it means finding the courage to say: we do not need perfect learners; we need engaged ones. We need learners who can challenge us, who can surprise us, who can help us grow—not just in knowledge, but in wisdom (Stanley & Clinton, 2024). This does not imply chaos or anti-discipline. Participation is not the absence of structure; it is the presence of shared purpose. It is about co-constructing norms, rather than enforcing them unilaterally. It is about moving from a culture of fear—fear of falling behind, fear of losing control—to a culture of trust. Trust in learners. Trust in educators. Trust in the process of learning as a messy, relational, and transformative journey.

Of course, such a vision will not come easily. It requires not just pedagogical shifts, but political ones. It demands that we challenge the systems that have commodified education, reduced it to competitive rankings, and hollowed out its democratic soul. It means advocating for policies that prioritize inclusion over efficiency, equity over standardization, and meaning over metrics (Kayyali, 2022 b). But more than that, it requires a reawakening of belief—in education as a public good, in learners as agents of change, in classrooms as spaces of possibility. It asks us to remember what education can be: not a quiet rehearsal for obedience, but a bold rehearsal for democracy. Not a performance of compliance, but a practice of becoming. In the end, the shift from participation to compliance is not just a shift in educational policy—it is a shift in worldview. It is a move away from believing that people, when empowered, can shape a better world, and toward believing that people must be shaped to fit the world as it is. Resisting that shift is the task before us—not just as educators, but as citizens. It is the task of reclaiming education not as a tool of conformity, but as a space of courage, connection, and co-creation. Let us remember that the classroom is not only a place where futures are prepared—it is a place where they are imagined. If we are to build futures worth living in, we must ensure that education remains a space where participation is not just permitted, but protected; not just tolerated, but treasured.

DIGITAL PEDAGOGIES IN CONTROLLED SOCIETIES: OPPORTUNITIES AND LIMITATIONS

There is something inherently paradoxical about the rise of digital pedagogies in societies where control—not freedom—is the dominant political and social currency. On the surface, digital education appears to be a democratizing force, promising open access, borderless classrooms, flexible learning, and the empowerment of voices that might otherwise go unheard (Borgos et al, 2023). But beneath this progressive veneer lies a more complicated, and often more troubling, reality. In contexts where surveillance is normalized, dissent is discouraged, and ideological conformity is a prerequisite for stability, the very tools that promise emancipation may be the ones that quietly deepen control. Digital pedagogy, broadly defined, refers to the use of digital tools, platforms, and methodologies to support teaching and learning. At its best, it is innovative, responsive, and inclusive—offering students new ways to engage with knowledge and with one another. In open societies, digital pedagogy has been linked to critical thinking, collaboration, creativity, and global awareness. But in more controlled environments—whether under authoritarian regimes or highly centralized bureaucratic systems—these same tools take on a different character. They are not just vessels of learning; they are instruments of governance, surveillance, and behavioral calibration. They are not always designed to provoke thought—they are often structured to constrain it.

This dual nature of digital pedagogy in controlled societies is not accidental—it is structural. Every digital interaction leaves a trace. Every keystroke, click, submission, attendance record, or forum post is logged, stored, and potentially analyzed. In contexts where state or institutional control is prioritized over individual autonomy, this data becomes a powerful mechanism—not just to improve learning, as proponents claim, but to monitor and manage students and educators alike. Participation is encouraged, but only within tightly regulated frames. Critical engagement is tolerated, so long as it does not transgress established boundaries (Sobkowiak et al, 2023). The result is a form of learning that is both expansive and restrictive—technologically rich, yet intellectually cautious. It is essential to be honest about both sides of this equation. There are, undeniably, opportunities. Digital pedagogy, even in controlled societies, can expand access to education in regions where physical infrastructure is lacking or movement is restricted. It can connect isolated learners with global conversations, expose students to resources that were previously unavailable, and offer flexible models of learning that accommodate different needs and schedules. For some students—particularly those in remote or underserved areas—online platforms may be the only viable educational lifeline. In this sense, digital learning can act as a bridge: connecting the disconnected, amplifying marginal voices, and offering glimpses of a world beyond immediate constraints.

Moreover, digital tools can empower creative pedagogies that bypass rigid curricula (Oyedotun, 2024). A teacher who cannot speak openly in a classroom may be able to use storytelling platforms, coded metaphors, or artistic media to raise critical questions indirectly. Students may use private forums or encrypted communication channels to collaborate on projects that challenge dominant narratives—quietly building communities of resistance and imagination. In the hands of thoughtful educators and courageous learners, digital pedagogy can thus become a subtle form of dissent: not loud or defiant, but persistent and deeply human. Yet, these opportunities exist in tension with profound limitations. The most immediate is the limitation of expression. In controlled societies, online speech is monitored. Platform providers may be legally required to report "suspicious" content. Educational institutions may restrict certain topics from being discussed. Self-censorship becomes a survival skill, and both students and teachers learn quickly what can be said and what must be avoided. Over time, this leads to a narrowing of intellectual horizons—not because people lack curiosity, but because they have been conditioned to suppress it.

Another limitation is the architecture of the platforms themselves. Many of the most widely used digital education tools are not built for critical pedagogy—they are built for compliance, for data collection, for performance management. Learning management systems are often designed around the logic of tracking and grading, rather than exploration and dialogue. The design of a platform shapes the experience of learning. If the platform is built to reward speed over depth, correctness over creativity, conformity over curiosity, then the pedagogical possibilities become constrained—regardless of the teacher's intentions. There is also the issue of digital inequality. Even in societies that invest heavily in digital infrastructure, access is not evenly distributed. Rural communities, ethnic minorities, economically disadvantaged groups, and individuals with disabilities often face significant barriers to participation (Belando-Montoro et al, 2022). In controlled societies, these disparities may be compounded by political marginalization. Certain regions may have more restricted access to the internet. Certain topics may be firewalled. Certain educators may be denied access to resources because of their perceived dissent. In such contexts, digital pedagogy risks reproducing—not reducing—existing inequalities.

Moreover, the widespread use of educational technology creates new vulnerabilities. Students' data may be used to profile them, not just for academic support, but for ideological loyalty or behavioral risk. Teachers may be evaluated not by peers or pedagogical experts, but by opaque algorithms that assess their compliance with institutional norms. Educational content may be shaped by centralized authorities or corporate interests, limiting curricular diversity and reinforcing dominant ideologies. In these conditions, digital pedagogy becomes a tool of normalization—a way of aligning thought and behavior with prescribed norms, under the guise of modernization. This convergence of pedagogy and surveillance poses deep ethical questions. Can learning truly be free in a space where every move is monitored? Can dialogue flourish when the consequences of speech are unpredictable? Can teachers foster critical thinking when they themselves are subject to invisible forms of control? These are not hypothetical concerns. In many controlled societies today, educators have been disciplined, silenced, or dismissed for encouraging students to question authority. Students have been flagged for expressing dissenting views online. Institutions have adopted technologies that automate discipline, filter content, and enforce ideological boundaries—all in the name of educational progress (Gilliard, 2021). And yet, despite these constraints, learning persists. Human beings are inherently meaning-seeking creatures. Even under conditions of constraint, they find ways to connect, to wonder, to challenge, to grow. In controlled societies, this learning often happens in the margins: in informal conversations, in coded texts, in subversive assignments that hide critique behind creativity. Digital pedagogy, when used with care, can support this kind of learning—not by pretending the constraints do not exist, but by acknowledging them, navigating them, and sometimes, cleverly subverting them.

For educators in these contexts, the task is delicate. They must balance safety with honesty, innovation with caution, ambition with survival. They must find ways to teach that invite students to think without putting them at risk. They must choose platforms and methodologies that support dialogue without exposing participants to unnecessary harm. They must model resilience—not by ignoring fear, but by moving forward in spite of it. This is not an easy path. It requires deep contextual knowledge, emotional intelligence, and moral clarity. It also requires solidarity. No educator should have to navigate these challenges alone. International networks of support, communities of practice, and ethical guidelines for digital pedagogy can all help sustain those working in difficult environments. But ultimately, the work must be grounded in local realities. Solutions must be developed not for, but with, those who live and teach in controlled societies. It is also vital to remember that control is never total (Hirsh-Pasek et al, 2022). Even in the most tightly managed systems, there are cracks—spaces where freedom flickers. A creative writing assignment that invites personal reflection. A science lesson that opens questions about environmental justice. A digital art project that explores identity. These are not overtly political acts—but they can be deeply pedagogical. They create space for voice, for thought, for possibility. And in controlled societies, that space is precious.

In the end, the question is not whether digital pedagogy is possible in controlled societies—it is how it can be made meaningful. This requires more than technology; it requires vision, courage, and a commitment to human dignity. It requires a pedagogy that is aware of its context, responsive to its risks, and attuned to its possibilities. A pedagogy that does not just transmit content, but nurtures capacity—the capacity to think, to question, to connect, to hope. Digital tools will not liberate education on their own (Smeplass, 2023). They are shaped by the systems in which they operate. But used wisely, and with care, they can help build the conditions for more expansive futures. They can support dialogue where silence is expected, creativity where conformity is imposed, connection where isolation is enforced. They can open windows—even if just a crack—into what education can still be: a shared human endeavor toward understanding, agency, and meaning. And so, in controlled societies, digital pedagogy is not a neutral endeavor—it is a profoundly ethical one. It is not just about using technology; it is about reclaiming education as a space of possibility. This does not mean ignoring danger. It means facing it with integrity. It means asking: what can we still do, even here, even now, to foster the habits of thought, care, and community that learning demands? What can we create, within constraints, that still honors the learner's right to know, to speak, to become? These are not easy questions. But they are the questions that matter. And in asking them, educators keep alive the hope that, even in the shadow of control, learning can remain a form of light.

BUILDING RESILIENCE: HUMAN AGENCY, ETHICS, AND EDUCATION FUTURES

In an age increasingly defined by technological acceleration, political uncertainty, and societal fragmentation, the future of education is no longer a question of reform—it is a question of resilience. But what do we mean by resilience in education? Too often, the word is thrown around in policy documents and mission statements as if it were self-evident: a kind of institutional toughness, a system's ability to "bounce back" after disruption. Yet, if we take the concept seriously—if we strip it of its corporate polish and return it to the lived experiences of learners, teachers, and communities—we begin to see that resilience is not about bouncing back to what was, but about enduring through what is, and imagining what might still be possible. Resilience in education is deeply human (Kayyali, 2025 b). It is not a system function or a dashboard metric. It resides in people—in the quiet courage of a teacher who continues to teach despite censorship, in the defiance of a student who dares to question, in the community that gathers around learning even when its legitimacy is under attack. It is rooted in agency—the capacity to act, to choose, to create meaning, even in contexts of constraint. It is inseparable from ethics—the ability to reflect, to care, and to take responsibility for the worlds we are shaping through education. And ultimately, it is inseparable from hope—not blind optimism, but the stubborn belief that other futures are still imaginable, and still worth striving for.

To speak of resilience today is also to acknowledge that many educational systems around the world are under pressure—some collapsing slowly under the weight of bureaucracy, others being hollowed out by authoritarianism, austerity, or technocratic overreach. In such environments, education becomes a terrain of contestation. It is not only about knowledge transmission but about power, ideology, and survival. The values that guide it—democracy, inclusion, critical thinking, compassion—are not just ideals. They become acts of resistance (Kayyali, 2024). And within this resistance, resilience takes shape. But resilience is not just about weathering storms. It is about preparing for the long journey ahead. It is about developing the moral and intellectual capacity to navigate a world where truth is contested, where technologies outpace regulation, where ecological and political crises loom large, and where the very idea of a shared future is under strain. In such a world, resilience cannot be taught as a fixed trait—it must be cultivated as a living practice. This practice begins with human agency. In educational contexts, agency is often misunderstood as individual autonomy—making choices, setting goals, managing one's own learning. While these are important, agency is more than self-direction. It is relational, situated, and political. It involves recognizing the conditions that shape our choices, challenging the structures that limit our possibilities, and imagining alternatives to what seems inevitable. Agency is not something we possess—it is something we exercise, often against the grain of dominant systems.

In post-democratic contexts, where participation is constrained and conformity is rewarded, fostering agency becomes both more difficult and more necessary. Students raised in environments of surveillance and scripted learning may struggle to see themselves as active contributors to society. Educators working under ideological scrutiny may feel unable to depart from official narratives (Ellis et al, 2024). And yet, even in these conditions, agency can take root—in a question that refuses to disappear, in a story that reveals complexity, in a project that opens space for voice. These moments may be small, even fleeting. But they matter. They signal that the capacity for thought, for choice, for resistance is still alive. And it is through these moments that ethical reflection enters the scene. Ethics in education is not a code of conduct or a checklist of responsibilities. It is a continuous process of asking: What kind of people are we becoming through our educational practices? What values are we reinforcing? Whose voices are we amplifying or silencing? Whose lives are we preparing students to see as worthy, and whose are we training them to overlook?

In a world increasingly governed by algorithms, efficiency metrics, and behavioral optimization, ethical questions often get sidelined (Tsamados et al, 2021). The logic of "what works" replaces the question of "what's right." But education without ethics is mere training. It may produce skilled workers, but it will not produce thoughtful citizens. It may prepare students to perform roles, but not to interrogate them. It may cultivate intelligence, but not wisdom. To build ethical resilience, then, we must return to the core questions of education. Not just what should be taught, but why. Not just how students should learn, but who they are becoming in the process. We must create space for moral inquiry, for philosophical discomfort, for complexity that resists easy resolution. This requires pedagogies that are not afraid of contradiction, that welcome dissent, that see vulnerability as a strength rather than a weakness. It also requires institutional courage. Educational institutions must resist the drift toward technocratic governance. They must push back against policies that reduce students to data points, teachers to content delivery agents, and curricula to market instruments. They must defend spaces for experimentation, for dialogue, for critical reflection. This is not easy. It may involve risk—political, financial, reputational. But without such courage, education cannot be transformative. It can only be transactional.

Resilience also demands that we attend to the emotional and psychological dimensions of learning. In times of uncertainty and upheaval, students carry more than academic burdens. They bring with them fears, traumas, dislocations. If education is to be resilient, it must be humane. It must create environments where students feel seen, heard, and held. Where relationships matter as much as results. Where care is not an add-on, but a foundation. This ethic of care extends beyond the classroom. It involves reimagining education as a collective endeavor—one that includes families, communities, and broader social networks. Resilience is not an individual trait—it is a communal capacity (Salas-Vallina et al, 2022). It grows in environments where people look out for one another, where knowledge is shared, where interdependence is recognized. In fractured societies, education can be a thread that weaves connection. It can offer not just skills or credentials, but a sense of belonging. And yet, resilience is not about simply enduring what is. It is also about envisioning what could be. If education is to have a future, it must be a future worth building. A future that moves beyond survival and towards flourishing. This means not just adapting to the world as it is, but daring to shape the world as it should be.

To do this, we need educational futures that are rooted in justice. Not the abstract justice of policy frameworks, but the lived justice of equitable access, of epistemic inclusion, of historical reckoning. We need curricula that center diverse voices, that challenge dominant narratives, that equip students to grapple with the realities of climate change, migration, inequality, and systemic oppression (Salinas et al, 2023). We need technologies that serve human dignity, not market interests. We need policies that are responsive to local contexts, and not dictated by global metrics. And above all, we need imagination. The imagination to see beyond the present constraints. To ask: What if education were truly emancipatory? What if schools were places where democracy was not just taught, but practiced? What if digital tools were used to foster collaboration rather than compliance? What if learning was not a preparation for life, but a practice of living more fully, more justly, more joyfully?

This is not a utopian fantasy. It is a necessary reorientation. In a time when despair is easy and cynicism is fashionable, hope is a radical act. But it is a hope grounded in action, in community, in an unshakable belief in the human capacity to learn, to grow, to care. Resilience is not the absence of struggle—it is the refusal to let struggle define what is possible (Neenan & Dryden, 2021). In this sense, education is not just one of many domains that must adapt to the future—it is the very ground on which the future will be built. It is where values are tested, where identities are shaped, where the next generation learns not just how to survive, but how to imagine and enact better ways of being. The futures of education will not be written solely by policymakers, technologists, or market forces (Kayyali, 2022 a). They will be co-written by all of us—by teachers who take risks, by students who ask hard questions, by communities that insist that education remain a public good. To build that future, we must start now—by cultivating the habits of resilience that education demands. The habit of asking why. The habit of listening deeply. The habit of acting with care. The habit of resisting what dehumanizes, and embracing what uplifts. Let us remember: the future is not something that happens to us. It is something we participate in shaping. And in education, as in life, that shaping begins with the everyday acts of courage, imagination, and ethics. It begins with choosing, again and again, to teach in ways that honor our shared humanity. That is resilience. That is agency. That is the work ahead.

CONCLUSION

As we close this inquiry into the future of education in a post-democratic era, one thing becomes clear: the questions we must ask are no longer just about pedagogy, curriculum, or access—they are about power, agency, and the role of education in shaping the human condition. The convergence of technological acceleration with the erosion of democratic norms has given rise to new educational realities—ones in which the language of efficiency often eclipses that of equity, and where systems once meant to liberate now risk becoming mechanisms of control. This chapter has aimed to confront this reality head-on, not with alarmism, but with critical urgency. We have explored how ICTs—when embedded in environments of weak democratic oversight—can amplify existing inequalities, automate compliance, and obscure accountability. What begins as an innovation for student engagement or institutional efficiency can, in the absence of transparent governance and ethical frameworks, morph into an infrastructure of surveillance and conformity. Educational technology, in such settings, is no longer neutral—it is profoundly political. Yet, we must resist the temptation to frame this transformation as inevitable. There is no technological determinism here. The tools we use do not dictate our futures; rather, it is how we choose to deploy them, under what values, and toward what goals that will shape the outcome. In post-democratic contexts, the challenge is not just to "use technology wisely" but to protect—and if necessary, reassert—the foundational human values that have long animated educational purpose: critical inquiry, collective dialogue, creative dissent, and the right to imagine alternative worlds.

Education must not become the quiet casualty of political drift. It must remain a contested space, where difficult questions can be asked, where multiple truths can be examined, and where learners are not reduced to data but recognized as whole persons with context, history, and voice. This is particularly crucial as systems increasingly prioritize performance metrics, behavioral analytics, and standardized outputs. In this environment, the role of educators as ethical stewards—those who can resist reductive tendencies and advocate for pedagogical justice—becomes more important than ever. We must cultivate resilience at both the institutional and individual level. This means designing policies that protect teacher autonomy, investing in digital literacy that goes beyond mere technical skills, and supporting communities of practice that promote ethical reflection on the use of technology in learning. It also means preserving space—within and beyond the classroom—for the kinds of conversations that defy easy answers, that value slowness, ambiguity, and dissent as essential to meaningful education. In the end, the future of education in a post-democratic world will depend less on the sophistication of our technologies and more on the strength of our convictions. Will we accept a future of docility and algorithmic governance, or will we fight for an education that fosters curiosity, courage, and care? This is not merely a pedagogical choice—it is a moral and political one. As educators, researchers, and citizens, we are called to shape this future—not as passive recipients of change, but as its authors.

REFERENCES

Abdou, E. D. (2023). *Education, civics, and citizenship in Egypt: Towards more inclusive curricular representations and teaching*. Springer Nature. DOI: 10.1007/978-3-031-33346-0

Ahmad, C. V. (2021). Causes of students' reluctance to participate in classroom discussions. *ASEAN Journal of Science and Engineering Education, 1*(1), 47–62. DOI: 10.17509/ajsee.v1i1.32407

BAMIDELE, E. F. (2024). EDUCATION FOR SOCIO-ECONOMIC AND POLITICAL RECONSTRUCTION IN THE 21ST CENTURY NIGERIA. *Journal of Education Innovation and Practice, 8*(1), 1–11.

Belando-Montoro, M. R., Naranjo-Crespo, M., & Carrasco-Temino, M. A. (2022). Barriers and facilitators to the retention and participation of socially, economically, and culturally disadvantaged university students. An international systematic review. *International Journal of Educational Research, 113*, 101968. DOI: 10.1016/j.ijer.2022.101968

Bickerton, C. J., & Accetti, C. I. (2021). *Technopopulism: The new logic of democratic politics*. Oxford University Press. DOI: 10.1093/oso/9780198807766.001.0001

Borgos, J., Kinser, K., & Kline, L. (2023). The borderless market for open, distance, and digital education. In *Handbook of Open, Distance and Digital Education* (pp. 355–369). Springer Nature Singapore. DOI: 10.1007/978-981-19-2080-6_22

Bruun, M. H. (2024). Algorithmic Governance, Public Participation and Trust: Citizen–State Relations in a Smart City Project. Social Anthropology. *Anthropologie et Sociétés, 32*(4), 13–30.

Casey, C. (2024). *Governing the Firm in the Social Interest: Corporate Governance Reimagined*. Taylor & Francis.

Elliott, A. (2022). *Algorithmic intimacy: The digital revolution in personal relationships*. John Wiley & Sons.

Ellis, V., Gatti, L., & Mansell, W. (2024). *The new political economy of teacher education: The Enterprise narrative and the shadow state*. Policy Press.

Facer, K. (2021 forthcoming). *Futures in education: Towards an ethical practice. Paper commissioned for the UNESCO Futures of Education report*. UNESCO.

Fawns, T., Mulherin, T., Hounsell, D., & Aitken, G. (2021). Seamful learning and professional education. *Studies in Continuing Education*, *43*(3), 360–376. DOI: 10.1080/0158037X.2021.1920383

Gilliard, C. (2021). Tagged surveillance. The Journal of Interactive Technology and Pedagogy, 20.

Giroux, H. A. (2024). *Teachers as intellectuals: Toward a critical pedagogy of learning*. Bloomsbury Publishing. DOI: 10.5040/9781350458635

Haggard, S., & Kaufman, R. (2021). *Backsliding: Democratic regress in the contemporary world*. Cambridge University Press. DOI: 10.1017/9781108957809

Hirsh-Pasek, K., Zosh, J. M., Hadani, H. S., Golinkoff, R. M., Clark, K., Donohue, C., & Wartella, E. (2022). *A Whole New World: Education Meets the Metaverse. Policy Brief.* Center for Universal Education at The Brookings Institution.

Hunter, A. (2023). Transformative learning in international education. In *Developing intercultural competence and transformation* (pp. 92–107). Routledge. DOI: 10.4324/9781003444169-7

Kayyali, M. (2022 a). *Universities of the Future: HEIs that are Oriented to Accreditation*. Quality Assurance, Research, and Ranking.

Kayyali, M. (2022b). Equity, Equality, Diversity, and Inclusion as Key Performance Indicators in Higher Education. *Online Submission*, *3*(5), 1–3.

Kayyali, M. (Ed.). (2024). *Building Resiliency in Higher Education: Globalization, Digital Skills, and Student Wellness: Globalization, Digital Skills, and Student Wellness*. IGI Global.

Kayyali, M. (2025 a). Leadership's Role in Fostering a Resilient Organizational Culture. In *Enhancing Resilience in Business Continuity Management* (pp. 267–294). IGI Global Scientific Publishing. DOI: 10.4018/979-8-3693-8809-9.ch009

Kayyali, M. (2025 b). E-Learning Platforms and Their Influence on Lifelong Learning. In Impact of Digitalization on Education and Social Sustainability (pp. 155-176). IGI Global.

Lacelle-Webster, A. (2023). A Democratic Theory of Hope: Collective Agency in Uncertain Times (Doctoral dissertation, UNIVERSITY OF BRITISH COLUMBIA (Vancouver).

Mendonça dos Santos, R. (2024). Teachers' narratives of resisting deprofessionalization (Doctoral dissertation, University of British Columbia).

Miles, J. (2024). Guilt, complicity, and responsibility for historical injustice: Towards a pedagogy of complex implication. *Pedagogy, Culture & Society*, *32*(3), 619–635. DOI: 10.1080/14681366.2022.2064537

Millefiorini, A. (Ed.). (2025). *Democracies on the Move: Contributions to a Social Theory of Contemporary Democracies*. Taylor & Francis. DOI: 10.4324/9781032722528

Neenan, M., & Dryden, W. (2021). Understanding and developing resilience. In *Cognitive Behavioural Coaching in Practice* (pp. 199–220). Routledge. DOI: 10.4324/9781003027164-9

Oyedotun, T. D. (2024). Digital pedagogy: The role of e-learning tools in facilitating 21st century skills. *International Journal of Innovation in Education*, *9*(1), 32–46. DOI: 10.1504/IJIIE.2024.144155

Ozalp, H., Ozcan, P., Dinckol, D., Zachariadis, M., & Gawer, A. (2022). "Digital colonization" of highly regulated industries: An analysis of big tech platforms' entry into health care and education. *California Management Review*, *64*(4), 78–107. DOI: 10.1177/00081256221094307

Salas-Vallina, A., Rofcanin, Y., & Las Heras, M. (2022). Building resilience and performance in turbulent times: The influence of shared leadership and passion at work across levels. *Business Research Quarterly*, *25*(1), 8–27. DOI: 10.1177/23409444211035138

Salinas, I., Fernández, M. B., Johnson, D., & Bastías, N. (2023). Freire's hope in radically changing times: A dialogue for curriculum integration from science education to face the climate crisis. *Cultural Studies of Science Education*, *18*(1), 21–39. DOI: 10.1007/s11422-023-10157-4 PMID: 36845560

Sharma, P., Joshi, S., Gautam, S., Maharjan, S., Khanal, S. R., Reis, M. C., . . . de Jesus Filipe, V. M. (2022, August). Student engagement detection using emotion analysis, eye tracking and head movement with machine learning. In International conference on technology and innovation in learning, teaching and education (pp. 52-68). Cham: Springer Nature Switzerland. DOI: 10.1007/978-3-031-22918-3_5

Singh, A. (2024). The Future of Learning: AI-Driven Personalized Education. Available at *SSRN* 5076438. DOI: 10.56557/ajocr/2024/v9i49018

Smeplass, E. (2023). Nurturing inclusivity and professional growth among vocational teachers through communities of practice. *Pedagogy, Culture & Society*, , 1–20.

Sobkowiak, M., Senn, J., & Vollmer, H. (2023). Rethinking planetary boundaries: Accounting for ecological limits. *Social and Environmental Accountability Journal*, *43*(3), 259–272. DOI: 10.1080/0969160X.2023.2283019

Stanley, P. D., & Clinton, J. R. (2024). *Connecting: The mentoring relationships you need to succeed in life*. NavPress.

Tariq, M. U. (2024). Navigating the personalization pathway: Implementing adaptive learning technologies in higher education. In *Adaptive Learning Technologies for Higher Education* (pp. 265–291). IGI Global. DOI: 10.4018/979-8-3693-3641-0.ch012

Trabelsi, Z., Alnajjar, F., Parambil, M. M. A., Gochoo, M., & Ali, L. (2023). Real-time attention monitoring system for classroom: A deep learning approach for student's behavior recognition. *Big Data and Cognitive Computing*, *7*(1), 48. DOI: 10.3390/bdcc7010048

Tsamados, A., Aggarwal, N., Cowls, J., Morley, J., Roberts, H., Taddeo, M., & Floridi, L. (2021). The ethics of algorithms: key problems and solutions. Ethics, governance, and policies in artificial intelligence, 97-123.

Tuhuteru, L. (2023). The role of citizenship education in efforts to instill democratic values. International journal of humanities education and social sciences, 2(4).

Wagner, P., & Kössler, T. (2022). Moulding democratic citizens: democracy and education in modern European history–an introduction. European Review of History: Revue européenne d'histoire, 29(6), 859-883.

Wilson, A., & Brezicha, K. F. (2023). Good citizenship and the "True and Inspiring Story of America": A critical policy analysis of the South Dakota civics and history initiative. *Peabody Journal of Education*, *98*(5), 500–515. DOI: 10.1080/0161956X.2023.2261311

Chapter 5
Post-Globalization and Education:
Redefining Access and Equity in a Technological Era

Mustafa Kayyali
http://orcid.org/0000-0003-3300-262X
Maaref University of Applied Sciences, Syria

ABSTRACT

In the wake of globalization's waning influence, a new era is emerging—one where educational access and equity must be reconsidered through the lens of technological transformation. This chapter explores the post-globalization landscape and the evolving role of Information and Communication Technologies (ICTs) in shaping the future of education. While digital tools have expanded access to knowledge, they have also introduced new forms of exclusion and inequality, particularly in underrepresented and marginalized communities. By interrogating how access is defined, who benefits from educational technologies, and what forms of equity are truly being pursued, this chapter offers a nuanced analysis of the intersection between technology and justice in education. Drawing on global examples and critical theory, it argues for a human-centered approach that reclaims education as a public good and not merely a digital service. Ultimately, it calls for systemic reform that aligns technological advancement with inclusive, ethical, and equitable educational futures.

DOI: 10.4018/979-8-3373-1190-6.ch005

INTRODUCTION

The promise of globalization once gleamed with optimism. It envisioned a world interlinked by open markets, global citizenship, and unprecedented flows of knowledge. In the realm of education, globalization ushered in a period of rapid transformation—cross-border collaborations, the digitization of curricula, the expansion of English as a lingua franca of academia, and the hopeful rhetoric of democratized access (Voronkova et al., 2023). But behind that optimism, beneath the sheen of policy papers and development reports, cracks began to show. The benefits of globalization were unevenly distributed. The myth of borderless learning masked deep-rooted inequities. And then, somewhere between rising geopolitical rifts, economic realignments, and a global pandemic that redefined the meaning of distance and proximity, we entered what many now describe as the post-globalization era. This shift—more tectonic than transitional—is not merely political or economic. It is epistemological. It challenges how we conceptualize access, belonging, and participation in knowledge systems. In education, this demands a radical rethinking of the frameworks that have guided our assumptions for decades. The post-globalization moment is not just about the end of something—it is also the beginning of a complex negotiation with new forms of technological mediation, socio-cultural fragmentation, and the reassertion of national and regional identities in knowledge production.

At the heart of this negotiation lies the role of Information and Communication Technologies (ICTs). No longer confined to the realm of innovation, ICTs are now woven into the very fabric of how education is delivered, assessed, and imagined (Roy & Swargiary, 2023). Yet, despite their ubiquity, these technologies are far from neutral. They are imbued with cultural assumptions, political agendas, and economic interests that shape who can access them, how they are used, and what kinds of futures they make possible. For all the talk of digital inclusion, the reality remains jarring: while some students learn in fully immersive virtual environments with real-time feedback and adaptive AI tutors, others struggle with spotty mobile data, power outages, or a lack of culturally relevant content. Access, therefore, cannot be reduced to a question of infrastructure—it is fundamentally a question of equity, justice, and voice. This chapter emerges from this tension. It seeks to explore how the collapse—or reconfiguration—of globalization impacts the landscape of educational access and equity in a technological era. It asks hard questions: What does it mean to have access to education when the very definition of education is in flux? How do ICTs amplify existing inequalities under the guise of innovation? And can we envision a post-globalization ethic of education that foregrounds human dignity, social justice, and epistemic plurality over digital uniformity? Drawing on a range of global contexts, this chapter argues that educational equity in the post-globalization era must move beyond simplistic metrics of device distribution or broadband penetration. True equity demands a deeper reckoning with the cultural, structural, and ideological barriers that shape who is included and who is marginalized in digital education spaces. It also requires us to confront the limits of globalized thinking—the idea that one model of education, one language, or one technological solution can serve all. In response, this chapter proposes a redefinition of access: one that is not merely about being online, but about being seen, heard, and meaningfully engaged. It calls for an approach to educational technology that is rooted in local knowledge, inclusive design, and ethical accountability. Post-globalization, then, is not an end—it is a challenge. A call to reimagine education not as a race toward digital perfection, but as a shared, human journey toward justice, inclusion, and collective flourishing.

FROM GLOBALIZATION TO POST-GLOBALIZATION: SHIFTING EDUCATIONAL PARADIGMS

It is difficult to pinpoint the precise moment when globalization began to lose its grip on the world's imagination. For decades, it stood as an almost unquestionable force—driving policy, shaping economies, influencing culture, and perhaps most profoundly, redefining education. It was the invisible engine behind university internationalization, the explosion of cross-border partnerships, the global migration of students and academics, and the sweeping digitization of learning platforms (Woodman et al., 2023). The very notion of "21st-century skills" was born in the womb of globalization: adaptability, digital fluency, intercultural competence. We were no longer educating for the local or the national—we were educating for the world. But if globalization's rise was gradual, its unraveling came like a wave pulling back from a shoreline, revealing the sharp rocks that had always been beneath the surface. Nationalism resurged. Borders were reasserted. The pandemic served as a brutal reminder of the limits of global interdependence, especially when it came to equitable vaccine access, online learning capabilities, and international cooperation. Supply chains fractured. Academic mobility was frozen (Kayyali, 2020). Even the internet, once heralded as a universal common, became increasingly territorial. Surveillance, censorship, data nationalism—all grew in the shadow of what once seemed like limitless connectivity (Samier, 2022). This, perhaps, is the beginning of what we might call *post-globalization*. Not a clean departure from the global, but a slow, uneven reorientation of priorities (Zamfir et al., 2023). A time when global ambitions are tempered by local realities. When the idea of one world, one market, one educational model feels not only unrealistic but, in many contexts, unjust. Post-globalization is less a theory than a condition—fluid, emergent, difficult to define. And yet its implications for education are profound. It forces us to confront what we thought education was supposed to do in the global age, and what it must now become.

In the heyday of globalization, educational systems were rapidly aligning themselves to a set of dominant values: competitiveness, efficiency, innovation, mobility. Schools and universities across the globe rushed to adapt to international benchmarks, accreditation standards, and testing regimes that were often rooted in Euro-American paradigms. English-language instruction proliferated. International student recruitment turned into a billion-dollar industry. Ranking systems became the currency of academic prestige (Kayyali, 2023). Curricula were restructured to reflect a supposedly global perspective, often at the expense of indigenous knowledge, local history, and critical pedagogies. There was a prevailing sense that education should prepare students to become "global citizens"—an idea that, while noble in theory, often translated in practice to an uncritical embrace of neoliberal values: self-optimization, entrepreneurialism, digital professionalism. The cosmopolitan ideal masked a growing homogenization. Students in Nairobi, Mumbai, São Paulo, and Jakarta were learning from the same platforms, reading the same "world-class" authors, taking the same MOOCs, applying for the same internships at multinational companies. Diversity was celebrated rhetorically, but structurally, educational systems were converging around a narrow vision of success. At the same time, glaring contradictions persisted. Even as digital access expanded, the global digital divide remained stubbornly wide. While some students had access to virtual labs, simulation tools, and personalized learning analytics, others were attending WhatsApp-based classes under candlelight (García-Martín & García-Sánchez, 2022). The promise of global learning was not shared equally. And as the technological sophistication of global education increased, so too did its detachment from the lived realities of marginalized communities. Then came the disruptions—some slow, others sudden. The rise of right-wing populism. Trade wars. Visa restrictions. Brexit. The pandemic. Each event chipped away at the infrastructure of global education. Cross-border mobility, once the lifeblood of international learning, was abruptly halted. Universities that had built their financial models around international students were forced to reckon with vulnerabilities they had long ignored (Kayyali, 2022). Teachers scrambled to translate classroom intimacy into remote formats. Families struggled to provide digital access in households where internet connectivity was a luxury, not a given. Education's global project—so meticulously constructed—was shown to be far more brittle than we had believed. And so, we find ourselves in the uneasy terrain of post-globalization. In this terrain, the old narratives no longer hold. Globalization, once positioned as a natural and inevitable progression, now seems exposed as partial and uneven. And if globalization created educational systems oriented toward mobility, uniformity, and global competitiveness, then post-globalization invites us—perhaps compels us—to craft something different.

One of the first shifts required in this era is a reconsideration of *where* knowledge comes from. For too long, the global education discourse has privileged knowledge produced in elite institutions in the Global North. These institutions set the tone, defined the standards, and dictated the metrics of excellence. Academic journals, citation indexes, textbook publishers—all disproportionately reflected Eurocentric frameworks (Ledgerwood et al., 2024). But as globalization recedes, there is a new opportunity to value local epistemologies, to revive indigenous pedagogies, and to re-center voices that have long been peripheralized. This is not a call for isolationism or academic protectionism. Rather, it is a call for epistemic justice—for an education that recognizes the plural origins of knowledge, and that treats difference not as a deficit to be remediated, but as a resource to be cultivated. It means moving beyond tokenistic inclusion toward structural transformation. It means challenging the assumption that success in education looks the same everywhere. A second major shift concerns the role of technology. If globalization brought us the first wave of educational technology—MOOCs, virtual exchanges, collaborative tools—post-globalization forces us to grapple with their limitations. The pandemic was a powerful litmus test (Hussain, 2023). On the one hand, it showcased the potential of digital learning. Teachers around the world showed remarkable adaptability. Students demonstrated resilience. Educational access, at least in theory, expanded. On the other hand, the crisis laid bare the systemic inequities that had always been present: disparities in access to devices, stable internet, supportive home environments, and accessible digital content.

In the post-global era, simply scaling up technology is no longer enough. What matters now is *how* technology is used, *who* it serves, and *whose* values it encodes. Are our digital tools fostering critical thinking or passive consumption? Are they reinforcing dominant languages and perspectives, or are they enabling multilingual, multicultural learning? Are they accessible to students with disabilities, students in rural areas, students who are undocumented or displaced? If not, then they are not instruments of equity—they are extensions of inequality. Technology must become more than a delivery mechanism. It must become a site of pedagogy, a space of resistance, a tool for community-building and storytelling. In this sense, the educational paradigm of post-globalization is not anti-technology—it is ethically and socially *aware* technology. Tools designed not just for scalability, but for solidarity. Platforms that are responsive to local needs, that are co-created with communities, that reflect the lived experiences of learners (Kayyali, 2025 a). Another crucial reorientation lies in the purpose of education itself. Under globalization, education was frequently positioned as a means to an economic end—skills for the knowledge economy, credentials for mobility, degrees for jobs. While these aims are not inherently wrong, they are incomplete. They risk reducing learners to human capital, and learning to a transactional process. In the post-globalization moment, we are being asked to imagine education not just as preparation for work, but as preparation for life—for democratic participation, ecological stewardship, intercultural empathy, and collective flourishing.

This shift requires courage. It means decentering standardized testing as the primary measure of learning. It means valuing the arts, humanities, and civic education, even when they do not yield immediate economic returns. It means listening to students not just as consumers of content, but as co-creators of knowledge. It means acknowledging that learning is as much about identity, belonging, and meaning as it is about technical proficiency. And perhaps most importantly, post-globalization asks us to rethink *where* education happens. During globalization's ascent, the emphasis was on global networks, elite universities, and flagship institutions (Menon et al., 2023). But the future of education may not lie in sprawling campuses or transnational mega-deals. It may lie in local community centers, in culturally sustaining classrooms, in alternative learning spaces that prioritize care, relationality, and mutual aid. It may lie in smaller, slower, more grounded forms of learning that honor the complexity of place and the dignity of each learner. This is not to romanticize the local or to abandon the global. It is to insist that education must be rooted before it can be connected. That true global understanding arises not from abstraction, but from deep, contextual engagement. And that equity cannot be achieved through uniformity—it must be pursued through pluralism. In sum, the transition from globalization to post-globalization is not just a geopolitical shift. It is an invitation to reimagine education at every level—from curriculum design to pedagogical values, from technological implementation to institutional mission. It is an opportunity to confront the unfinished business of globalization: the inequities it obscured, the voices it sidelined, the knowledges it overlooked. To navigate this new era, we need more than innovation. We need imagination. We need ethical leadership, critical reflexivity, and genuine humility. We need educators who are willing to ask not just what works, but *for whom*, *in what context*, and *to what end*. And we need systems that are willing to evolve—not just to keep pace with change, but to shape change in ways that are just, inclusive, and sustainable. The post-globalization era is already here. The question is not whether we will respond, but *how*.

TECHNOLOGY AS A DOUBLE-EDGED SWORD: ACCESS AND ALIENATION IN THE DIGITAL AGE

There's a particular kind of silence that fills a classroom when technology fails. A screen freezes mid-lecture, a platform crashes during an exam, a student's voice flickers in and out over an unstable connection. And in that pause—often awkward, sometimes frustrating—something deeper is revealed: a fragile dependency on a system we're still learning to understand. In recent years, educational technology has surged from supplementary tool to foundational infrastructure (Lamb & Weiner, 2021). The classroom, once defined by walls and chalkboards, now extends into cloud servers, algorithmic tutors, and asynchronous forums. But as this transition accelerates, we must ask: at what cost? For whom does this digital expansion signify empowerment, and for whom does it deepen disconnection?

Technology in education is often introduced with the best of intentions (Kayyali, 2021). It promises to widen access, personalize learning, break down barriers of geography and socioeconomic status. And in many ways, it has. Online courses have reached students in conflict zones. Open-source materials have brought world-class content to underfunded schools (Özmacun, 2023). Assistive technologies have given students with disabilities a more equitable seat at the learning table. For educators, digital tools have unlocked new pedagogical possibilities—interactive simulations, multimedia storytelling, real-time feedback loops. These gains are real. They matter. And yet, they exist alongside an unsettling truth: the very technologies designed to connect us can also estrange us—from one another, from ourselves, and from the deeper purposes of education. One of the most visible tensions in this landscape is the notion of "access." In policy circles and development discourse, access has become a kind of shorthand for progress. More devices, more bandwidth, more digital tools—these are the metrics by which success is often measured. But access is not a monolith. A student with a smartphone and a data plan in a rural village does not inhabit the same learning environment as a student with fiber-optic broadband, a laptop, and a quiet room (Pregowska et al., 2021). The former may technically be online but is often navigating a maze of limitations: low-quality interfaces, platform incompatibilities, competing family needs, or unreliable electricity. To treat both situations as equal is to flatten lived experience into infrastructure statistics. It is to mistake connectivity for capacity. True access is not just about logging on—it is about logging in with dignity, with support, with a sense of inclusion. And for that, we need to reckon with the vast disparities that exist beneath the surface of digital rollouts. We need to understand that "digital divide" is not just a phrase—it is a lived reality, felt in moments of exclusion, frustration, and quiet resignation.

Equally complex is the psychological landscape of digital learning. While technology enables flexibility, it can also foster isolation. The communal rituals of learning—the informal conversations before class, the shared glances during group discussions, the subtle cues of body language—are difficult to replicate online. In the absence of these moments, many students report a sense of disconnection. They attend class, submit assignments, join forums, and yet feel unmoored—like avatars moving through content rather than embodied learners engaged in a shared journey.

This is especially acute for first-generation students, those navigating language barriers, or students in culturally alienating platforms. Alienation, in this sense, is not just emotional. It is structural. It emerges when the design of educational technology assumes a one-size-fits-all learner (Fitriyah, 2025). Platforms often reflect the values and assumptions of their developers—typically based in the Global North, predominantly English-speaking, and embedded in particular cultural norms. As a result, learners from other contexts may find the interface unfamiliar, the content irrelevant, or the pedagogy unintuitive. The promise of a "universal" education is quietly undercut by a cultural asymmetry that leaves many students feeling like guests in someone else's house. Even more subtle, perhaps, is the way technology reshapes our very relationship to knowledge. In analog classrooms, learning unfolds through dialogue, contemplation, and sustained inquiry. Online, it is often fragmented—divided into modules, quizzes, and checkboxes. Efficiency becomes a virtue; speed, a measure of mastery (Bhuyan & Chakraborty, 2024). The algorithms that power adaptive learning platforms prioritize what can be quantified—completion rates, click-throughs, time-on-task. But what of curiosity? What of uncertainty, or the slow, meandering process by which deep understanding emerges?

This shift toward data-driven learning risks narrowing the soul of education. Students begin to see knowledge not as a horizon to explore, but as a set of tasks to complete. Reflection gives way to repetition. Meaning is displaced by measurement. And in this new terrain, it is not only students who feel the pressure—teachers, too, find themselves caught in systems that prioritize analytics over relationships. Their roles become part technician, part monitor, part content-delivery interface. The affective, intuitive, deeply human dimensions of teaching are easily lost in dashboards and deadlines. There is also the issue of surveillance. As educational platforms collect more data—on student performance, attendance, engagement, and even keystroke behavior—a quiet normalization of monitoring has taken root (Preuveneers et al., 2021). Students are tracked in ways that would have been unimaginable a generation ago. Proctoring software scans their eye movements, learning management systems log every click. Ostensibly, this is to ensure fairness and optimize learning. But it raises urgent questions: Who owns this data? How is it used? What assumptions are embedded in the algorithms? And how do these dynamics affect students' sense of autonomy and trust?

Surveillance can erode the foundation of a learning community. When students feel watched, they often perform rather than explore. They internalize a sense of being constantly evaluated, which can stifle risk-taking and intellectual vulnerability. Moreover, surveillance tends to reproduce existing inequalities. Students with neurodivergent traits, different cultural norms around eye contact, or non-standardized home environments may be disproportionately flagged by automated systems (Kennedy et al., 2025). What is framed as objective monitoring is, in practice, shaped by bias and blind spots. And yet, technology remains indispensable. It is not something we can, or should, abandon. The challenge, then, is not to reject digital tools but to humanize them. To design with empathy. To move from a paradigm of *scaling up* to one of *digging deep*. This means involving educators and learners in the design of the platforms they use (Aldhafeeri & Alotaibi, 2023). It means investing in digital literacy—not just in how to use tools, but in how to critique them, question them, shape them. It means recognizing that the most powerful technologies are those that amplify human connection, not replace it. Consider, for instance, a rural teacher who uses WhatsApp to deliver poetry prompts to her students each week—not because it's the most sophisticated tool, but because it's the one they all have. Or the university that creates a low-bandwidth platform in multiple languages to reach displaced learners across borders. Or the community learning center that blends face-to-face dialogue with locally curated digital resources. These are not examples of high-tech innovation—they are examples of context-sensitive care. They reflect a shift from asking "what's possible?" to asking "what's needed?"

That question—*what's needed?*—must become the ethical compass of digital education. Because what's needed is not always more features, more automation, more platforms. Sometimes what's needed is slowness. Or simplicity. Or the humility to listen. Technology can be dazzling, but education is not a spectacle. It is a relationship. And relationships require presence, patience, and trust.

There is also a role for resistance. In a world where educational platforms are increasingly owned by tech conglomerates, we must be wary of the commodification of learning. When students become users and knowledge becomes content, education risks becoming a marketplace rather than a commons. In this climate, open-source tools, community-driven platforms, and non-commercial alternatives are acts of defiance—and hope. They remind us that education can still be a public good, not a private service. To resist, however, is not to retreat into nostalgia. It is to claim agency over the future. It is to imagine a digital ecosystem that is plural, porous, and participatory. An ecosystem where learners are not just consumers of content, but creators of knowledge (Paskevicius, 2021). Where teachers are not data-input supervisors, but mentors, facilitators, provocateurs. Where technology is not the center of the classroom, but a bridge—linking ideas, people, and worlds. None of this is easy. It requires rethinking funding models, training programs, institutional priorities. It requires admitting that some of our most celebrated innovations may be exacerbating rather than solving problems. It requires the courage to step back, even as the world pushes us to speed up. And perhaps, above all, it requires storytelling. Because in the end, technology is only as powerful as the stories we tell with it. Stories that invite students to see themselves in the curriculum. Stories that honor the multiplicity of their lives. Stories that resist the flattening logic of standardization and instead celebrate the vibrant, messy, unfinished process of becoming. In this sense, the double edge of technology is not a dilemma to resolve—it is a tension to hold. Access and alienation exist in tandem. Innovation and inequity are entangled. But in acknowledging this complexity, we move closer to an educational practice that is both honest and hopeful. The digital age is not destiny. It is design. And we are its architects.

DIGITAL LITERACIES, POWER, AND THE POLITICS OF EDUCATIONAL REFORM

It is no longer enough to know how to click, swipe, type, or even code. In today's educational landscape, the concept of digital literacy has outgrown its technical roots. It is no longer merely about operating devices or navigating interfaces—it is about understanding the power structures that underpin them. It is about asking who designs the tools we use, whose interests they serve, and what ideologies they carry. To be digitally literate today is to be politically aware. It is to read not only the screen, but the systems behind it. This shift in definition matters. Because if we fail to interrogate the deeper layers of digital literacy, we risk producing generations of students who are proficient but powerless—students who know how to use technology, but not how to critique it; who can consume digital content, but not question its origins or consequences; who adapt to systems without realizing they can also reshape them. This is not just an educational concern—it is a democratic one.

Educational reform efforts across the world are increasingly centered on digital skills (Wang et al., 2024). Curricula are being rewritten to include coding, media literacy, information evaluation. Governments are investing in STEM hubs, robotics labs, and digital classrooms (Kayyali, 2025 b). While these shifts are necessary and often overdue, they are also incomplete. Because beneath the surface of these reforms lies a deeper question: What kind of citizens are we trying to cultivate? Are we preparing students merely to fit into a digital economy, or are we preparing them to challenge it? Are we training them to be agile workers in algorithmic systems, or are we nurturing them to be critical participants in digital democracies? Are we teaching them to ask who profits from their clicks, their data, their attention?

To engage meaningfully with these questions, we must first expand our understanding of what literacy really means in the digital age. Literacy, at its core, has always been about power. To read and write is to participate in meaning-making, to shape how reality is understood and shared. In colonial contexts, literacy was a tool of both control and resistance. In feminist movements, it was a way to reclaim silenced narratives. In working-class communities, it was a path to agency. Digital literacy, then, is no different—it is the new terrain on which the politics of knowledge, identity, and voice are negotiated (Knight et al., 2023). Yet, much of the current digital literacy discourse remains technocratic. It emphasizes tools over theory, proficiency over perspective. Students learn how to navigate platforms but not how those platforms navigate them. They are taught how to evaluate sources but not how to trace the political economies of information. They may understand what a data breach is, but not how data extraction shapes their online behavior, or how predictive algorithms reinforce bias.

To build true digital literacy, we need to teach students to ask: Who made this app? What data does it collect? What behaviors does it incentivize? What norms does it reinforce? What does it silence? These are not just technical inquiries—they are ethical and political ones. And yet, such questions are rarely included in mainstream educational reform. Why? Because to teach this kind of literacy is to challenge power. It is to push back against the narrative that technology is neutral, that platforms are benign, that innovation is always good. It is to expose the ways in which digital infrastructures mirror and amplify existing inequalities—racial, gendered, economic, linguistic. And that is uncomfortable. It unsettles institutions. It unsettles markets. But that discomfort is necessary. Because the cost of silence is high. When we do not teach critical digital literacy, we leave students vulnerable—to misinformation, to manipulation, to surveillance. We also leave society vulnerable—to the erosion of democratic norms, to the concentration of power in the hands of a few tech giants, to a future where agency is replaced by automation.

Educational reform must therefore be about more than integrating tablets into classrooms or shifting exams online (Christopoulos, A., & Sprangers, 2021). It must be about cultivating critical consciousness. About helping students understand not only how to function in a digital world, but how to question it, reshape it, humanize it. That means embedding critical media studies into the core curriculum. It means teaching students about digital labor, about platform capitalism, about algorithmic bias. It means recognizing that literacy is not just about decoding texts, but about decoding systems. This kind of literacy is deeply relational. It cannot be taught through checklists or modules alone. It requires dialogue. Reflection. Trust. It requires classrooms that make space for discomfort, for complexity, for dissent. It also requires teachers who are themselves supported in developing these literacies—not through top-down directives, but through collaborative inquiry and professional freedom. But here lies another layer of the problem: educational systems are increasingly shaped by the very forces they are meant to interrogate. Ed-tech companies sponsor curricula. Standardized testing platforms collect vast amounts of student data (Adeoye & Otemuyiwa, 2024). Learning analytics are used to make decisions about funding, advancement, and discipline. In such a climate, it becomes difficult to separate pedagogy from profit. Reform becomes entangled with market interests. And critical literacy becomes a risk rather than a requirement. We must therefore be vigilant. Not only about what we teach, but about how reforms are shaped, who is shaping them, and what agendas they carry. A curriculum that emphasizes coding while ignoring critical theory is not neutral—it reflects a political choice about what counts as knowledge. A platform that tracks student behavior while claiming to enhance learning is not innocent—it participates in a broader culture of surveillance.

True reform is not about digital enhancement—it is about epistemic justice. It is about reclaiming education as a space where students can explore not only how the world works, but how it might be otherwise. A space where they can imagine alternative futures, not just optimize for the present. A space where they learn not only how to adapt to systems, but how to resist and reimagine them. Digital literacies, then, must be rooted in context. What it means to be literate in a refugee camp is different from what it means in a Silicon Valley classroom. What it means in a postcolonial society is different from what it means in a settler one. A universal approach to digital literacy is a form of epistemic violence—it erases difference, history, and power. Instead, we must adopt a pluralist approach (Sugeng et al., 2024). One that recognizes multiple literacies, multiple truths, multiple ways of knowing. This includes oral traditions, visual storytelling, embodied knowledge. It includes ancestral technologies and grassroots media. It includes literacies that are not yet recognized by the system, but that sustain communities every day. This also means honoring the literacies students already bring with them. The ability to navigate complex social dynamics online. The skill to remix media, create memes, mobilize hashtags. These are not trivial acts—they are forms of literacy, of authorship, of resistance. To build on them is to meet students where they are—not to pull them into our definitions of literacy, but to co-create new ones together. Of course, this work is slow. It does not lend itself easily to dashboards or KPIs. It cannot be reduced to quantifiable outcomes. But that is the point. Literacy is not a product—it is a process.

A lived, ongoing negotiation between self and world. Between the familiar and the unfamiliar. Between what is and what could be. And this process must extend beyond the classroom. Families, communities, libraries, grassroots organizations—all have a role to play. Reform must not only be top-down but also bottom-up. Participatory. Responsive. Grounded in lived experience. This means involving parents in digital literacy workshops. Supporting community-based educators. Funding local media initiatives. It means recognizing that learning happens everywhere, not just within formal institutions. Ultimately, the politics of educational reform are inseparable from the politics of the digital. Who controls the tools, the data, the narratives? Who sets the standards, the priorities, the pace? These are questions of governance, of justice, of accountability. And they are questions we must answer together—not just as educators, but as citizens. Digital literacies are not luxuries. They are lifelines (Eden et al., 2024). They are the foundation upon which democratic participation, personal agency, and collective flourishing rest. To treat them as optional, or to frame them narrowly, is to betray the very purpose of education. So let us teach our students not only to scroll, but to scrutinize. Not only to code, but to critique. Not only to connect, but to care. Let us reform not by replacing old tools with new ones, but by reimagining what it means to learn, to know, to act. And let us do so with humility. Because the digital age is still young. Its contours are still being shaped. And the classroom—however virtual it may become—remains one of the last places where we can choose the direction of that shaping. Let's not waste that opportunity.

TOWARD A POST-GLOBALIZATION ETHIC: HUMAN-CENTERED ICT INTEGRATION IN EDUCATION

There is a quiet reckoning unfolding beneath the surface of today's conversations about educational technology—a slow, often unspoken acknowledgment that somewhere along the path of innovation, we lost something. In the rush to digitize, optimize, and globalize, we have built remarkable systems. But we have also surrendered parts of education that cannot be reduced to code or measured by dashboards. We have privileged speed over reflection, reach over resonance, and uniformity over contextual understanding. As we drift further into a post-globalization reality, where the cracks in the old models of connection and exchange are becoming too visible to ignore, we find ourselves at a crossroads—not of technological capability, but of ethical direction. What would it mean to imagine a truly human-centered approach to ICT (Information and Communication Technologies) in education—one not governed by the logic of markets or efficiency, but by the moral imperatives of dignity, justice, care, and inclusion? What does it look like to integrate technology not *into* education, but *with* education—organically, critically, and with deep respect for the lives and contexts of learners?

The post-globalization moment invites us to ask these questions with new urgency. For decades, globalization framed educational technology as a liberating force: a way to level the playing field, to democratize knowledge, to bring learning across borders and barriers (Roudometof, 2024). And to an extent, that vision held promise. Many lives have been transformed by distance learning, open educational resources, and digital platforms that connect classrooms across continents. But the same era that opened doors for some quietly reinforced walls for others. The very mechanisms of globalization—standardization, centralization, commodification—often reproduced the very inequities they claimed to dismantle. In a global educational economy driven by scalability, local nuance became noise. Cultural specificity was flattened into generic content. "World-class" often meant Western-centric. Platforms were built for the average user, who was rarely average and never universally defined. Even well-intentioned interventions frequently arrived in communities as solutions in search of problems—technologically elegant, but socially tone-deaf.

Now, as the grand narratives of globalization recede—disrupted by pandemic, political fragmentation, economic inequality, and environmental collapse—we are being asked to reimagine our foundational assumptions. The question is no longer how to scale faster or connect more widely. The question is how to deepen, to root, to reconnect education with the lived humanity of those it serves. This is the heart of a post-globalization ethic: not anti-technology, but *post-illusion*. It is a commitment to restoring complexity in a system that too often seeks simplicity. At the core of this shift is a reframing of purpose. If the old narrative told us that education must prepare learners for the global market, the emerging narrative asks what kind of world we are preparing them to shape. This is not merely a semantic difference—it is a moral one. Under a human-centered framework, the goal of ICT integration is not to produce digital workers, but digital citizens. Not passive consumers of content, but active co-creators of knowledge. Not compliant users of platforms, but critical participants in a shared future.

Such a vision begins with listening. Listening not to metrics, but to people. To students who say they feel invisible in online classrooms. To teachers who struggle to make prescribed platforms work in chaotic, real-life environments. To parents who see their children disengage despite all the tools in front of them. To communities that ask, again and again, for education that reflects their histories, languages, and hopes (Sato & Loewen, 2022). In a truly human-centered ICT ecosystem, these voices are not anecdotes—they are architects. This requires more than empathy; it requires co-creation. Too often, technology is designed *for* users rather than *with* them. A post-globalization ethic insists on shared authorship. It understands that innovation rooted in extractive logic—where ideas are mined from communities but decisions are made elsewhere—is not innovation at all. It is replication of a colonial mindset, dressed in digital clothing. Human-centered design, in its most ethical form, is not about usability alone. It is about mutuality, reciprocity, accountability. This accountability must extend to the entire life cycle of educational technology—from funding and design to implementation and evaluation. Who funds the platforms? What are their motives? Who owns the data? Who benefits from the analytics? When platforms collect vast amounts of behavioral data under the guise of personalization, we must ask: What are we personalizing for? Whose interests are being served? And at what cost to autonomy, privacy, and trust?

A human-centered ethic recognizes that data is not just abstract information—it is intimate. It reflects lives, struggles, dreams. To treat it casually or purely commercially is to dehumanize. Transparent data practices, ethical data governance, and learner consent must be non-negotiable principles in any future-facing ICT policy. But even as we talk about rights and agency, we must not forget that humanity is not only rational—it is also emotional, embodied, relational. Education is not merely the transmission of knowledge; it is a web of connections: between teacher and learner, between peers, between one's inner world and the wider society. Technology, no matter how sophisticated, cannot replace these connections (Fawns, 2022). But it can either erode them or support them, depending on how it is conceived and applied. Consider the difference between a learning management system that tracks engagement through clicks and one that allows students to express how they are feeling before class begins. Between a digital tool that auto-grades essays and one that offers scaffolding for collaborative writing. Between a curriculum that centers "universal competencies" and one that invites students to explore their local histories through multimedia storytelling. The difference is not just technical—it is philosophical. It is a matter of values. And it is here that the post-globalization ethic finds its most radical challenge: slowness. In a world obsessed with acceleration, where technologies are measured by their ability to "disrupt," choosing to slow down is a deeply countercultural act. But it is essential. Human-centered ICT integration cannot be rushed. It takes time to build trust. Time to adapt tools to context. Time to unlearn assumptions and co-create meaning. Speed, in this sense, is not always a virtue. Sometimes it is a violence.

Let us not forget that many of the most marginalized communities have already experienced centuries of educational imposition—curricula that erased, languages that shamed, systems that excluded. If we now show up with tablets and platforms, expecting immediate adoption, we repeat a pattern of dominance. Respectful ICT integration requires not just technical adaptation, but historical humility. And it requires imagination (Klosi, 2024). A human-centered approach is not about rejecting technology—it is about reimagining it. What if platforms prioritized care over control? What if digital spaces were designed to facilitate community-led learning, intergenerational dialogue, creative risk-taking? What if we shifted our design questions from "what is efficient?" to "what is meaningful?" From "how many users?" to "how much belonging?"

There are already seeds of this future around us. In the grassroots learning collectives using open-source tools to reclaim indigenous knowledge. In the teachers who blend radio broadcasts with community discussion circles. In the youth-led digital campaigns that turn social media into spaces for justice education. These are not anomalies. They are blueprints. But to scale such practices ethically, we must rethink what "scale" means. The dominant model views scale as uniformity—one tool, many users. But a post-globalization ethic envisions scale as resonance—many tools, each rooted in place, connected by shared principles of justice and care. This is a messier, slower, but ultimately more resilient vision of change. To support it, policymakers must become facilitators, not just regulators. Funders must be willing to back process, not just product. Technologists must sit in circles, not just boardrooms. And educators must be honored not as adopters of tools, but as thinkers, leaders, creators in their own right.

In the end, a post-globalization ethic is not a roadmap. It is a stance—a way of approaching the intersection of technology and education with humility, integrity, and hope (Konrad, 2021). It resists the cynicism that says systems cannot change. It resists the reductionism that says equity is bandwidth and devices. It resists the abstraction that sees learners as data points rather than full human beings. Instead, it invites us to begin again. To rebuild educational technology from the inside out—from story, from relationship, from place. To ask, again and again, not just *what* we are building, but *why*, *for whom*, and *at what cost*. Because the future is not found in our tools. It is found in our choices. And the most human choice we can make is to center people—not platforms—in everything we do.

CONCLUSION

As we close this chapter, it is important to recognize that the post-globalization moment does not offer easy solutions. It is not a clean break from the past, nor is it a utopian reinvention of the future. Rather, it is a liminal space—uncertain, messy, full of contradictions—where new possibilities are forged through struggle, reflection, and collective effort. And in that space, education finds itself at a pivotal crossroads. We have seen how ICTs—once heralded as great equalizers—can both illuminate and intensify the disparities they were meant to address. The proliferation of online learning, virtual classrooms, and algorithm-driven platforms has undoubtedly expanded educational reach. But reach is not the same as inclusion. Bandwidth does not equal belonging. And content delivery alone does not guarantee critical engagement or learner empowerment. The digital age, with all its seductive promise, risks becoming yet another mechanism through which privilege is reproduced and difference is flattened—unless we choose otherwise.

Choosing otherwise requires a fundamental shift in our understanding of access and equity. We must move beyond narrow definitions that equate access with devices or connectivity, and instead embrace a multidimensional view—one that considers cultural relevance, linguistic diversity, socio-economic context, learner agency, and emotional safety. Equity is not achieved by handing everyone the same tablet or logging everyone into the same learning management system. It is achieved by asking: Who is this education for? Who gets to define what is worth learning? Whose voices are centered, and whose are excluded?

This chapter has argued that the post-globalization context offers a chance—a rare one—to pause and recalibrate. To question whether the frameworks we inherited from the globalization era truly serve the needs of today's learners, especially those at the margins. It calls on educators, policymakers, technologists, and communities to co-create systems of learning that are not only technologically advanced but also ethically grounded, socially responsive, and culturally situated. Moreover, we must begin to imagine education not merely as preparation for economic productivity, but as a space for meaning-making, healing, resistance, and solidarity. ICTs, in this vision, are not ends in themselves but tools—tools that must be wielded with care, humility, and a deep sense of responsibility. The goal is not to digitize the status quo but to challenge it—to build digital ecosystems that support pluralism, promote justice, and restore the dignity of learners whose needs have long been sidelined. Post-globalization is not a retreat from the world—it is an invitation to engage with it differently. It is a call to re-center the human in an era increasingly defined by algorithms and data. And it is a reminder that in our pursuit of educational reform, we must not lose sight of the simple, powerful truth that every learner deserves not just access to education, but access to a future that is equitable, empowering, and their own.

REFERENCES

Adeoye, M. A., & Otemuyiwa, B. I. (2024). Navigating the Future: Strategies of EdTech Companies in Driving Educational Transformation. JERIT. *Journal of Educational Research and Innovation Technology*, *1*(1), 43–50. DOI: 10.34125/jerit.v1i1.10

Aldhafeeri, F. M., & Alotaibi, A. A. (2023). Reimagining education for successful and sustainable digital shifting. *SAGE Open*, *13*(1), 21582440231154474. DOI: 10.1177/21582440231154474

Bhuyan, N., & Chakraborty, A. (2024). The Problem of Efficiency: Redefining the Relation Between Success & Excellence in Business Ethics. *Philosophy of Management*, *23*(1), 17–39. DOI: 10.1007/s40926-023-00234-7

Christopoulos, A., & Sprangers, P. (2021). Integration of educational technology during the Covid-19 pandemic: An analysis of teacher and student receptions. *Cogent Education*, *8*(1), 1964690. DOI: 10.1080/2331186X.2021.1964690

Eden, C. A., Chisom, O. N., & Adeniyi, I. S. (2024). Promoting digital literacy and social equity in education: Lessons from successful initiatives. *International Journal of Management & Entrepreneurship Research*, *6*(3), 687–696. DOI: 10.51594/ijmer.v6i3.880

Fawns, T. (2022). An entangled pedagogy: Looking beyond the pedagogy—technology dichotomy. *Postdigital Science and Education*, *4*(3), 711–728. DOI: 10.1007/s42438-022-00302-7

Fitriyah, A. (2025). Inclusive Policy Design for Transformative Education: A Global Synthesis of Teacher Identity Formation in Resource-Constrained Contexts. *Sinergi International Journal of Education*, *3*(2), 94–106. DOI: 10.61194/education.v3i2.703

García-Martín, J., & García-Sánchez, J. N. (2022). The digital divide of know-how and use of digital technologies in higher education: The case of a college in Latin America in the COVID-19 era. *International Journal of Environmental Research and Public Health*, *19*(6), 3358. DOI: 10.3390/ijerph19063358

Hussain, S. M. A. (2023). Actualising soft power through cultural diplomacy: A media, education, and communication perspective. *Dogo Rangsang Research Journal*, *13*(6), 196–205.

Kayyali, M. (2020). Post COVID-19: New era for higher education systems. *International Journal of Applied Science and Engineering*, *8*(2), 131–145. DOI: 10.30954/2322-0465.2.2020.6

Kayyali, M. (2021). Positive impact of high technology on higher education.

Kayyali, M. (2022). Post-COVID-19 Pandemic Main Challenges with e-Learning.

Kayyali, M. (2023). Importance of Rankings in Academia. *Online Submission*, *4*(1), 1–3.

Kayyali, M. (2025 a). E-Learning Platforms and Their Influence on Lifelong Learning. In Impact of Digitalization on Education and Social Sustainability (pp. 155-176). IGI Global.

Kayyali, M. (2025 b). The Impact of Personalized Learning on Student Engagement and Achievement in STEAM. In *Integrating Personalized Learning Methods Into STEAM Education* (pp. 51–78). IGI Global Scientific Publishing. DOI: 10.4018/979-8-3693-7718-5.ch003

Kennedy, L. J., Richdale, A. L., & Lawson, L. P. (2025). A comparison of neurodivergent and non-neurodivergent students' characteristics and higher education experiences in Australia and New Zealand. *Autism in Adulthood: Challenges and Management*, aut.2024.0159. DOI: 10.1089/aut.2024.0159

Klosi, I. (2024). THE TEACHER OF THE FUTURE: VISION AND SKILLS. *KNOWLEDGE-International Journal*, *64*(5), 609–613.

Knight, J., Dooly, M., & Barberà, E. (2023). Getting smart: Towards critical digital literacy pedagogies. *Social Semiotics*, *33*(2), 326–349. DOI: 10.1080/10350330.2020.1836815

Konrad, V. (2021). New directions at the post-globalization border. *Journal of Borderlands Studies*, *36*(5), 713–726. DOI: 10.1080/08865655.2021.1980733

Lamb, A. J., & Weiner, J. M. (2021). Technology as infrastructure for change: district leader understandings of 1: 1 educational technology initiatives and educational change. *Journal of Educational Administration*, *59*(3), 335–351. DOI: 10.1108/JEA-10-2020-0220

Ledgerwood, A., Lawson, K. M., Kraus, M. W., Vollhardt, J. R., Remedios, J. D., Westberg, D. W., Uskul, A. K., Adetula, A., Leach, C. W., Martinez, J. E., Naumann, L. P., Reddy, G., Chucky Tate, C., Todd, A. R., Weltzien, K., Buchanan, N. C., González, R., Montilla Doble, L. J., Romero-Canyas, R., & Zou, L. X. (2024). Disrupting racism and global exclusion in academic publishing: Recommendations and resources for authors, reviewers, and editors. *Collabra. Psychology*, *10*(1), 121394. DOI: 10.1525/collabra.121394

Menon, P., Samaddar, K., & Aparna, K. H. (2023). Towards globalisation of B-schools: Best practices for sustainable higher education. *International Journal of Business and Globalisation, 33*(1-2), 130–146. DOI: 10.1504/IJBG.2023.128326

Özmacun, T. E. (2023). A critical perspective on intellectual property rights through the lens of the free and open-source software experience (Doctoral dissertation, Middle East Technical University (Turkey)).

Paskevicius, M. (2021). Educators as content creators in a diverse digital media landscape.

Pregowska, A., Masztalerz, K., Garlińska, M., & Osial, M. (2021). A worldwide journey through distance education—From the post office to virtual, augmented and mixed realities, and education during the COVID-19 pandemic. *Education Sciences, 11*(3), 118. DOI: 10.3390/educsci11030118

Preuveneers, D., Garofalo, G., & Joosen, W. (2021). Cloud and edge based data analytics for privacy-preserving multi-modal engagement monitoring in the classroom. *Information Systems Frontiers, 23*(1), 151–164. DOI: 10.1007/s10796-020-09993-4

Roudometof, V. (2024). How should we think about globalization in a post-globalization era?. Dialogues in Sociology, 29768667241293053.

Roy, K., & Swargiary, K. (2023). *ICT Revolutionalizing Education*. Scholar's Press.

Samier, E. A. (2022). To surveil or not to surveil: Educational surveillance in populist and nationalist times. In Populism and Educational Leadership, Administration and Policy (pp. 188-205). Routledge.

Sato, M., & Loewen, S. (2022). The research–practice dialogue in second language learning and teaching: Past, present, and future. *Modern Language Journal, 106*(3), 509–527. DOI: 10.1111/modl.12791

Sugeng, S., Fitria, A., & Bastomi, S. (2024). Epistemic Injustice and Digital Disinformation: Addressing Knowledge Inequities in the Digital Age. *DISKURSUS-JURNAL FILSAFAT DAN TEOLOGI STF DRIYARKARA, 20*(1), 135–169. DOI: 10.36383/diskursus.v20i1.483

Voronkova, V., Vasyl'chuk, G., Nikitenko, V., Kaganov, Y., & Metelenko, N. (2023). Transformation of digital education in the era of the fourth industrial revolution and globalization.

Wang, C., Chen, X., Yu, T., Liu, Y., & Jing, Y. (2024). Education reform and change driven by digital technology: A bibliometric study from a global perspective. *Humanities & Social Sciences Communications*, *11*(1), 1–17. DOI: 10.1057/s41599-024-02717-y

Woodman, T. C., Whatley, M., & Glass, C. R. (Eds.). (2023). *Digital internationalization in higher education: Beyond virtual exchange*. Taylor & Francis. DOI: 10.4324/9781003444237

Zamfir, A. M., Davidescu, A. A., & Năstasă, A. (2023, June). Implications of Educational Poverty in a Post-Globalization World. In Griffiths School of Management and IT Annual Conference on Business, Entrepreneurship and Ethics (pp. 265-284). Cham: Springer Nature Switzerland.

Chapter 6
Unveiling AI's Dual Impact:
Analyzing Creativity, Application, and Decision-Making Skills in Education

Varanasi Rahul
http://orcid.org/0000-0003-2407-7654
Jain (Deemed to be) University, India

Y. Fathima
Jain (Deemed to be) University, India

L. B. Muralidhar
http://orcid.org/0000-0003-3453-613X
Jain (Deemed to be) University, India

R. Manju Priya
http://orcid.org/0000-0001-9144-5301
Jain (Deemed to be) University, India

N. Sathyanarayana
http://orcid.org/0000-0002-4185-7751
Jain (Deemed to be) University, India

G. Srividhya
http://orcid.org/0000-0002-5216-546X
Jain (Deemed to be) University, India

ABSTRACT

Integrating Artificial Intelligence (AI) tools, such as ChatGPT, into education is transforming learning paradigms. This chapter investigates AI's dual impact on students' creativity, application, and decision-making skills through an experimental study involving control and experimental groups. Results highlight that while AI can significantly enhance learning outcomes, improper usage risks cognitive laziness and dependence. This chapter provides an in-depth discussion on fostering responsible AI engagement, ethical considerations, and strategies to harness its potential for education. This study contributes to the discourse on sustainable AI integration in academia by bridging theoretical frameworks with empirical findings.

DOI: 10.4018/979-8-3373-1190-6.ch006

1. INTRODUCTION

The integration of Artificial Intelligence (AI) in education has significantly transformed teaching, learning, and assessment. Key themes emerging from the literature include intelligent tutoring, adaptive learning systems, and ethical considerations. Intelligent Tutoring Systems, initially proposed by Anderson et al. (1995) and later supported by VanLehn (2011) and Koedinger et al. (2012), tailor instruction to individual learner needs. Similarly, adaptive learning systems personalize learning pathways, though they raise concerns around data privacy and educational inequality (Gasevic et al., 2017). Ethical issues—particularly bias and data protection—are underscored by Diakopoulos (2016), Selbst and Barocas (2018), and Tene and Polonetsky (2014). Technologies like Natural Language Processing enhance human-computer interaction (Graesser et al., 2018), while the combination of AI and gamification, explored by Hamari & Koivisto (2014) and Deterding et al. (2011), shows promise for increasing student engagement. Successful AI implementation requires balancing personalization with ethical safeguards.

AI also influences creativity in education by supporting ideation, problem-solving, and alternative thinking strategies. Generative models and intelligent tutors promote creative workflows (Chittaro & Buttussi, 2015), but assessing creativity in AI-supported contexts remains challenging (McCosker & Wilken, 2012). Ethical implementation demands careful attention to the interplay between technological augmentation and learner agency.

AI enhances students' application of knowledge in realistic contexts. Adaptive systems promote deeper understanding and skill transfer (VanLehn, 2011), while virtual simulations foster experiential learning (Ferguson, 2012; Sitzmann et al., 2008). However, aligning AI assessments with authentic applications remains a challenge. AI tools also facilitate critical thinking and problem-solving (Aleven et al., 2009; Van den Heuvel-Panhuizen et al., 2019), while ethical concerns call for fairness and transparency in AI-driven evaluations (Bull & Kay, 2016). AI plays a growing role in shaping students' decision-making abilities, a crucial aspect of cognitive development. AI supports information analysis (Baker et al., 2008; Gobert et al., 2019), simulates decision-making environments, and guides learners through complex scenarios. Yet, overreliance on AI raises concerns about diminishing critical thinking. Integrating ethical frameworks into AI tools—emphasizing value-sensitive decision-making—is essential. By analyzing learning patterns, AI fosters personalized decision-making pathways, enhancing autonomy and skill development (Khan et al., 2021).

2. PREVIOUS STUDIES

Integrating artificial intelligence (AI) in education has attracted extensive literature attention from scholars, teachers, and policymakers who examine both the potential for change and the multifaceted challenge in AI application. As technology continues to redefine where and how information is processed and transmitted, several studies have sought to examine AI's impact on learning achievements, student interest, and moral practices in institutions. The following offers comprehensive literature on AI's benefits and limitations in education, explaining how AI is used to advance individualized learning and skill development while raising questions on cognitive dependency, diminished creativity, and moral questions such as data protection and bias.

AI's penetration in education is multifaceted, including intelligent systems, adaptive learning technologies, and simulation technologies. These technologies have reshaped traditional teaching models and have given birth to innovative dynamics in teachers and students' interactions with content. As powerful tools for differentiated instruction and timely feedback, AI offers substantial benefits and serious drawbacks that must be carefully weighed. In this overview, the literature is arranged into three major sections: AI's benefits in education, potential adverse effects of using AI, and moral imperatives for AI's implementation.

2.1 Positive Impact of AI in Education

2.1.1 Intelligent Tutoring System and Customized Instruction

One of AI's most impactful educational contributions is designing intelligent tutoring systems (ITS) to emulate the individual attention of experienced teachers. VanLehn (2011) describes how ITS can observe students' learning behaviors and patterns in real time and offer individualized feedback based on their needs. By continuously monitoring their performance, these systems adapt the challenge of exercises and instructional pace, such that every student is appropriately challenged and assisted where necessary.

Beyond simply mapping instructional content to student ability, ITS systems include advanced prediction facilities for learning trajectories. Koedinger et al. (2012) propose that such prediction facilities provide interventions such as supplementary practice on topics where students have proved to have struggles. Such a procedure does not only ensure individualized learning but also bridges learning gaps, which otherwise may slow up academic progress. For instance, in mathematics, an ITS might recognize that a student is struggling to grasp fractions and, in response, offer a chain of scaffolded exercises and in-time hints, hence enabling mastery before transitioning to advanced content.

Recent studies have expanded on such initial evidence, focusing on the role of ITS in serving students who have multiple learning profiles. Through data analytics, such systems can adapt to variability in learning pace and style, something of considerable application in diverse classrooms. Furthermore, implementing natural language processing (NLP) strategies in ITS makes more affluent, more natural conversation patterns possible in response, yielding more significant and personal interaction. Such personalization, combined with greater understanding and retention, facilitates greater student autonomy by enabling students to take on responsibility for their learning.

Examples for Intelligent Tutoring System and Customized Instruction

Claude for Education – Anthropic's AI Assistant in Universities: Anthropic has introduced "Claude for Education," an AI assistant tailored for higher education. Unlike traditional AI tools, Claude emphasizes ethical usage by employing a "Learning Mode" that utilizes the Socratic method to pose reflective, guiding questions, promoting critical thinking over rote answers. Institutions like Northeastern University and the London School of Economics have adopted Claude to support learning through interaction, including study guide generation from uploaded materials.

Wharton School's AI-Integrated Curriculum: The Wharton School at the University of Pennsylvania has overhauled its curriculum to include a new MBA major and undergraduate concentration in Artificial Intelligence. This initiative balances technical training with ethical, economic, and social considerations, preparing students for a business world increasingly shaped by AI. The curriculum includes courses like "Foundations of Deep Learning" and "Big Data, Big Responsibilities," integrating AI into various aspects of business education.

Duolingo's AI-Powered Language Learning: Duolingo has leveraged AI to transform language education by generating content faster and offering personalized learning experiences. Features like AI-driven video calls allow users to practice conversations, enhancing engagement and providing a more interactive learning environment. Duolingo's CEO emphasizes AI's role in extending personalized education beyond languages to subjects like math and music.

2.1.2 AI-Powered Simulation and Practical Skills Acquisition

AI-powered simulations have evolved to offer an irreplaceable platform for linking theoretical knowledge and application in life. In experiential-learning-oriented fields like medicine, engineering, and natural sciences, such simulations offer an irreplaceable platform for skill development. Ferguson (2012) notes that AI-powered virtual environments offer students a platform to test challenging scenarios in a safe environment. For example, medicine students might undertake simulated surgery, which enhances their technical skill and prepares them for the stressful context of treating actual patients.

These simulations often incorporate immersive technologies including augmented reality (AR) and virtual reality (VR), amplifying sensory impact and encouraging deeper engagement. Students can better transfer their learning to everyday settings as they interact in realistic scenarios. Furthermore, AI-driven simulations often include game-like features such as leveling, scoring, and immediate response, encouraging students to overcome challenging exercises. Such gameified learning is perfect for maintaining students' attention and keeping them entirely focused, effectively eliminating some boredom in standard rote learning practices.

Moreover, AI-driven simulation is by no means limited to scientific studies. In business and social studies, for instance, virtual role-plays offer students access to replicating negotiating, crisis management, and moral decision-making scenarios. These systems might include unpredictable parameters, hence copying the intricacies of realistic scenarios and preparing students for constantly changing scenarios in their professional lives. Thus, pragmatic values of AI-powered simulation lie way beyond content dissemination, such as challenging students to think critically and develop analytical, strategically oriented thinking for professional development.

2.1.3 Adaptive Learning Technologies and Student Engagement

Adaptive learning technologies offer a building block for AI's influence on education. Adaptive systems, through machine learning, constantly assess student performance and adapt instructional content accordingly. Koedinger et al. (2012) observe that adaptive technologies offer students content that is never too challenging or easy, thus crucial to maintaining students on an optimal learning curve. Such frequent calibration of challenge is necessary to sustain student interest and keep students from cognitive load and boredom.

The ability to provide individualized learning routes is most prized in large, heterogenous classrooms where individual attention is in short supply. Adaptive learning systems can identify patterns of mastery or challenge and modify the course in real time, so each student is offered an individualized learning opportunity. In addition to content adaptability, such systems tend to include AI-driven chatbots and virtual assistants who offer on-demand help. These digital assistants can offer answers, clarify thorny questions, and even propose supplementary materials—ensuring an ongoing learning process that continues after school hours.

Studies have shown timely, constructive feedback from such adaptive systems suits student motivation and learning. Students who receive constructive, timely feedback on their mistakes or questions develop reflective learning habits. Furthermore, data gathered by adaptive learning technologies can provide teachers with insights on student progress and course quality and thus better instructional decisions. Such a loop creates a symbiotic relationship between teaching and technology, where teachers and students both benefit through an augmented, responsive, and dynamic learning environment.

Artificial Intelligence (AI) is revolutionizing education by offering personalized learning experiences, real-time feedback, and adaptive content delivery. Tools like Intelligent Tutoring Systems (ITS) and AI-driven simulations enable students to engage with material at their own pace, fostering deeper understanding and retention. For instance, platforms such as Duolingo utilize AI to tailor language lessons to individual learners, enhancing engagement and efficacy. Moreover, AI assists educators by automating administrative tasks, allowing them to focus more on interactive teaching and student support. When thoughtfully integrated, AI serves as a powerful ally in creating dynamic, student-centered learning environments.

2.2 Negative Impact of AI in Education

While AI's educational benefits have multiple facets, ongoing literature also reveals potential drawbacks to AI in education that must be critically examined. Critics have shown concern regarding the long-run effects of excessive AI system use, such as their influence on cognition, creativity, and the overall role of human teachers.

2.2.1 Cognitive Slowness and Impairment in Critical Thinking Skills

One of the most quoted issues in literature is laziness in cognition, where students excessively rely on AI tools to do their problems and handle information. Baron (2023) believes that if students continue to use AI systems to provide them with direct answers and automated solutions, they lose their motivation to think critically and in detail. Such dependence will likely weaken their ability to analyze information critically, synthesize diverse views, and build independent analytical ability.

Several studies have revealed that students who frequently interact with AI tutorial systems lose their capability to solve problems. The convenience of having available answers might discourage students from undertaking challenging problems and thinking them through in various ways. The effect is most pronounced in subject areas where iterative thinking and innovative problem-solving are required, and trial and error is integral to learning.

Moreover, the reception of information passively, rather than active engagement, may lead to lower intellectual inquisitiveness. Without needing to research and weigh up different positions laboriously, students may take shallow understanding on trust. Their education and professional development may take a knock in the long term. Teachers have found that while AI tools operate to standardize content presentation, they need to complement teaching practices, fostering thinking and constructive problem-solving.

2.2.2 Impact on Creativity and Originality in Man

Another significant concern is whether AI can suppress creative thinking and originality. McCosker and Wilken (2012) suggest that AI systems, and content-generating and solving systems in general, have the potential to stifle creativity by giving rise to predictable, standard outputs. If students have AI do essays, problems, or provide suggestions, their creative muscles might atrophy. AI content tends to copy patterns and trends derived from data in their training, and such may fail always to nurture innovative or novel thinking.

The risk of homogeneous learning is most evident in contexts where AI tools are used as content creation tools. For instance, students might begin to copy AI-generated styles and formats, taking their thoughts and approaches to a homogeneous position and eliminating heterogeneity in thinking. Such standardization is worst in disciplines where innovative thinking is crucial, such as literature, fine and performing arts, and social sciences, where individual thinking is to be cultivated.

The application of AI-provided feedback may also discourage students from crafting their individual voices and personal analytical models. AI may suggest and correct quickly, but it cannot substitute for sensitive understanding and empathy on the part of instructors who cherish individuality in students. The challenge, therefore, is to reconcile AI's ability to provide efficient solutions to teaching practices that promote independent, innovative thinking and distinct intellectual identities.

While AI offers numerous benefits, it also presents challenges that educators must address. Over-reliance on AI tools can lead to diminished critical thinking skills, as students may accept AI-generated information without question. Additionally, issues such as data privacy, algorithmic bias, and unequal access to technology can exacerbate existing educational disparities. It's crucial for educators to implement AI thoughtfully, ensuring that it complements rather than replaces traditional teaching methods, and that students are guided to use AI as a tool for enhancement, not a crutch.

2.3 Ethical Considerations in AI-Enhanced Education

As AI technologies assume increasing prominence in education, the imperative for ethical oversight and robust regulative structures demands ever-more considerable attention. The use of AI in education is no technical issue; scholars have argued that it raises questions of fairness, transparency, and social justice.

2.3.1 Algorithmic Bias and Fairness

A recurring argument in the moral debate is the issue of algorithmic bias. AI education tools tend to tap massive datasets to predict and suggest. Diakopoulos (2016) recognizes that if such datasets do not have adequate heterogeneity or have been shaped by prior biases, resultant algorithms can inadvertently deepen current disparities. For example, if data for training is representative of predominantly one student population, AI may favorably promote favored learning strategies or assessment practices to disadvantage minority students. These sorts of biases can increase education disparities by omitting students who do not have the predominant profile.

Addressing algorithmic bias requires a conscious push towards more excellent AI model and implementation transparency. Researchers suggest representative and diverse datasets and regular auditing of AI systems to identify and remedy discriminatory patterns. Furthermore, greater demands for collaboration between teachers, technologists, and ethicists to develop guidelines focused on fairness and equity are rising. These actions are imperative for protecting individual student rights and establishing an inclusive and diverse learning environment.

2.3.2 Data Protection and Security Problems

The widespread use of AI in education also raises serious data protection and security issues. AI learning systems tend to collect and analyze vast amounts of personal data, such as student behavior and performance. Tene and Polonetsky (2014) argue that safeguarding such data is imperative today, where digital intrusions have become rampant. Infringement or malicious manipulation of student data can have serious ramifications, affecting student records, individual privacy, and job prospects.

Therefore, Educational institutions need robust data protection systems to safeguard sensitive data in storage and transit. These include data encryption, access restrictions, and regular security scans to check for vulnerabilities. In addition to technical protection, there is also an ethical imperative to fully disclose data collection practices to students and teachers and obtain their consent for data use. Being open about data practices not only inspires trust but also empowers individuals to take their digital footprints in their own hands.

2.3.3 The Role of Human Teachers in AI-Powered Learning

Despite the several benefits of AI education tools, there is consensus in academe today that such technologies have to complement—not replace—the vital role of teachers. Whereas AI can transmit content and provide data-driven insights, AI lacks the emotional intelligence and context sensitivity to provide an integrated learning experience. Teachers are uniquely capable of inspiring creativity, analytical thinking, and moral judgment—capacities not so easily automated by computers.

Thus, AI integration in education must have an optimal mix of machine and human interaction. In most studies, AI tools' effects are optimal if employed as an adjunct to standard teaching. For instance, teachers might employ AI systems' real-time analytics to identify learning holes and thereafter develop individual interventions to plug them. Teachers, in parallel, provide personal touches of mentorship, empathy, and fine-grained feedback to address the nuanced, socio-emotional learning components.

There is greater attention to enabling teachers to interact with AI systems effectively. Educator professional development in digital literacy and the moral use of technology can prepare teachers to use AI tools to enhance learning while keeping adverse effects to a minimum. If AI is accepted and integrated in such a manner, education institutions have the potential to create learning environments that are both technologically advanced and deeply human.

In AI-enhanced education environments, the role of the teacher evolves from a traditional knowledge deliverer to a facilitator of inquiry, reflection, and ethical engagement with technology. Teachers are no longer solely responsible for transmitting knowledge but are critical in shaping the conditions under which students engage meaningfully with AI tools. While AI systems can provide personalized feedback and content, they lack contextual awareness, emotional intelligence, and ethical reasoning—traits intrinsic to human educators.

Educators must leverage AI insights to enhance pedagogical design, detect learning gaps, and offer differentiated interventions. For instance, an economics instructor might use analytics from an AI system to identify students struggling with data interpretation, and then design specific activities to improve that skill. Simultaneously, teachers guide students in evaluating AI-generated suggestions, ensuring learners remain active agents in the cognitive process.

Professional development programs are essential for preparing teachers to integrate AI meaningfully. These programs should emphasize digital literacy, data ethics, and design thinking to help educators co-create learning environments where technology enhances human agency rather than replaces it.

The ethical integration of AI in education necessitates a balance between innovation and responsibility. Educators must ensure transparency in AI applications, safeguard student data, and promote fairness in algorithmic decision-making. Pedagogically, AI should augment human instruction, not supplant it. Teachers play a vital role in guiding students to critically assess AI-generated content, fostering digital literacy and ethical awareness. By maintaining a human-centric approach, educators can harness AI's potential while upholding the integrity of the learning experience.

Examples for the Role of Human Teachers in AI-Powered Learning:

Khan Academy's Khanmigo – AI as a Learning Companion: Khan Academy has developed "Khanmigo," an AI tool functioning as a tutor to provide personalized support across various subjects. It aids students in learning and assists teachers with lesson planning and data analysis. Partnered with Microsoft, Khanmigo is available for teachers in 49 English-speaking countries, aiming to enhance learning outcomes significantly.

QANDA – AI-Based Math Problem Solving: QANDA is an AI-based learning platform developed by Mathpresso Inc., specializing in math problem-solving and tutoring. Using optical character recognition technology, it scans problems and provides step-by-step solutions, aiding students in understanding complex mathematical concepts. As of March 2024, QANDA has solved over 6.3 billion questions, demonstrating its widespread impact.

2.3.4 Future Perspectives and Final Reflections

The education and AI literature is diverse, multifaceted, encouraging and cautionary. AI's potential to re-engineer learning through experiential simulation, individualized instruction, and adaptive technologies to adapt to various student needs is encouraging. On the cautionary, AI's threat of cognitive dependency, curtailed creativity, algorithmic bias, and data protection underscores the imperative for responsible, thoughtful application.

Looking forward, future studies ought to prioritize building integrative models, striking a balance between AI's strength and irreplaceable aspects of teaching by humans. Collaboration between researchers, technologists, teachers, and policymakers is necessary for designing strategies for maximally optimizing learning, while safeguarding values of equity, creativity, and inquiry. Longitudinal studies evaluating AI incorporation, especially in diverse classrooms, will yield insights on optimizing such technologies for successful, long-run implementation. The evolution of AI in education requires continued oversight and reassessment. As AI technologies advance and evolve rapidly, so do the regulative and moral frameworks in position to direct their implementation. Ensuring future studies do not only assess direct benefits to education but also general society effects of AI-enabled learning systems is imperative. Remaining watchful and agile, the education sector can ensure AI continues to serve as an instrument for empowerment, and never for dependency or inequality.

Generative AI's interactive nature empowers learners to explore multiple knowledge pathways. Student agency refers to learners' ability to set goals, make learning choices, and reflect on outcomes. In this study, ChatGPT served as a thinking partner—generating prompts, offering alternatives, and sparking critical dialogue.

Student agency is not guaranteed by mere access. Our findings suggest students with higher self-regulation and curiosity exhibited significant skill gains, while passive users showed minimal change. This dichotomy supports the view that AI's effectiveness is contingent upon students' epistemic engagement. Therefore, cultivating reflective practices and metacognitive skills should accompany AI use.

2.4 Research Gap and Theoretical Framework

Despite a growing body of literature highlighting the potential and pitfalls of Artificial Intelligence in educational contexts, a significant research gap remains unaddressed. Most existing studies are based on self-reported data such as surveys, interviews, or perception studies, which are susceptible to bias and fail to capture actual behavioral or cognitive change. These methods often overlook how learners actively interact with generative AI tools in real-world educational environments. To bridge this gap, the present study adopts a robust experimental design—specifically a pretest-posttest control group method—to offer empirical evidence of AI's influence on critical educational outcomes like creativity, application, and decision-making.

The theoretical framework underpinning this research is rooted in Constructivist Learning Theory, which emphasizes the learner's role in actively constructing knowledge through experience and social interaction. This framework aligns closely with the integration of generative AI, which can serve as both a scaffold and a stimulus for deeper learning. In addition, the study draws upon Epistemic Agency Theory, which focuses on students' capacity to take ownership of their learning process, critically assess knowledge sources, and generate original insights. The convergence of these theories helps contextualize the findings and supports the interpretation that effective AI integration fosters not only knowledge acquisition but also reflective and autonomous learning behaviors.

3. RESEARCH GOALS

The overarching objective of this study is to extensively analyze and explore in detail the multiple effects of AI adoption in education, and most importantly, how AI influences students' choice-making, cognition, and creativity. In their extensive analysis, through systematic inquiry, the study tries to reconcile technology and teaching practices so that AI is used to enhance learning while curbing potential dangers.

3.1 Assessing AI Tools' Impact on Creativity, Application, and Decision-Making Abilities in Students

This objective evaluates how AI tools, including conversational tools like ChatGPT, affect students' ability to think innovatively and apply their knowledge in everyday contexts. From empirical evidence and in-depth case studies, the study will evaluate whether interactions between AI systems stimulate novel thinking or inadvertently lead to a dependency that dampens creative expression. The study will explore various dimensions of creativity—such as solving and combining novel ideas—to evaluate whether AI enhances deeper understanding or bypasses education. The study will also analyze AI's influence on decision-making ability by evaluating how students synthesize AI-provided insights and their analytical thinking. The evaluation will incorporate quantitative and qualitative methods, including surveys, observational studies, and testing, to capture AI's impact on thinking and creative ability.

3.2 Identifying The Promoting and Inhibiting Conditions for Cognitive Development Using AI

Recognizing that contextuality is vital in establishing AI's benefits and limitations in education contexts, this objective is to identify under what conditions AI boosts cognition and under what conditions AI might undermine development. The studies analyze factors such as instructional design, type of student activity, and built-in constraints of AI systems. Through comparative evaluation of various education contexts and teaching strategies, the studies strive to identify optimal practices to enhance cognition. The studies search for how course design and teaching facilitation complement or contrast AI-enhanced tools. In addition, the studies evaluate how extrinsic factors such as prior student knowledge and learning style interact with AI tools. The multifactor evaluation is expected to provide prescriptive insights on optimizing AI application to ensure AI incorporation boosts independent learning, thinking, and solving, and does not reinforce mere passivity in receiving and passing on information.

3.3 Providing Actionable Suggestions for Educators and Policy-Makers

The final objective aims to transfer research insights to concrete, evidence-based recommendations for direct implementation in classrooms. In integrating evidence on AI's impact on cognition, application, and creativity, the research is poised to propose comprehensive guidelines for curricula and policy formulation. The recommendations will include AI's ethical dimensions, such as fairness, data protection, and transparency, while concentrating on strategies for using AI to complement, and not replace, traditional teaching practices. Policy makers and school administrators will be given tools to assess and adopt AI tools efficiently and effectively, such that technology enhancements translate to better education while avoiding potential drawbacks. In so doing, the research will provide an optimal approach to AI benefits and, in advance, overcome constraints, to result in better, integrated AI adoption in education today.

4. METHODS AND RESEARCH MODEL

This study used a robust experimental design, like a pretest-posttest control group design. The reason for such a design is to systematically assess the impact of ChatGPT on students' ability to think, apply, and make decisions in analyzing economics case studies. The students' performance in an experimental group that used ChatGPT was compared to that of students in a control group that did not, in an attempt to isolate treatment effects from extraneous influences.

4.1 Experimental Design

Participants were randomly assigned to either the experimental or the control group. Random assignment minimized potential biases and ensured internal validity by spreading extraneous variables equally between groups. Both groups were pretested to obtain baseline measurements of participants' skill levels in the three skill areas to be developed, including creativity, application, and decision-making.

The experimental group received an intervention where they used the ChatGPT instrument to assist them on their in-class exercises on solving economics case studies. The intervention ran for seven days, where students were directed to use ChatGPT to analyze case study questions, develop hypotheses, and suggest solutions. The control group, however, performed the same exercises but did not have access to using the ChatGPT instrument. Aside from the intervention instrument, all other aspects of the learning process were kept identical between groups. The instructional style of the lecturer, teaching materials, and classroom arrangement were kept identical to ensure no extraneous influences affected the outcome.

This design allowed for a direct comparison of posttest and pretest scores between the two groups. Focusing only on the effect of ChatGPT, any noticeable improvement in the experimental group's performance could directly be attributed to the treatment and not to differential teaching quality or extraneous influences.

4.2 Data Collection

Data collection in the current study was structured around three significant tools designed to evaluate the focused skills:

4.2.1. Torrance Test of Creativity (TTCT) – Verbal Test

The TTCT Verbal Test was used to assess students' creative thinking ability. The instrument comprises six distinct creative thinking exercises representing types of creative thinking in daily life and business. These exercises include:

Asking Questions: The students must answer questions designed around a case study, challenging them to identify gaps and analyze underlying problems.

Guessing Causes: The students must make inferences about the possible reasons behind the case, relying on their analytical and inference skills.

Guessing Consequences: Here, participants predict the ramifications or effects of decisions made in the case situation, emphasizing forecasting and planning.

Product Improvement: In case studies, the students suggest ways to improve the situation, challenging them to think critically and innovatively.

Unusual Uses: In this activity, students must recognize novel or unexpected uses for the case study context, which encourages them to think in ways that are different from standard thinking.

Just Suppose: In their last activity, students imagined and evaluated potential scenarios by answering what might have happened if an unexpected situation had occurred in the case study.

Each was rated using rubrics developed a priori, which numerically rated students' performance on every creative activity.

4.2.2. Analytical Scoring Skills

An analytical grading system was employed to evaluate students' ability to apply in case analysis. The grading system broke up the analysis process into several primary components, including:

Problem Identification: The student's ability to identify correctly the root issues in the case study.

Data Interpretation: The ability to accurately interpret data and information provided.

Application of Economic Principles: The student's capability to properly apply economic theory and principles in analyzing.

Each component was also graded qualitatively using structured rubrics, providing clear criteria for assessment.

4.2.3. Problem Solving and Decision-Making Competency Skills

The decision-making skill of students was measured using a special test to evaluate their ability to solve problems and make decisions. The instrument is meant to concentrate on:

Critical Thinking: Evaluation of students' capability to analyze and think critically through multiple-problem scenarios.

Analytical Skills: Assesses the logical structure of arguments and inferences drawn from case study data.

Application of Knowledge: Measurement of ability to effectively use acquired concepts to make decisions.

Similar to the remaining instruments, all components of the decision-making scale were rated numerically on structured rubrics to ensure consistent and objective marking.

4.3 Procedure

The overall procedure of the study was divided into three phases in order to ensure systematic data collection and analysis:

4.3.1. Pretest

Before treatment, students in both treatment and control conditions performed baseline testing on the aforementioned three tools. The pretest obtained initial baseline values for students' ability to apply, decide, and think creatively. The initial measurement played an important role in facilitating comparison of subsequent improvement in performance and in keeping in check any potential initial between-group differences.

4.3.2. Intervention

The intervention ran for seven days, during which the experimental group used ChatGPT to think through their thoughts, generate potential solutions, and critically evaluate AI-generated content. The students in the experimental group were encouraged to use the tool to think through their thoughts, to generate potential solutions, and to evaluate AI-generated content critically. The control group, by contrast, approached identical case studies but in complete absence of ChatGPT, relying on standard learning and teaching by instructors. The teaching methodology and in-room activity consistency between classes were maintained to ensure only ChatGPT's influence impacted the result.

4.3.3. Posttest

Following the treatment, both classes were administered a posttest, using identical format and content to the pretest. The posttest sought to evaluate any improvement or advancement in students' ability in the three skill areas under focus. By analyzing posttest and pretest scores, the study determined whether ChatGPT enhanced creativity, application, and decision-making.

4.4 Ensuring Validity and Reliability

Several steps were taken throughout the study to maintain result validity and reliability. The participants' random assignment to either of two groups helped to limit selection bias and provide equivalence between the two. In addition, using standard teaching practices and identical classrooms for both groups helped limit confounding variables that could otherwise affect the results. The data collection tools used—the TTCT, analytical scoring scale, and problem-solving and decision-making competency test—were selected for their documented reliability and validity in measuring subject skill. The student performance score for every instrument was supplemented by carefully developed rubrics, to provide score-objectivity and reproducibility.

In essence, the approach to this research was designed in such a manner to isolate the effect of the Chat GPT instrument on student economics case study performance. Applying a pretest-posttest control group design, valid measurement tools, and standard teaching practices, the research offers precise, reliable measurement of how AI can affect application, creativity, and decision-making ability. The systematic implementation and design of the experimental procedure ensure any resultant disparities in performance may confidently be attributed to treatment, yielding valuable insights on AI's influence on education.

With the methodological framework established, the following section presents the empirical findings derived from our study, highlighting the impact of generative AI on student creativity, application, and decision-making.

5. DATA, FINDINGS, AND RESULTS

5.1. Data Analysis

Using the pretest-posttest experimental design using the control group, this study aimed to compare experimental and control group students' scores derived through the scales before and after the experiment. A T-test is used to review and check whether there is any relationship between the independent and dependent variables. Standard deviation is used to compare the deviation of the three measurements' pre and post test results to check the more significant deviation. The difference between the means of the pre- and post-test results of the individual experimental group was computed to check for students who could not develop their creative skills, application skills, or decision-making skills even after using the AI tool (Chat GPT).

5.2. Findings and Results

5.2.1 T Test Results

Result of Torrance Test of Creativity

Table 1 shows the T test results for the response of Torrence test of creativity. The posttest results (mean = 3.89, Variance = 0.17) are significantly greater than those of the pretest (mean = 3.01, Variance = 0.14). There is a positive correlation between pre and post-tests in six assessments of the Torrance test of creativity (Pearson's correlation = 0.08). An independent samples T-test indicated a significant difference between the pre and post-tests of the experimental group with a p–value = 0.00, which results in rejection of the null hypothesis and acceptance of the Alternative hypothesis. AI has a positive impact on the creativity skills of economics students when solving case studies.

Table 1. T-test results of Torrence test of creativity

t-Test: Paired Two Sample for Means		
	Variable 1	*Variable 2*
Mean	3.01	3.89
Variance	0.14	0.17
Observations	30.00	30.00
Pearson Correlation	0.08	
Hypothesized Mean Difference	0.00	
Df	29.00	
t Stat	-9.09	
P(T<=t) one-tail	0.00	
t Critical one-tail	1.70	
P(T<=t) two-tail	0.00	
t Critical two-tail	2.05	

Result of the Analytical Scoring Skills

Table 2 shows the T-test results for the response of Torrence test of creativity. The posttest results (mean = 3.97, Variance = 0.18) are significantly greater than those of the pretest (mean = 2.87, Variance = 0.20). There is a positive correlation between pre and posttests in six assessments of the Torrance test of creativity (Pearson's correlation = 0.14). An independent samples T-test indicated a significant difference between the pre and posttests of the experimental group with a p–value = 0.00, which results in rejection of the null hypothesis and acceptance of the Alternative hypothesis. AI positively impacts the application skills of economics students when solving case studies.

Table 2. T – test results of Analytic scoring scales

t-Test: Paired Two Sample for Means		
	Variable 1	*Variable 2*
Mean	2.87	3.97
Variance	0.20	0.18
Observations	30.00	30.00
Pearson Correlation	0.14	
Hypothesized Mean Difference	0.00	
Df	29.00	
t Stat	-10.48	
P(T<=t) one-tail	0.00	
t Critical one-tail	1.70	
P(T<=t) two-tail	0.00	
t Critical two-tail	2.05	

Result of the Problem Solving and Decision-Making Competency Skills

Table 3 shows the T-test results for the response of Torrence test of creativity. The posttest results (mean = 4.02, Variance = 0.16) are significantly greater than those of the pretest (mean = 3.01, Variance = 0.21). There is a positive correlation between pre and posttests in six assessments of the Torrance test of creativity (Pearson's correlation = 0.08). An independent samples T-test indicated a significant difference between the pre and posttests of the experimental group with a p–value = 0.00, which results in rejection of the null hypothesis and acceptance of the Alternative hypothesis. AI positively impacts the decision-making skills of economics students when solving case studies.

Table 3. T – test results of Problem solving and decision-making competency skills

t-Test: Paired Two Sample for Means		
	Variable 1	*Variable 2*
Mean	3.01	4.02
Variance	0.16	0.21
Observations	30.00	30.00
Pearson Correlation	-0.11	
Hypothesized Mean Difference	0.00	
Df	29.00	
t Stat	-8.65	
P(T<=t) one-tail	0.00	
t Critical one-tail	1.70	
P(T<=t) two-tail	0.00	
t Critical two-tail	2.05	

5.2.2 Standard Deviation results

Table 4 shows that the Torrence Test of Creativity has an SD of 0.53, the Analytical Scoring Scale has an SD of 0.58, and problem-solving and decision-making skills have an SD of 0.64 for the pre- and post-tests of the experimental group. The SD results show that problem-solving and decision-making skills have more deviation with a score of 0.64.

Table 4. Standard Deviations of experimental group Pre and post test results

Tests	SD
Torrance Test of Creativity	0.53
Analytic Scoring scales	0.58
Problem Solving and Decision-making Competency Scales	0.64

Empirical studies indicate that students engaging with AI tools like ChatGPT exhibit notable improvements in creativity, application, and decision-making skills. However, the data also reveals a subset of students who show minimal progress, highlighting the importance of active engagement and proper guidance in AI utilization. These findings underscore the necessity for educators to provide structured support, ensuring that all students can effectively leverage AI to enhance their learning outcomes.

The subsequent discussion interprets these findings, exploring their implications within the broader context of AI integration in educational settings.

6. DISCUSSION

The findings of this study contribute to an ongoing and nuanced discourse about the dual impact of Artificial Intelligence (AI) in education. Our analysis reinforces the notion that AI is not inherently beneficial or harmful; rather, its outcomes are shaped by the manner in which it is integrated and used. While some previous studies emphasize AI's detrimental effects—such as decreased motivation and cognitive engagement—others highlight its transformative potential to enhance creativity, critical thinking, and problem-solving. This duality prompts a critical examination of the conditions under which AI supports or undermines educational objectives.

In recent years, researchers have increasingly cautioned against the overuse of AI tools in learning environments. One of the most frequently cited drawbacks is the tendency for AI to encourage intellectual passivity. As Nikita (2023) notes, AI systems that quickly and mechanically solve routine problems may discourage learners from developing critical thinking or memorization skills. When answers are easily available, students may gravitate toward convenience, bypassing the mental effort required for meaningful learning. Over time, this may foster a culture of cognitive dependency, where both students and educators become reliant on AI, abandoning traditional methods of analytical exploration and self-driven discovery.

This concern is echoed by Baron (2023), who highlights the erosion of decision-making skills due to excessive automation. When AI tools are consistently used to guide actions and decisions, learners may become disengaged from the process of evaluating alternatives or understanding trade-offs. In such scenarios, the analytical rigor historically cultivated by educational systems is at risk of being diminished. Teachers, too, may gradually delegate core teaching tasks to AI, thereby limiting opportunities for meaningful interaction and pedagogical reflection.

Recognizing these risks, scholars such as Posner and Fei-Fei (2020) have emphasized the need for a recalibrated approach to AI integration—one that augments rather than replaces human intellectual labor. Their argument is not against the use of AI, but against its uncritical adoption. They propose that educational technologies must be designed and deployed in ways that stimulate human cognition, inquiry, and creativity, ensuring that learners are not mere recipients of AI-generated content but active participants in the learning process.

Contrasting these concerns, our own experimental study offers compelling evidence of AI's constructive role in educational settings, particularly when students engage with the tools in active, thoughtful, and creative ways. The majority of student participants reported that tools like ChatGPT enhanced their understanding, broadened their knowledge base, and inspired innovative approaches to problem-solving. Rather than replacing their thinking, AI served as a catalyst, allowing students to tackle complex economic case studies with increased confidence and depth.

A key outcome observed was the emergence of what is known as **epistemic agency**—the learner's ability to take ownership of knowledge construction. By interacting with AI, students were prompted to critically evaluate the accuracy, relevance, and logic of AI-generated content. They learned to compare, contrast, and validate information, thus strengthening their analytical and decision-making capabilities. In this way, AI served not as a shortcut to answers, but as a scaffold for deeper thinking.

Creativity also appeared to benefit from AI interaction. For capable and motivated students, AI became an ideational partner, helping them view problems from multiple angles and generate novel solutions. It stimulated brainstorming, encouraged divergent thinking, and prompted iterative problem-solving—hallmarks of creative engagement. These students were not passively consuming AI output but were interrogating, refining, and expanding upon it, resulting in more original and effective solutions.

These findings underline a crucial insight: **access to AI tools alone is insufficient** to produce positive learning outcomes. The effectiveness of AI in education is mediated by various factors, including students' initial cognitive skills, their willingness to engage deeply with the content, and the learning environment's design. Teachers play an essential role in shaping these factors—by guiding, scaffolding, and motivating students to use AI as a tool for exploration rather than automation. Ayanwale, et al (2025)

This duality demands a balanced and context-sensitive approach to AI implementation. On one side, educators must design learning experiences that maximize AI's potential to foster creativity, inquiry, and critical thinking. On the other, they must guard against the tendency of some learners to misuse AI as a passive resource. Curriculum designers and policymakers should work collaboratively to embed AI literacy into educational frameworks—ensuring students are not just users of technology but are trained to understand its capabilities, limitations, and ethical dimensions.

AI's role in education is multifaceted, offering opportunities for personalized learning and skill development. When students actively engage with AI tools, they can experience enhanced autonomy and deeper understanding. Conversely, passive use may lead to dependency and superficial learning. Educators must therefore foster environments that encourage critical interaction with AI, integrating it as a complementary resource within a broader pedagogical framework. This balanced approach ensures that AI serves as a catalyst for educational growth rather than a substitute for human instruction.

6.1 Ethical Considerations: Bias, Privacy, and Accountability in AI-Enhanced Education

6.1.1 Algorithmic Bias in Educational AI Systems

AI systems in education can inadvertently perpetuate existing societal biases present in their training data, leading to unfair or discriminatory outcomes for certain student groups. For instance, if an AI tool is trained on data that underrepresents students from specific backgrounds, it may not provide equitable support or assessments for those students. Addressing algorithmic bias requires careful curation of training data, ongoing monitoring, and the inclusion of diverse perspectives in AI development processes.

6.1.2 Data Privacy and Student Information Security

The integration of AI in education often involves the collection and analysis of extensive student data, including personal details and learning behaviors. This raises significant concerns regarding data privacy and the potential for misuse or unauthorized access. Educational institutions must implement robust data protection policies, ensure compliance with relevant regulations, and maintain transparency with students and parents about data usage.

6.1.3 Accountability in AI-Driven Educational Decisions

Determining accountability when AI systems influence educational outcomes is a complex challenge. If an AI tool provides incorrect guidance or assessments, it is essential to establish clear responsibility—whether it lies with the developers, educators, or the institution. Establishing transparent governance structures and accountability frameworks is crucial to address errors and maintain trust in AI-driven educational tools.

The study also identified a subgroup of students—termed the **"no impact" group**—who showed no measurable improvement in creativity, application, or decision-making. Detailed analysis revealed that these students either lacked the motivation to interact meaningfully with AI or defaulted to using it as a simple answer-providing machine. For instance, several participants showed no improvement in the Torrance test scores or decision-making assessments. Their engagement with AI was shallow, marked by copy-paste behavior and minimal critical evaluation. This resulted in stagnant performance, despite having access to the same tools as their peers.

Synthesizing the key insights from our analysis, the concluding section reflects on the broader implications of our findings and suggests directions for future research in the realm of AI-enhanced education.

7. CONCLUSION

The integration of Artificial Intelligence (AI) into educational settings has sparked significant scholarly interest. While earlier studies present a mixed outlook—highlighting both the transformative potential of AI and its unintended consequences—our research offers a more affirming perspective. In particular, our experimental approach focusing on economics education revealed that when used thoughtfully, AI significantly enhances student creativity, application of knowledge, and decision-making. The key insight from our findings is that limitations attributed to AI are often the result of passive user engagement rather than inherent technological flaws.

7.1 A Dual Landscape in Prior Research

Previous literature paints a dual landscape regarding AI's role in education. On the one hand, numerous studies showcase AI as a driver of personalization, creativity, and deeper understanding through tools such as intelligent tutoring systems, adaptive learning platforms, and immersive simulations. On the other hand, there are critical voices raising concerns about the detrimental effects of over-reliance on AI. Scholars like Nikita (2023) have warned of "learned laziness," a phenomenon where students, when frequently exposed to AI-generated answers and problem-solving aids, gradually lose the motivation to engage critically with content. Similarly, Baron (2023) argued that the constant availability of AI-generated content might discourage students from putting in the cognitive effort required to explore complex ideas.

A deeper analysis of these concerns reveals that such negative outcomes are not due to the AI tools themselves, but rather how users engage with them. Our experimental design allowed us to test this hypothesis directly, confirming that active, reflective use of AI tools enhances cognitive outcomes, while passive use yields negligible benefits. This challenges the assumption that AI inherently diminishes cognitive engagement.

7.2 Key Contributions of Our Experimental Study

Our study aimed to move beyond theoretical speculation by directly observing students' interactions with AI in a controlled, academic context. The focus was on three high-order cognitive domains—creativity, application, and decision-making—

within the framework of economics case study analysis. By equipping students with advanced AI tools, including large language models and economic data simulators, we sought to evaluate whether meaningful learning improvements could be observed. Our approach incorporated both qualitative and quantitative measurement tools to ensure a robust and multifaceted analysis.

7.3 Enhancing Creativity

One of the standout results from our study was AI's role in improving students' creative capacity. In this context, creativity refers not merely to artistic output but to the ability to generate novel solutions, consider alternative viewpoints, and develop innovative approaches to problem-solving. Rather than accepting AI-generated content uncritically, students used it as a springboard for further exploration. They would critique suggestions, adapt models, and iterate upon initial insights—thus engaging in a creative process enriched, but not replaced, by AI. Youssef, et.al (2024)

The use of AI as a creative collaborator allowed students to engage in higher-order thinking activities. Brainstorming with AI helped students frame problems differently and encouraged lateral thinking. As students worked through scenarios, they began to see AI not as a crutch, but as a co-participant in the ideation process. This partnership led to the production of diverse solutions, greater originality, and a noticeable improvement in creative expression, particularly in developing economic strategies and policy evaluations.

7.4 Boosting Application Skills

A fundamental aspect of economics education is the ability to apply theoretical concepts to real-world scenarios. In our experimental setting, AI tools played a crucial role in fostering this application skill. Students used AI to interpret data sets, simulate market responses, and evaluate policy alternatives. These simulations served as dynamic learning environments where students could test hypotheses and view outcomes in real-time. Borji, A., & Zhao, Y. (2024)

By engaging in this form of experiential learning, students were able to internalize theoretical principles more effectively. The ability to visualize economic models and examine the consequences of policy shifts helped bridge the gap between abstract theory and practical understanding. Students reported increased confidence in applying their knowledge and demonstrated improved performance in assessments that required applied thinking.

7.5 Strengthening Decision-Making Abilities

Decision-making is a cognitively demanding process that involves weighing options, assessing risks, and making informed judgments. In our study, AI provided the scaffolding necessary for students to develop these skills. By presenting multiple decision pathways and offering predictive insights, AI tools encouraged students to think critically and evaluate the implications of each choice.

Rather than promoting reliance on AI recommendations, the process cultivated epistemic agency—the students' ability to take ownership of their learning and knowledge construction. Many students began questioning AI suggestions, proposing alternative approaches, and justifying their decisions with evidence. This reflective process led to improved analytical depth and greater coherence in their strategic thinking, particularly evident during post-intervention testing.

7.6 The "No Impact" Subgroup

Interestingly, not all participants benefited equally. A subgroup—referred to as the "no impact" cluster—showed little or no improvement across the three skill areas. Upon further investigation, it became clear that the issue stemmed from passive usage patterns. These students relied on AI outputs without engaging critically or interactively. They tended to copy solutions, skipped reflective steps, and treated AI as a shortcut rather than a learning partner. Carver, L. B., & Nouri, J. (2024)

This finding underscores the importance of instructional context. Effective AI use in education is not automatic; it depends heavily on how students are guided to use the technology. The lack of engagement seen in the "no impact" subgroup signals a need for stronger scaffolding, clearer expectations, and structured opportunities to foster active learning behaviors.

7.7 Methodological Strengths

A major strength of our study lies in its methodological design. While many studies rely on surveys and self-reported data, our approach focused on direct observation and empirical measurement. We used standardized tools such as the Torrance Test of Creative Thinking and application rubrics to objectively assess outcomes.

This method allowed us to capture nuanced interactions and real-time decision-making processes that are often missed in traditional research. We were able to document not only whether students used AI, but how they used it, how often they interacted with feedback, and how their strategies evolved over time. This fine-grained analysis contributed to the reliability and depth of our findings. Wang, Y., & Zhang, H. (2024)

7.8 Implications for Educators and Policymakers

Our results have significant implications for both educators and policymakers. For educators, the key takeaway is that AI can be a powerful tool for enhancing learning—but only when integrated thoughtfully into pedagogy. Teachers must design learning activities that encourage students to critique, reflect, and expand on AI outputs. Additionally, professional development programs should equip educators with the skills to guide effective AI usage in classrooms.

For policymakers, supporting the responsible integration of AI means more than supplying hardware or software. It involves creating an ecosystem that includes curriculum alignment, teacher training, and ongoing student support. Addressing motivational issues—such as those observed in the "no impact" group—is also essential. This can be achieved through incentive structures, awareness programs, and targeted interventions aimed at cultivating active learning behaviors.

REFERENCES

Aleven, V., McLaren, B., & Sewall, J. (2009, April). Scaling Up Programming by Demonstration for Intelligent Tutoring Systems Development: An Open-Access Web Site for Middle School Mathematics Learning. *IEEE Transactions on Learning Technologies*, *2*(2), 64–78. DOI: 10.1109/TLT.2009.22

Anderson, J. R., Corbett, A. T., Koedinger, K. R., & Pelletier, R. (1995, April). Cognitive Tutors: Lessons Learned. *Journal of the Learning Sciences*, *4*(2), 167–207. DOI: 10.1207/s15327809jls0402_2

Ayanwale, O. R., Ayeni, F. O., & Adekunle, O. M. (2025). Generative artificial intelligence in pedagogical practices: A systematic review of the literature. *Cogent Education*, *12*(1), 2485499. DOI: 10.1080/2331186X.2025.2485499

BARON NS. (2023) Even kids are worried ChatGPT will make them lazy plagiarists, says a linguist who studies tech's effect on reading, writing and thinking. Fortune. https://fortune.com/2023/01/19/what-is-chatgpt-ai-effect-cheating-plagiarism-laziness-education-kids-students/

Borji, A., & Zhao, Y. (2024). *Generative AI and its impact on personalized intelligent tutoring systems: A roadmap for education.* arXiv preprint arXiv:2410.10650. https://arxiv.org/abs/2410.10650

Bull, S., & Kay, J. (2016, January 12). SMILI: A Framework for Interfaces to Learning Data in Open Learner Models, Learning Analytics and Related Fields. *International Journal of Artificial Intelligence in Education*, *26*(1), 293–331. DOI: 10.1007/s40593-015-0090-8

Carver, L. B., & Nouri, J. (2024). *Generative AI and agency in education: A critical scoping review and thematic analysis.* arXiv preprint arXiv:2411.00631. https://arxiv.org/abs/2411.00631

Chittaro, L., & Buttussi, F. (2015, April 18). Assessing Knowledge Retention of an Immersive Serious Game vs. a Traditional Education Method in Aviation Safety. *IEEE Transactions on Visualization and Computer Graphics*, *21*(4), 529–538. DOI: 10.1109/TVCG.2015.2391853

Diakopoulos, N. (2016, January 25). Accountability in algorithmic decision making. *Communications of the ACM*, *59*(2), 56–62. DOI: 10.1145/2844110

Ferguson, R. (2012). Learning analytics: Drivers, developments and challenges. *International Journal of Technology Enhanced Learning*, *4*(5/6), 304. DOI: 10.1504/IJTEL.2012.051816

Gasevic, D., Siemens, G., & Rose, C. P. (2017, January 1). Guest Editorial: Special Section on Learning Analytics. *IEEE Transactions on Learning Technologies, 10*(1), 3–5. DOI: 10.1109/TLT.2017.2670999

Hamari, J., & Koivisto, J. (2014, November). Measuring flow in gamification: Dispositional Flow Scale-2. *Computers in Human Behavior, 40*, 133–143. DOI: 10.1016/j.chb.2014.07.048

Koedinger, K. R., Corbett, A. T., & Perfetti, C. (2012, April 9). The Knowledge-Learning-Instruction Framework: Bridging the Science-Practice Chasm to Enhance Robust Student Learning. *Cognitive Science, 36*(5), 757–798. DOI: 10.1111/j.1551-6709.2012.01245.x

McCosker, A., & Wilken, R. (2012, May 2). Café Space, Communication, Creativity, and Materialism. *M/C Journal, 15*(2). https://doi.org/DOI: 10.5204/mcj.459

Nikita (2023) Advantages and Disadvantages of Artificial Intelligence. Simplilearn. https://www.simplilearn.com/advantages-and-disadvantages-of-artificial-intelligence-article

Polonetsky, J., & Tene, O. (2014, July 1). The Ethics of Student Privacy: Building Trust for Ed Tech. *International Journal of Information Ethics, 21*, 25–34. DOI: 10.29173/irie367

Posner, T., & Fei-Fei, L. (2020). AI will change the world, so it's time to change A. *Nature, 588*(7837), S118–S118. DOI: 10.1038/d41586-020-03412-z

Selbst, A. D., & Barocas, S. (2018). The Intuitive Appeal of Explainable Machines. SSRN *Electronic Journal*. https://doi.org/DOI: 10.2139/ssrn.3126971

Sitzmann, T., Kraiger, K., Stewart, D., & Wisher, R.SITZMANN. (2006, September). THE COMPARATIVE EFFECTIVENESS OF WEB-BASED AND CLASSROOM INSTRUCTION: A META-ANALYSIS. *Personnel Psychology, 59*(3), 623–664. DOI: 10.1111/j.1744-6570.2006.00049.x

Vanlehn, K. (2011, October). The Relative Effectiveness of Human Tutoring, Intelligent Tutoring Systems, and Other Tutoring Systems. *Educational Psychologist, 46*(4), 197–221. DOI: 10.1080/00461520.2011.611369

Wang, Y., & Zhang, H. (2024). Generative AI in education and research: A systematic mapping review. *Review of Education, 12*(2), e3489. DOI: 10.1002/rev3.3489

Youssef, A. B., Jain, P., & Al-Fadhli, S. (2024). Generative AI in higher education: A global perspective of adoption strategies. *Discover Education, 2*, 15. Advance online publication. DOI: 10.1016/j.discove.2024.1001516

Chapter 7

Leveraging Cloud Computing for Enhanced Efficiency and Collaboration in Education Institutions

Pooja Dehankar
Ajeenkya D.Y. Patil School of Engineering, India

Kashvi Chaturvedi
http://orcid.org/0009-0005-3192-3400
Ajeenkya D.Y. Patil University, India

Susanta Das
http://orcid.org/0000-0002-9314-3988
Ajeenkya D.Y. Patil University, India

ABSTRACT

The desire for increased productivity, teamwork, & creativity is driving a major revolution in the education industry. Cloud computing, which offers institutions an inexpensive, flexible, & scalable platform for delivering excellent instruction, has played a significant role in enabling this shift. The potential of cloud computing to improve communication & collaboration between administrators, professors, & students is one of its main advantages for institutions. Cloud computing can improve educational outcomes by giving students access to top-notch digital resources. However, there are many issues with cloud computing adoption in institutions. The chapter discusses these issues & offers recommendations on how institutions might

DOI: 10.4018/979-8-3373-1190-6.ch007

reduce these risks. It offers a thorough overview of the potential of cloud computing in institutions, highlighting its advantages, difficulties, & best practices. By utilizing cloud computing, institutions can lower costs, improve collaboration & communication, and outcomes—all of which will ultimately increase their competitiveness in the digital age.

1. INTRODUCTION

Cloud computing is being used by many businesses because of its dynamic scalability and utilization of virtualized resources as a service through the Internet. The future of education will probably be significantly impacted by it. In order to run their information systems efficiently without having to invest more funds in computers and network equipment, educational institutions, particularly those with tight budgets, can benefit greatly from cloud computing. Utilizing cloud-based software made available by service providers, universities allow their users/students to complete academic and business-related tasks (Ercan, 2010).

From textbooks to the internet today, the availability and accessibility of academic resources have changed over time. The shift from in-person to online instruction has also been seen in tuition. This is in addition to the scholarly materials offered in a variety of formats and by different organizations. Both opportunities and problems are presented by these phenomena. This phenomenon has been facilitated by cloud computing and information and communication technology (ICT). Communication dynamics have been brought about by the availability and accessibility of academic resources by different entities and in different media. Therefore, in order to provide their customers with better service, organizations and people should base their activities on the availability and accessibility of academic materials (Odhiambo, 2024).

Higher education greatly benefits from the use of virtual desktops in cloud environments. These days, it appears that transferring labs to the cloud is required in order to keep students' dedication in this novel situation. The aim is to empirically use the technology acceptance model and the system usability scale in a population of students who use a virtual desktop platform that has been tailored to present the laboratories of a computer science programming course in a bachelor's degree program in computer science. Lera et al. (2021) performed a thorough examination of their platform's system usability scale (SUS) and the technology acceptance model (TAM). The results of this study shed light on the direct relationship between perceived usefulness, convenience of use, and attitude toward technology.

Active learning approaches encourage students to actively participate in their own education, foster collaborative learning, and create a shared understanding of the subject as a whole. Cloud computing both enhances and changes the learning environment by acting as a link between active methodology and students' educational activities (Sein-Echaluce et al., 2021). In order to identify and handle the individual, group, and collective evidence of the students' activities in this research, an active approach is combined with a cloud computing system. A popular pedagogy in educational institutions generally, and especially during and after the COVID-19 pandemic, is e-learning by using a cloud-based learning management system with its many extra benefits. Cloud e-learning uptake and deployment appear to be challenging without high service quality (Naveed et al., 2021). For Higher Education Institutions (HEIs), resource optimization is a major challenge. As the newest computing technology of the fourth industrial revolution, cloud computing has become the primary norm for resource and service delivery. Retaining clients for this cutting-edge technology has become difficult for cloud service providers as cloud computing has developed into a mature technology that is being quickly implemented in many HEIs worldwide. Few studies have been conducted on the ongoing usage of cloud computing in an organizational context, despite current research trends focusing on the acceptance or adoption of technology (Qasem et al., 2020).

Nearly every nation has been severely impacted by coronavirus pandemic (COVID19), which had an effect on how education is delivered worldwide. Numerous colleges, universities, and schools were either forced to offer online courses or were under lockdown. Educational institutions became more effective at providing high-quality teaching services virtually in order to handle the problems brought on by COVID-19. In that situation, cloud computing gave teachers a great platform to enhance their productivity and methods of instruction. Cloud computing offers a perfect opportunity because it not only saves money but also power because multiple stakeholders may use the infrastructure at the same time for research, teaching, and learning (Agrawal, 2021).

Cloud computing technologies have being used in education more and more in recent years. A systematic mapping study (SMS) on the interest in innovative approaches, tools, and techniques used in cloud-based educational activities was presented by Scalera et al. (2020). The SMS that was being delivered, was created by utilizing a standardized, trustworthy, and exacting technique and executed in a way that can be replicated and verified. The selection of articles from recognized online digital libraries forms the basis of the adopted procedure.

The following chapter begins by going over how cloud computing is integrated into educational institutions and the three main models that determine its structure, i.e. IaaS, PaaS, and SaaS. The adoption of cloud computing in educational and administrative contexts, its effects on teaching, learning, and institutional efficiency, and the growing role of cloud-based IoT and digital learning technologies in improving accessibility and engagement are the topics covered in following subsections. The chapter also examines security issues, mitigating techniques, collaborative learning frameworks, and upcoming developments, such as the incorporation of blockchain, Artificial Intelligence (AI), and quantum computing. The chapter concludes with bibliometric analysis that maps global research trends, topic clusters, and collaborative networks. The following flowchart, Figure 1, will visually depict the flow of the chapter.

Figure 1. Flowchart Presenting Chapter Flow

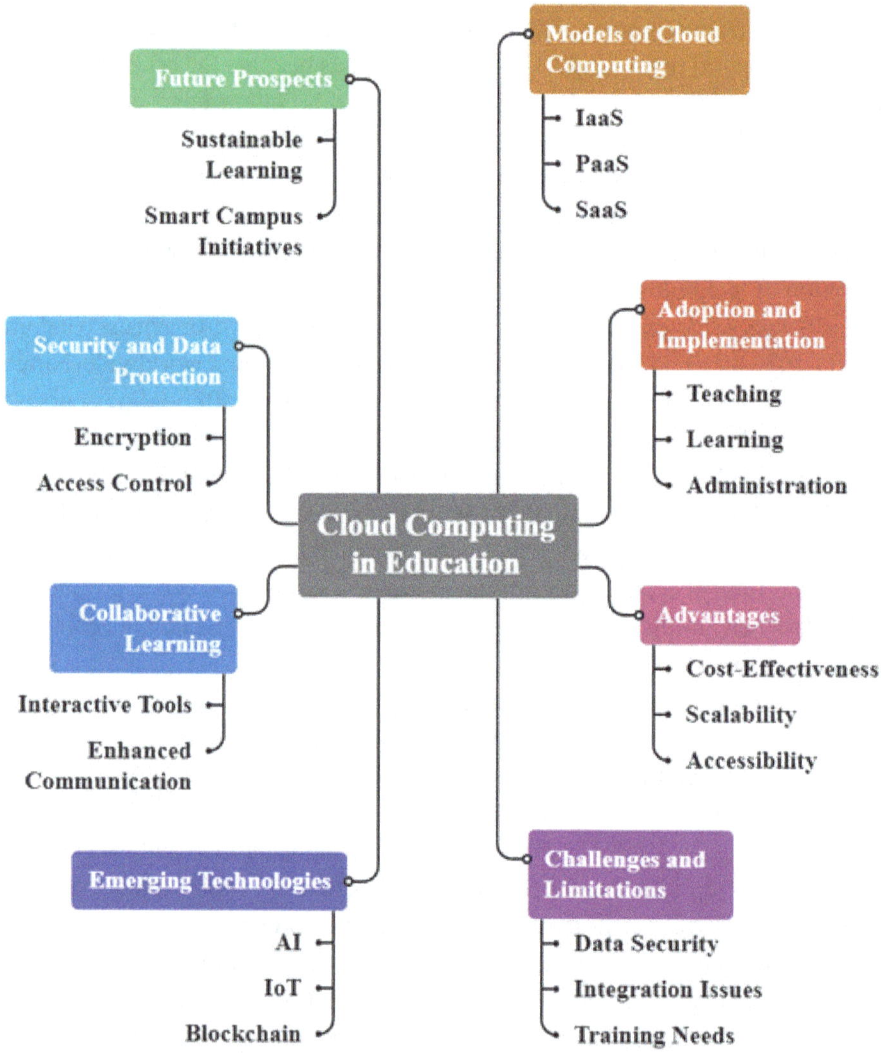

2. CLOUD COMPUTING IN EDUCATION INSTITUTIONS

Today's education has a greater connection with information technology (IT) when it comes to communication, collaboration, and content transmission. Universities, colleges, and schools have extremely high requirements for servers, storage, and software. Cloud computing refers to Internet-based computing where computers and other machines, including the electrical grid, are provided with access to shared resources, software, and data on demand. The three business models for cloud computing that are now in use are IaaS (Infrastructure as a Service), PaaS (Platform as a Service), and SaaS (Software as a Service) (Lakshminarayanan et al., 2013).

Cloud computing will probably have a big effect on the future of schooling. For educational institutions, particularly those with limited funding, cloud computing is a great way to run their information systems efficiently without having to invest additional funds in computers and network equipment. In order to facilitate commercial and academic duties for their own users and students, universities utilize the cloud-based solutions that are available from service providers. The educational sector will benefit from the cloud computing infrastructure, particularly in universities where computer use is more prevalent (Ercan, 2010).

Many businesses are beginning to employ cloud computing because of its dynamic scalability and utilization of virtualized resources as a service via the Internet. With applications in practically every field, including education, cloud computing is expanding quickly. These days, e-learning is likewise growing in popularity and strength as a broad trend. Many hardware and software resources are often needed for e-learning systems. Cloud computing is the ideal option for many educational institutions that cannot afford such expenditures, particularly in universities where computer use is increasingly prevalent and there is potential to improve the advantages of shared applications for instructors and students (Bora & Ahmed, 2013).

Integration of cloud computing into educational systems is revolutionizing the manner in which academics and industry collaborate in the rapidly evolving technology environment of the present. The researchers are analyzing the degree to which the cloud technology will enhance the process of learning and facilitate the new business-academia collaboration model. The utilization of cloud resources by the educational institutions can lead to improvement in the management of resources, collaborative research, and provision of the practical skills that the modern-day businesses are looking for in the students (Mehta et al., 2024).

3. ADOPTION OF CLOUD COMPUTING IN EDUCATIONAL SETTINGS

The education sector is characterized by the inherent ambiguity of data analytics, the different priorities of stakeholders, and the varied student outcomes. The present research has been conducted with the aim of finding out how the school management can use the data storage, analytics, and collaboration features of cloud computing to handle these risks. To achieve this, we introduce a neutrosophic model that employs "truth," "falsity," and "indeterminacy" values to characterize these complexities in educational data. Real-world case studies showcase the positive impact of cloud computing on decision-making and uncertainty reduction. The neutrosophic framework, despite its effectiveness in dealing with the "indeterminacy" that is embedded in educational data, cloud computing has a considerable potential to be the navigator through the "fog" of educational administration. The presentation of the application of cloud computing to support and clarify school administration decision-making from the unique neutrosophic standpoint is based on the report of Yasser et al. (2025).

Cloud computing has completely revolutionized the educational landscape due to the fact that it facilitates collaboration, improves learning, and grants access to educational content at any time. Adaptive technologies and analytics make it possible to customize learning experiences and increase learner engagement by tailoring education to each student's needs. Cloud solutions allow teachers to focus more on teaching by simplifying administrative tasks like attendance and grading. Educational institutions handle a great deal of sensitive data, including grades, learner records, academic information, and intellectual property. When this data is kept in the cloud, problems like data breaches, unauthorized access, and regulatory compliance occur. Attacks are more likely to target organizations that use cloud-based services to host their infrastructure and applications. This makes cloud security crucial in education (Ponde, 2024).

Based on the assessment of its economic impact (EI), cloud computing has been recognized as a very strong technology with the ability to change business practices entirely and, at the same time, increase the economic growth of the regions. The main or primary ways in which cloud computing brings about these financial benefits are through better productivity, more creative work that leads to breakthrough inventions, and the lowering of expenses. Among the barriers that hinder the quantification of cloud computing's EI are the crises of data security, user privacy, the challenge of system integration, and the necessity of new laws. In the case of the cloud, one can point out the government, health care, and finance areas as the main industries that are being influenced by its varying economic impact both positively and negatively. It is very important to evaluate the EI of cloud computing using a comprehensive approach, which means taking both its pros and cons into account, so that one can make well-informed decisions about its adoption and usage (Mondal & Goswami, 2024).

Sharma et al. (2025) conducted an analysis of Tier 1 and Tier 2 cities in India to create a through adoption model for cloud services, examining both time periods (i.e., prior to and after COVID). They did this by combining the TAM with categorical data. The findings indicate a relationship between perceived utility and privacy, security, and knowledge sharing. The adoption of cloud computing does spur creativity, according to the results, and this boosts HEI's academic performance.

Adoption of cloud computing in e-government has many advantages, but it also has special difficulties, especially for developing nations. This study broadens the analysis to look into more factors that influence cloud adoption from a technical, organizational, environmental, and personal standpoint, whereas previous research has concentrated on technical elements. Forty primary papers were found after a thorough examination of the literature across six academic databases by Alkhasawneh et al. (2025). The results highlighted how crucial it is to take an integrated organizational-individual viewpoint in order to comprehend cloud adoption more fully. The report recommended more empirical support for its comprehensive conceptual framework for government agencies' adoption of cloud computing. That evaluation provides public sector managers and decision-makers with practical guidance for addressing adoption challenges, enhancing service quality, and increasing cloud computing adoption rates in government organizations.

Research on cloud computing adoption and usage at educational institutions, as well as how contextual factors can affect cloud computing dispersion and adoption, has been few. In order to compete on the global stage, universities in poor nations must contend with difficult socioeconomic and political limitations that prevent them from investing in pricey information systems. A model that considers cultural, economic, and technological factors in the perception and adoption of cloud computing at universities in sub-Saharan Africa is put forth using concepts from the technology acceptance model and the diffusion of innovation theory. Outsourcing corporate IT infrastructure to third-party data centers with a common pool of computing hardware, storage, and networking resources is the foundation of the cloud computing paradigm. Services are made quickly and on-demand via the Internet. The anticipated advantages include sharing of almost infinite CPU, bandwidth, and storage space, easier and automated IT infrastructure administration, and elastic and dynamic resource supply. With increased scalability and significant infrastructure management cost savings, resource virtualization makes this feasible (Sabi et al., 2016).

With cloud provisioning companies like IBM setting up cloud computing centers in numerous developing nations like China, India, Vietnam, Brazil, South Africa, and South Korea, the cloud computing industry in the developing world is starting to attract significant attention from international agencies, national governments, and local IT players. Other international cloud providers are aggressively looking for opportunities and setting up cloud computing centers in developing nations, including Microsoft, Amazon, VMware, Salesforce, Dell, and Parallels (Kshetri, 2010).

4. CLOUD-BASED IOT IN EDUCATION

The use of cloud-based Internet of Things (IoT) in education has ushered in a period of innovation and transformation. The topic of cloud-based IoT and its significant impact on education are examined. "Cloud-based IoT" refers to the way that sensors and devices are linked to the cloud so that data may be shared, analyzed, and communicated. Together, IoT and cloud computing have the power to fundamentally alter educational institutions' administrative procedures and teaching methods. It examines the applications of the IoT in student interaction, campus management, tailored learning, and classrooms. There are many benefits to using cloud-based IoT in the classroom. Teachers can learn about students' development in real time. Adapt their tactics appropriately to promote individualized learning paths. By maximizing resource allocation and improving campus security through data driven decision-making, administrative chores become more efficient. Real-world case studies illustrate how cloud-based IoT is being used in education while illuminating the difficulties encountered and the tactics used. It anticipates new developments that might affect the education industry. Edge computing and intelligence work together to expand the IoT's potential and offer educational opportunities (Gautam & Kotiyal, 2024).

5. EXAMPLES OF CLOUD-BASED EDUCATIONAL TOOLS

Cloud computing makes it easier for students and teachers to access IT resources, as well as the platform for delivering and consuming services. It is an immaterial paradigm shift that brings new, essential elements to learning environments and models that are not available in more traditional technical frameworks. This refers to the rental or shared usage of computer power on a dispersed network of servers that store data and execute applications. This challenge is faced by both teachers and students in today's classrooms. Currently, e-learning platforms like Moodle are being used to accomplish this. This one thing alone requires more storage space. Teachers who wish to contribute extra materials must adhere to file size and type constraints. Periodically doing updates is necessary to maintain storage servers in optimal operating condition. The usage of cloud-based technologies in education may result in lower maintenance costs and greater availability (Sharma, 2025).

The advent of digital learning dates back to the COVID-19 epidemic, and as the world has been more advanced, education has been impacted to shift to digital, including the learning process. Students become more independent when the flip classroom approach is implemented with a learning management system (LMS). The findings demonstrate that implementing a flip classroom with different LMS can boost student autonomy by encouraging students to be more engaged, critical thinkers, and creative thinkers (Runiasih et al., 2025).

LMSs have changed as a result of AI, which has also made distant learning easier and allowed for individualized adaption. The literature shows gaps in the efficacy and acceptance of virtual assistants in educational environments, despite the field's seeming quick advancement. It is possible to see the thematic progression from subjects like regression analysis to LMS and e-learning. The theme clusters emphasize the concepts of ant colony optimization, ontology, and e-learning. A temporal study shows that ideas like a league table and a Cartesian plane provide a thorough understanding of the development of important concepts. According to analysis, terms like "learning style" and "learning management systems," which are new and developing, merit more research (Cabrera et al., 2025).

6. COLLABORATIVE LEARNING

Cloud computing has significantly facilitated the usage of online collaborative learning activities. Although a particular online application or service has been specifically mentioned, it is unclear how various cloud computing tools have influenced the idea of collaborative learning and how easily available these resources are to students today (Al-Samarraie & Saeed, 2018).

One encouraging development in contemporary ICT pedagogical systems is the use of cloud computing services. The idea behind a cloud-based platform for open learning and research is taken into account. The benchmarking studies on the use of ICT tools for research and learning in higher education pedagogical systems are completed. AWS virtual desktop, IBM Box, WPadV4, and other cloud-based components are used to support a model of the cloud-based open learning and research collaboration environment. Analyzing official international documents, publications on the topic, observing, comparing, analyzing the experience of using scientific and educational network packages, applying cloud technologies, benchmarking studies, and experimental studies are some of the research methods (Bykov et al., 2020).

Collaborative learning solutions are software applications and sites that facilitate communication, collaboration, and mutual learning in work and educational environments. These tools allow users to collaborate asynchronously or synchronously, which fosters good communication, task management, idea exchange, and content sharing. Popular instances include Trello, Slack, Google Workspace, and Kahoot. Utilizing these technologies improves accessibility, refines problem-solving abilities, and increases participation in both physical and virtual environments. Their core characteristics, such as cross-platform compatibility, real-time update, and software interconnectivity, render them integral to contemporary workflows. Although it has advantages, there are still problems such as technical issues, resistance from users, and data security issues. Emerging developments including AI, immersive augmented reality (AR) and virtual reality (VR) technology, and improved analytics are influencing the direction of collaboration tools in the future, increasing their impact and inclusion (Ok et al., 2024).

Google Docs, a web-based word processor, is renowned for its collaborative and accessible features. Because it allows users to create, edit, and exchange documents in real time, it is suitable for peer review, group projects, and feedback. However, its formatting options are limited and may not fully support certain formatting requirements for assignments, such as those for MLA or APA styles. Furthermore, accessibility features might be improved, particularly by permitting closed captioning for videos meant for individuals with disabilities and making tables more accessible. The revision history function can be hard to use, and the commenting system can be cumbersome and confusing. Additionally, it would be easier to evaluate student work and group projects if it were easier to track changes and revisions, particularly in huge documents (Bikmohammadi, 2023).

7. CLOUD COMPUTING'S IMPACT ON STUDENT PERFORMANCE EXAMINING EMPIRICAL EVIDENCE

The most important Cs of education are teamwork, communication, creativity, and critical thinking. Currently, though, these Cs are connected to technology to increase its dependability and effectiveness. Higher education is heavily reliant on educational technology, which many people utilize every day. Students are excited to use these academic-supporting technology. The survey method is utilized to gather the data, and a five-point Likert scale is employed. Partial least square structural equation modeling and confirmatory factor analysis were the statistical methods used on the data set. Perceived usefulness and ease of use have been found to positively correlate with all dimensions (Raza & Khan, 2022).

The success of cloud computing is heavily influenced by end-user happiness, information quality, system quality, administrative IT capability, and technical IT capability. Additionally, the performance of businesses is strongly and favorably impacted by the success of cloud computing. The requirement for managers to focus more on system quality, information quality, user satisfaction, and technical IT capacity is further supported by an importance-performance map analysis (Khayer et al., 2020).

Because of its widespread adoption and creative use to meet students' academic demands, cloud computing is quickly becoming an essential component of the educational community. Using TAM as a theoretical foundation, the use of cloud computing improves student academic performance through individual traits and the knowledge management paradigm. Perceived utility has a positive correlation with learnability, knowledge application, and knowledge sharing. Perceived utility and ease of use also have a big impact on cloud computing adoption, which helps students accomplish better in school (Ali et al., 2018).

8. ANALYSES OF MITIGATION STRATEGIES

A revolutionary method of learning, mobile cloud computing (MCC) has a significant influence on the teaching and learning dynamics. MCC can be advantageous to educational institutions by facilitating easy storage of education data and allowing users to access this information at any time and from anywhere. To ensure the successful implementation and complete integration of MCC globally, several barriers need to be addressed. Data security is a very key among such challenges, threatening the successful implementation of MCC severely. Students and teachers are particularly not aware of such cloud data security threats, highlighting the necessity of ongoing training programs to enhance understanding of MCC-related data security issues. Risks such as data loss, data breaches, data location, backup and restoration, data segregation, and data scavenging also play a significant role in the effectiveness of MCC deployment in HEIs (Qashou et al., 2025).

One of the most important components of contemporary IT, cloud computing offers businesses a range of advantages, ranging from cost efficiencies to scalability and flexibility. However, besides the business advantages, businesses must secure their information, applications, and cloud networks to be able to leverage cloud services efficiently. A few of the most popular threats to cloud computing security are data breaches, malware infections, account hijacking, and distributed denial-of-service (DDoS) attacks. A variety of the mitigation options are examined, which include vulnerability management, identity and access management (IAM), security information and event management (SIEM), security awareness training, and encryption methods. The usefulness of the shared responsibility model and its associated difficulties are discussed, along with new developments in cloud security, including serverless computing, containerization, and the integration of AI and Machine Learning (ML) (Ahmadi, 2024).

To protect sensitive data, preserve operational integrity, and defend against a wide range of cyberthreats, security is of utmost importance. CloudGuard Analytics offers real-time vulnerability monitoring and AI-driven detection to improve cloud security. Future trends include hybrid approaches with improved performance that could be investigated to strengthen cloud security and emphasize the need for adaptive security strategies that incorporate strong defenses, continuous monitoring, and threat intelligence tools to handle the ever-changing nature of cyberthreats. By placing a strong emphasis on cyber security, businesses can successfully handle the difficulties posed by cloud computing while safeguarding the availability, confidentiality, and integrity of important data in the current digital landscape. These realizations serve as the foundation for creating a security architecture especially made for cloud-based applications (Sharma & Sharma, 2024).

9. TAM AND UTAUT2 FOR MODERN EDTECH

In order to advance educational equity, especially in rural China, digital education platforms are essential for combining digital resources, learning settings, and educational services. The adjusted extended Unified Theory of Acceptance and Use of Technology 2 (UTAUT2) model, which includes teacher innovativeness, Technological Pedagogical Content Knowledge (TPACK) Ability, and perceived trust as extended constructs, was utilized by researchers to formulate the study objective. Gender, teaching experience, and culture were used as moderators. The results show that the main factors influencing teachers' intention and actual use of digital education platforms are performance expectation, effort expectation, facilitating conditions, hedonic incentive, teacher innovativeness, and TPACK ability. The results of moderation studies show that the associations between UTAUT2 components and adoption behavior are significantly adjusted by individual variations (gender, teaching experience, and culture). These factors help develop decision-making models for resource allocation, inform a behavioral modeling framework for policy design, and provide quantitative parameters for digital infrastructure planning (e.g., prioritizing training resources or network upgrades based on TPACK gaps and facilitating conditions). These results confirm that the modified UTAUT2 model is applicable in rural learning environments and provide engineering-relevant recommendations for maximizing the deployment of digital education ecosystems (Huang et al., 2025).

10. LIMITATIONS OF CLOUD ADOPTION: VENDOR LOCK-IN

Organizations are worried about interoperability problems when using cloud solutions, despite the advantages of cloud computing. One of the things preventing cloud interoperability is vendor lock-in. Although integrating open-source technologies into siloed cloud solutions is one method of reducing vendor lock-in, little is known about the elements that should work together to avoid vendor lock-in problems: origins, causes, and drawbacks of vendor lock-in, with a focus on how open-source technologies like Kubernetes and containers might lessen vendor lock-in problems. Vendor lock-in is when cloud providers' infrastructure, APIs, and tools are proprietary and cannot be switched from by clients at will without paying huge switch costs. Workloads can be moved from one vendor to another by businesses using open-source technologies such as Kubernetes and containers, given that they have to be integrated into their current walled cloud solutions. Application containers are tiny programs that run as if they have their own complete operating system, memory, CPU, and network infrastructure. A cloud environment based on opensource technologies like Kubernetes and containers can promote integration between siloed cloud solutions and open-source technology. Using the concepts of UTAUT (Unified Theory of Acceptance and Use of Technology), performance and effort expectations, social influence, and facilitating conditions, this qualitative generic inquiry sought to identify the barriers that organizations face when implementing cloud solutions and to investigate the elements that will facilitate the development of synergies between open-source technologies and siloed cloud solutions (Jeyaseelan, 2025).

11. SECURITY CONCERNS

The sharing of resources is key security concerns with cloud computing concept. It is necessary for cloud service providers to let their clients know how secure their cloud is. Main fear with cloud computing is data security. A few other security issues are network and virtualization security vulnerabilities. Achieving end-to-end security will be challenging because of the cloud's complexity. To be compatible with the cloud architecture, new security methods must be created and outdated ones must be drastically modified (Padhy et al., 2011).

An architecture for service delivery and access is cloud computing that uses the Internet to deliver virtualized and dynamically scaled resources as a service. For businesses, this strategy opens up new possibilities. Instead of requiring large upfront investments, it offers new business and operating models that let users pay for the resources they actually utilize. The absence of a single architectural approach or de facto standard that can satisfy the demands of a corporate cloud approach is the largest obstacle in cloud computing (Rimal et al., 2011).

Cloud computing is a widely adopted paradigm because it provides superior computing services at a reduced cost and with improved performance. Cloud computing's user-friendliness, net-centric technique, and flexible architecture have made it more and more popular. However, the fact that the cloud computing paradigm is still unable to address security vulnerabilities, which could worsen consumer data privacy and quality of service, is reducing its widespread use (Khan, 2016).

Applications of IoT and integrated cloud include smart metering, smart cities, smart homes, video surveillance, healthcare, and agriculture, among others. The integration of IoT and cloud involves a number of difficulties and problems, such as standardization of M2M communication and interoperability, energy and power efficiency of devices for data processing and transmission, big data produced by multiple devices, security and privacy, integration methodology, pricing and billing, network communications, storage, etc. (Malik & Om, 2017).

In a social setting, trust is usually used to describe a circumstance in which one party is prepared to depend on the behavior of another. The former gives up control of the acts to the latter. In the end, it's unclear if the person you trust will act or fulfill their end of the bargain. For the privacy and security of their assets (i.e., their data and processes), clients in the cloud computing environment need to have faith in a cloud provider. The cloud service model determines how much control over data and procedures is lost. Lack of control over data and processes in cloud computing increases the danger of losing data availability, confidentiality, and integrity for clients (Khan & Malluhi, 2010).

12. FUTURE TRENDS

The Quantum Advantage, which promises exponential speedup for particular computer jobs, is introduced by quantum computing, which uses quantum bits (qubits) to process information in parallel. AI powered by ML algorithms, has proliferated and is transforming industries via data-driven insights and automation. The implementation of AI models is supported by cloud computing, which is renowned for its scalability and accessibility. The Quantum Machine Learning (QML) paradigm improves on traditional ML models by utilizing the special capabilities of quantum computing. Harnessing the full potential of this disruptive integration requires constant research, teamwork, and flexibility. Organizations must overcome obstacles and grasp opportunities as quantum technologies develop in order to create a future where cloud-based AI and quantum computing expand the realm of computational potential (Padmanaban, 2024).

A key component of serverless computing is the auto-scaling capability, which enables apps to grow or shrink automatically. This eliminates the need for direct server management by enabling applications to be set up to adjust to traffic and demands as they arise and acquire resources as needed. When creating serverless applications, auto-scaling is a crucial concept that both industry and academia are beginning to acknowledge. Despite the scientific and industrial community's keen interest in serverless computing auto-scaling, no thorough, systematic, and lucid study has been carried out (Tari et al., 2024).

The goal of 5G, the fifth-generation cellular network, is to significantly improve quality of service by achieving lower latency and better throughput. By bringing cloud capabilities closer to end users (or user equipment, or UEs), edge computing is a new technology that makes it possible to move toward 5G while addressing the inherent issues with traditional cloud computing, like high latency and a lack of security (Hassan et al., 2019).

Future directions and opportunities in cloud learning are particularly relevant to the applications of blockchain, AR, VR, AI, and IOT. AI and ML are required for adaptive learning and intelligent tutoring systems to tailor instruction to individual learners. For ongoing education and capacity building, education offers safe and validated certifications. AR and VR enable learning to take place across geographical locations along with experience learning, which is believed to be the best type of learning. A company can utilize real-time data gathering and analysis from IoT enabled smart classrooms to enhance learning environments and utilize resources. Through developing and updating the existing educational-learning environment, these technologies can help educational institutions provide a more diversified, flexible, and effective learning setup that can prepare students for the collaborative and complicated nature of today's work environment (Paul & Das, 2025). Key understandings about the long-term benefits and potential drawbacks of the technology will be obtained from a study of long-term impacts of MCC implementation on student achievement, academic attainment, and institutional effectiveness. Subsequent studies should investigate how cutting-edge tools for example AI and ML can maximize MCC's usability and functionality in educational settings. In addition, cross-national comparisons of studies from different regions and education systems may assist in determining the socioeconomic determinants of MCC use and adoption (Qashou et al., 2025).

13. BIBLIOMETRIC ANALYSIS

The following section presents bibliometric analysis on the topic "Cloud Computing in Education." The following figures were made by combination of various sources for different images. The data source for all the figures is Lens, from where the data on the respective topic was extracted for the years 2005 to 2025 since cloud computing was introduced in the year 1996 and saw its progress in education starting from the year 2005. The search string used for searching and extracting the data is "Cloud computing in education", same as the chapter title.

Figures 2 to 4 are made using VOSViewer (VOSViewer, 2025) which helps in creation of network clustering, timeline chart and density chart. This application combines all the keywords, as mentioned and provide by authors as well as the keywords marked by Lens, into different cluster and links them showing relation between different themes or fields. Timeline chart and Density chart helps in understanding how the research focus has shifted from some keywords to other group of keywords over period of time and understanding the importance placed on each keyword irrespective of time respectively. Figures 5 to 7 are made using Python code and Lens data (Lens, 2025). They help in understanding the trend in research on this topic, important keywords, and the countries' contribution in research on this field.

A co-occurrence network visualization of terms taken from the Lens dataset is shown in Figure 2. Every node represents a keyword; its closeness indicates how frequently it appears alongside other keywords, while its size indicates frequency. The colors indicate clusters made and segregated by VOSviewer's modularity-based clustering method, each of which represents a different thematic study subject within cloud computing in education. The thickness of connecting lines indicates the intensity of link.

The most prevalent cluster, the red one, is centered on implementation frameworks and technology infrastructure. Keywords like "cloud computing," "virtualization," "SaaS," "IaaS," "data centers," and "network architecture" are included. This cluster focuses on the fundamental technical foundations that allow educational institutions to implement computing solutions that are both scalable and affordable. Building strong, service-oriented architectures to enable digital learning environments is the focus of this early research.

The focus of the green cluster is educational applications and pedagogical integration. Terms like "e-learning," "distance education," "MOOCs," "virtual classrooms," "blended learning," and "curriculum development" are among them. This cluster demonstrates how scholars and educators have revolutionized the delivery of education by utilizing cloud infrastructure to facilitate flexible content access and cross-border collaborative learning.

Data security, privacy, and ethical governance in cloud-enabled education are the main concerns of the blue cluster. This field is dominated by keywords like "data encryption," "authentication," "privacy preservation," "user trust," and "secure access control." It highlights the significance of compliance, risk mitigation, and policy frameworks and reflects growing scholarly concern about the dangers of keeping private student and educational data on external servers.

Although it is smaller, the yellow cluster symbolizes new interdisciplinary developments, especially the combination of cloud computing, AI, ML, and learning analytics. Common terms used here include "AI," "deep learning," "personalized learning," "big data," and "predictive analytics." This cluster shows how research is moving toward data-driven, intelligent, and adaptable educational ecosystems that use real-time analytics to maximize learning results.

Lastly, however smaller, the orange and purple clusters draw attention to specific topics including "collaborative learning," "mobile learning," "IoT," "green computing," and "edge computing." These subjects show how cloud-based learning has spread to mobile platforms and environmentally friendly computing techniques. When taken as a whole, these clusters highlight the field's interdisciplinary nature by tying together data science, pedagogy, cloud infrastructure, and sustainability. They also indicate how technological advancements have consistently reshaped innovative educational practices.

Figure 2. Bibliometric Analysis on Cloud Computing in Education using VOSViewer and data from Lens: Cluster Networking

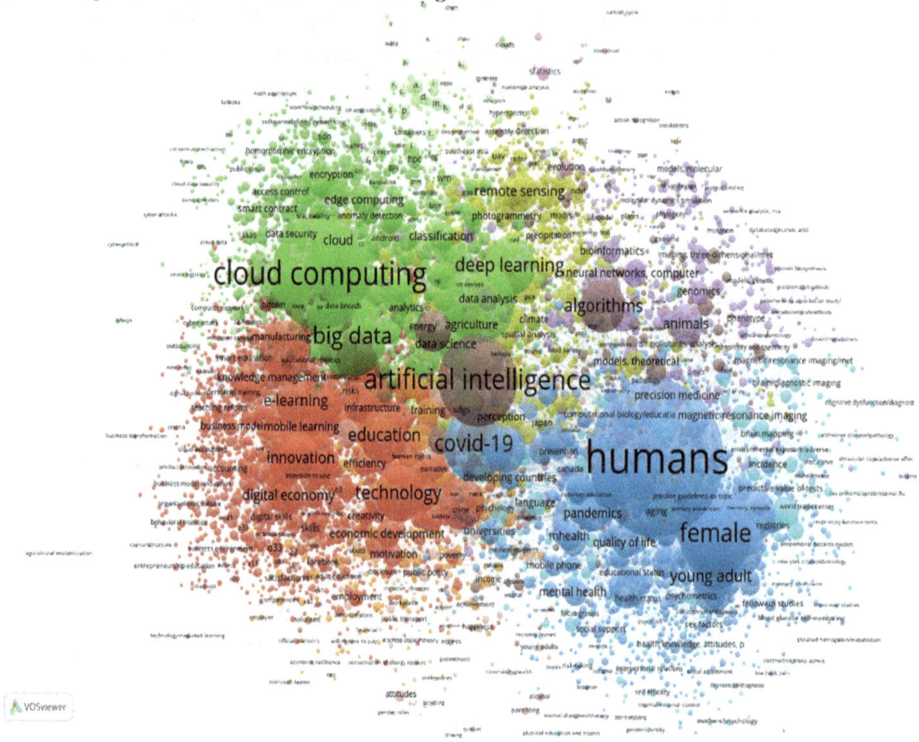

A temporal overlay map is shown in Figure 3, which uses color gradients from blue (previous years) to yellow (recent years) to illustrate the chronological development of study themes. One may see the development of the research focus over the course of 25 years with this visualization.

Blue-colored nodes predominate in the early period (2000–2008), reflecting foundational studies that concentrated on the technological and infrastructure aspects of cloud computing. This era is characterized by terms like "virtualization," "grid computing," "data centers," and "service architecture." During this phase, the majority of publications came from the fields of computer science and engineering, which established the basic technological foundation required for applications in education.

The green-shaded transitional period (2009–2016) signifies a significant change from the creation of new technologies to the application of those technologies in education. Terms like "e-learning," "distance learning," "virtual classrooms," "MOOCs," and "LMS (Learning Management Systems)" have become popular, indicating that educational institutions have started utilizing cloud platforms to support instruction, teamwork, and information exchange. This time frame reflects how the education sector used cloud solutions to improve accessibility and scalability, along with the growth of global online learning initiatives and massive open online courses.

The latest stage (2017–2025), which is indicated in yellow, reflects the continuous incorporation of ethics and intelligence into cloud-based learning platforms. This phase is dominated by keywords like "machine learning," "AI in education," "data privacy," "learning analytics," "adaptive learning," "blockchain," and "cybersecurity." The change from green to yellow illustrates how research has progressed beyond simple adoption to optimization, emphasizing intelligent, safe, and customized learning environments.

All things considered, the overlay map shows a distinct temporal shift from the development of infrastructure to innovations in education and, more lately, toward intelligent automation and moral leadership. This color shift effectively demonstrates how the scholarly conversation surrounding cloud computing in education has developed, broadened, and changed to accommodate changing demands in digital education.

Figure 3. Bibliometric Analysis on Cloud Computing in Education using VOSViewer and data from Lens: Timeline Chart

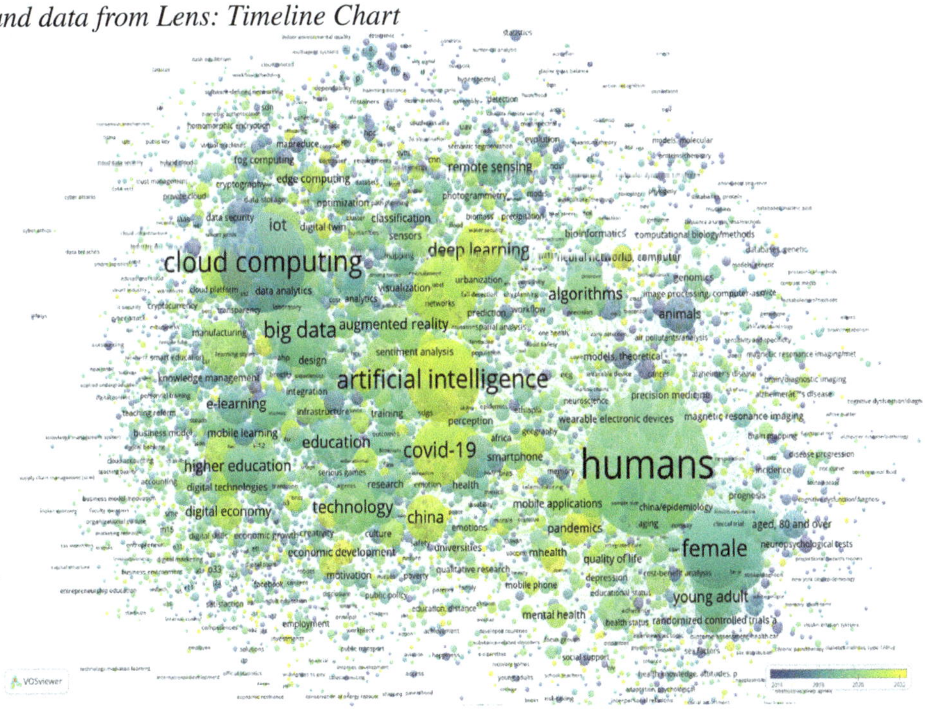

A density visualization map, which displays keyword concentration and connectedness in a heatmap-like fashion, is displayed in Figure 4. With green and blue parts signifying developing or less-studied subjects and yellow and red sections reflecting the most extensively explored themes, the color gradient illustrates the frequency and intensity of research activity.

The terms "cloud computing," "education," "e-learning," "learning management systems," and "virtual classrooms" are surrounded by a bright yellow zone at the center of the map. This dense area, which represents established study areas that have drawn consistent attention since the early 2010s, represents the field's intellectual core. It illustrates how the core of scholarly research is the integration of cloud infrastructure with learning systems.

Related yet specialized areas like "security," "data privacy," "authentication," "service reliability," "user trust," and "cost efficiency" are captured by moderately thick orange and green zones that go outward. These zones represent thematic extensions in which scholars have tackled issues related to preserving and scaling cloud platforms for education.

Emerging and multidisciplinary research horizons such as "AI-driven education," "learning analytics," "mobile cloud computing," "IoT in education," and "sustainability" are represented by the outermost blue and green peripheries. These less populated locations draw attention to areas of development and innovation that have gained popularity recently, especially since cloud-based schooling is starting to change due to technologies like edge computing and AI.

The research environment depicted in this picture is both developed and dynamic. The map's periphery is ablaze with new topics that indicate the field's transition toward intelligent, moral, and sustainable learning settings, while the core is still dominated by traditional pedagogical and infrastructure research. The domain's dual nature—a dynamic fringe of innovation and a stable core of known knowledge—is well captured by the density map.

Figure 4. Bibliometric Analysis on Cloud Computing in Education using VOSViewer and data from Lens: Density Visualization

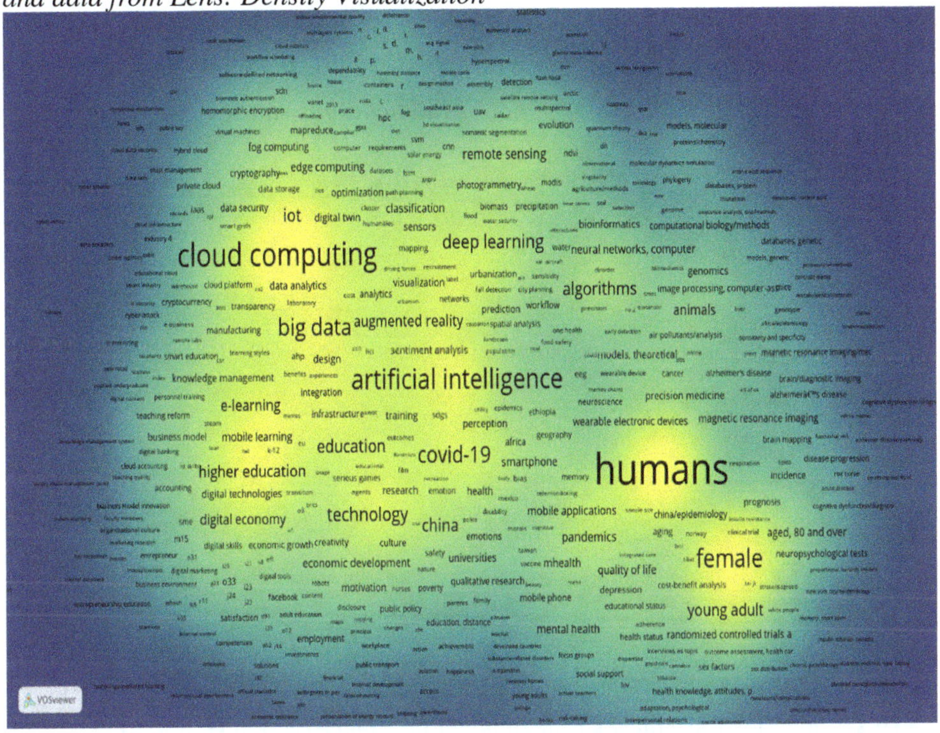

A longitudinal publishing trend is shown in Figure 5, which shows the yearly increase in cloud computing research output in education between 2000 and 2025. The vertical axis displays the total number of publications retrieved from Lens for each year, while the horizontal axis indicates the years.

The trendline shows that there was little research effort in the early years (2000–2007). The idea of cloud computing was still in its infancy, and its potential uses in education were only theoretical. Publications started to steadily increase between 2008 and 2012, in line with higher education institutions' growing use of cloud technology and internet-based learning.

Between 2013 and 2019, there is a significant increase that marks a significant stage of the field's growth. The widespread usage of cloud-hosted LMS platforms, the growth of online learning platforms like MOOCs, and the incorporation of collaborative cloud solutions like Google Workspace and Microsoft Azure Education all contribute to this boom. This time frame illustrates how cloud computing has becoming widely used in international educational systems.

A noticeable increase on the chart indicates a significant turning point during the COVID-19 pandemic (2020–2022). The increased use of remote and hybrid learning by educational institutions around the world has led to an unprecedented dependence on cloud-based solutions. Scalability, network dependability, fair access, and data ethics are just a few of the areas where research has expanded.

Figure 5. Line Chart depicting Number of Publications with respect to Time data from Lens

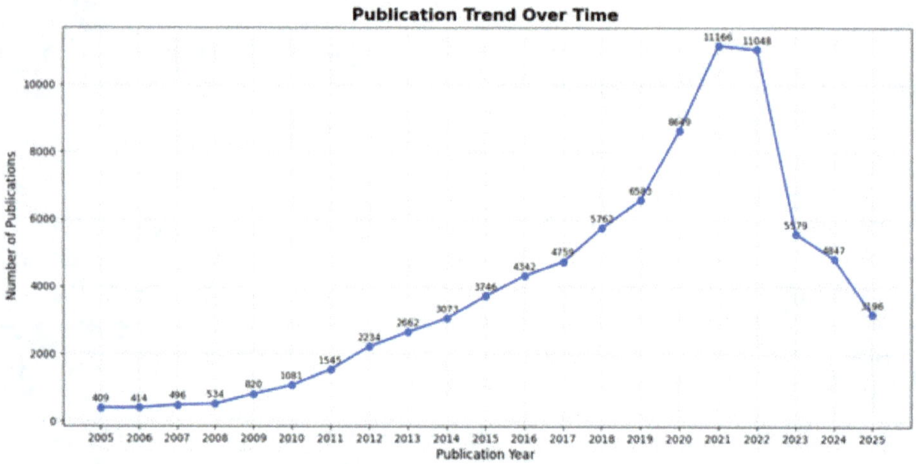

This graph illustrates the socio-technical importance of cloud computing during global upheavals in addition to its quantitative growth. The premise that educational technology research is driven by necessity, creativity, and accessibility is visually reinforced.

A Python-generated word cloud that provides a textual frequency depiction of the most common keywords, titles, and abstract concepts in the dataset is shown in Figure 6. Each word's size and prominence are correlated with its frequency, giving a brief yet thorough overview of the theme landscape.

Large terms like “cloud computing,” “education,” “e-learning,” “virtual classrooms,” “learning management systems,” “distance learning,” and “MOOCs” predominate at the center, reinforcing the field's emphasis on cloud-enabled digital learning platforms. These key words serve as the research area's fundamental conceptual framework.

Mid-sized phrases like “learning analytics,” “data security,” “privacy,” “collaboration,” “AI,” “remote learning,” and “machine learning” surround these key concepts. These highlight how cloud computing has expanded both technologically and analytically, with more recent research focusing on automation, personalization, and ethical data use.

New research directions centered on resource efficiency, trust mechanisms, and smart ecosystems are exemplified by peripheral but developing terminology like "blockchain," "green computing," "IoT," and "sustainability." The term "cloud" describes how the area has evolved from simple cloud deployment to whole digital learning ecosystems that include ethical, technological, and pedagogical factors.

Figure 6. Word Cloud Depicting Important Keywords

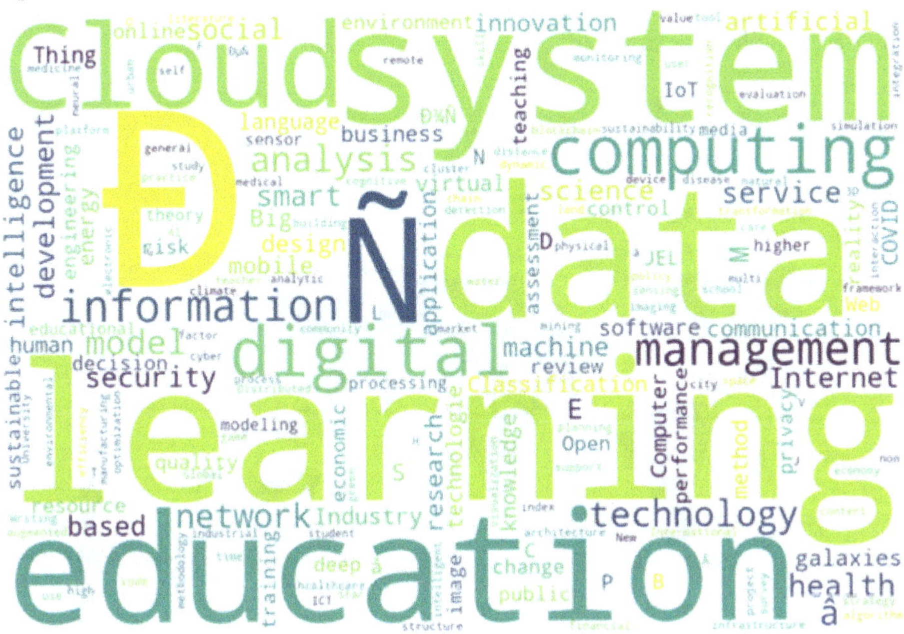

Based on co-authorship data, Figure 7 shows a geographical cooperation network that illustrates global research collaborations. Each node represents a nation, and the size of each node corresponds to the volume of publications. Collaborative linkages are indicated by edges, or lines; larger lines indicate stronger ties.

The most prominent nodes are the United States, China, the United Kingdom, and India, suggesting that these countries provide the most research on cloud computing in education worldwide. They serve as focal points, sustaining broad partnerships in North America, Europe, and Asia. The European cluster, which includes Germany, Spain, Italy, and the Netherlands, creates a highly integrated network that connects the research ecosystems in the East and the West.

Other notable contributors include South Korea, Japan, Australia, and Canada, all of which actively engage in technological experimentation and pedagogical innovation. Despite appearing as smaller nodes, emerging economies in the Middle East,

Africa, and Latin America are becoming more interconnected, which is indicative of more scholarly activity and global collaborations bolstered by cross-border funding and knowledge-sharing initiatives.

The web-like arrangement of connections depicts a globally integrated research community in which developing regions provide context-specific applications and adaptable models, while developed nations spearhead technological progress. This collaborative network's structure highlights how worldwide educational reform is, with cloud computing acting as a bridge for international academic cooperation as well as a technological equalizer.

Figure 7. World Graph Depicting Countries' Contribution

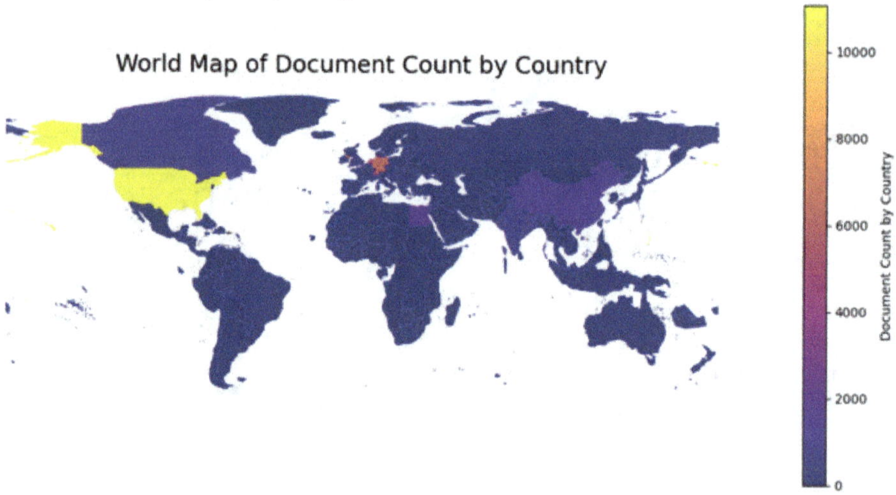

14. CONCLUSION

As cloud computing becomes more popular, institutions of learning can potentially enhance efficiency and collaboration across all levels, transforming the manner in which education is delivered and ushering in a more dynamic and future-proof learning platform. This will enable personalized learning experiences, increased involvement of students, easier operations, and flexible availability of resources. In order to make the most out of the potential of Cloud computing in the classroom, its integration must be approached with careful deliberation on data security, staff training, and good internet connectivity.

Teaching staff and students can access learning resources and tools anywhere, anytime, due to cloud platforms. Real-time communication between teachers, students, and campuses through cloud-based technology allows for a more interactive learning environment. Since cloud computing does not require enormous on-premise hardware infrastructure, it can largely lower operating expenses of educational institutions. The ease with which cloud services can scale down or up to accommodate changing demands allows for launching new educational programs and fine-tuning student enrollment. The high level of data protection and backup capabilities of cloud storage protects valuable student information. Providing that teachers and students get sufficient training to use cloud-based tools and resources. Installing strong security protocols to protect personal student data stored on the cloud. Providing reliable internet connectivity to enable easy access to cloud-based services.

Additionally, the bibliometric analysis provides quantitative insights into the development of cloud computing research in education between 2005 and 2025. It illustrates the shift from early technology frameworks to educational and data-centric innovations by looking at worldwide publication trends, theme clusters, and international collaborations.

All things considered, cloud computing offers schools a groundbreaking chance to increase productivity, enhance collaboration, and provide every student with a tailored learning experience ultimately ensuring that education is shaped by digital innovation.

REFERENCES

Agrawal, S. (2021). A survey on Recent Applications of Cloud Computing in Education: COVID-19 Perspective. *Journal of Physics: Conference Series, 1828*(1), 012076. DOI: 10.1088/1742-6596/1828/1/012076

Ahmadi, S. (2024). Systematic Literature Review on Cloud Computing Security: Threats and Mitigation Strategies. *Journal of Information Security, 15*(02), 148–167. DOI: 10.4236/jis.2024.152010

Al-Samarraie, H., & Saeed, N. (2018). A systematic review of cloud computing tools for collaborative learning: Opportunities and challenges to the blended-learning environment. *Computers & Education, 124*, 77–91. DOI: 10.1016/j.compedu.2018.05.016

Ali, Z., Gongbing, B., & Mehreen, A. (2018). Understanding and predicting academic performance through cloud computing adoption: A perspective of technology acceptance model. *Journal of Computers in Education, 5*(3), 297–327. DOI: 10.1007/s40692-018-0114-0

Alkhasawneh, R., Cob, Z. C., & Latif, A. B. A. (2025). Cloud computing adoption in government organizations in developing countries. *International Journal of Public Administration in the Digital Age, 12*(1), 1–27. DOI: 10.4018/IJPADA.368246

Bikmohammadi, M. (2023). *Incorporation of Google Docs in composition classes to Foster collaboration: An analysis of Google Docs' affordances.*

Bora, U. J., & Ahmed, M. (2013). E-Learning using Cloud Computing. In Blue Eyes Intelligence Engineering & Sciences Publication [IJISME]. *International Journal of Science and Modern Engineering, 1*(2), 9–10.

Bykov, V., Mikulowski, D., Moravcik, O., Svetsky, S., & Shyshkina, M. (2020). THE USE OF THE CLOUD-BASED OPEN LEARNING AND RESEARCH PLATFORM FOR COLLABORATION IN VIRTUAL TEAMS. In *Information Technologies and Learning Tools 76*(2), 304-305.

Cabrera, B. C. C., Leal, M. C., Martínez, J. S. A., Soto, A. L. R., Sepulveda, J., Valencia, J., Garcés-Giraldo, L. F., & Valencia-Arias, A. (2025). Artificial Intelligence (AI) and Learning Management Systems (LMS): A bibliometric analysis. *Journal of Infrastructure Policy and Development, 9*(1), 8029. DOI: 10.24294/jipd8029

Ercan, T. (2010). Effective use of cloud computing in educational institutions. *Procedia: Social and Behavioral Sciences, 2*(2), 938–942. DOI: 10.1016/j.sbspro.2010.03.130

Gautam, A., & Kotiyal, A. (2024). Connecting the future. In P. D. Singh & M. Angurala (Eds.), *Integration of Cloud Computing and IoT: Trends, Case Studies and Applications* (1st ed.), 274–308). Chapman and Hall/CRC eBooks. https://doi.org/DOI: 10.1201/9781032656694-15

Hassan, N., Yau, K. A., & Wu, C. (2019). Edge Computing in 5G: A review. *IEEE Access : Practical Innovations, Open Solutions*, *7*, 127276–127289. DOI: 10.1109/ACCESS.2019.2938534

Huang, J., Pinmanee, S., & Chaveesuk, S. (2025). Acceptance and Use of Digital Education Platform: An Application of UTAUT2. *Operational Research in Engineering Sciences: Theory and Applications*, *8*(1), 237–264.

Jeyaseelan, S. (2025). Vendor Lock-In Issues In Cloud Computing and How to Neutralize Them. In M. Muchnick, *A Dissertation Presented in Fulfillment of the Requirements for the Degree Doctor of Philosophy*.

Khan, K. M., & Malluhi, Q. (2010). Establishing trust in cloud computing. *IT Professional*, *12*(5), 20–27. DOI: 10.1109/MITP.2010.128

Khan, M. A. (2016). A survey of security issues for cloud computing. *Journal of Network and Computer Applications*, *71*, 11–29. DOI: 10.1016/j.jnca.2016.05.010

Khayer, A., Bao, Y., & Nguyen, B. (2020). Understanding cloud computing success and its impact on firm performance: An integrated approach. *Industrial Management & Data Systems*, *120*(5), 963–985. DOI: 10.1108/IMDS-06-2019-0327

Kshetri, N. (2010, January). Cloud computing in developing economies: drivers, effects, and policy measures. In *Proceedings of PTC 10*, 1-22.

Lakshminarayanan, R., Kumar, B., & Raju, M. (2013). Cloud computing benefits for educational institutions. *arXiv (Cornell University)*. https://doi.org//arxiv.1305.2616DOI: 10.48550

Lens. Lens. Retrieved October, 2025, from https://www.lens.org/

Lera, F. J. R., González, D. F., Rico, F. M., Guerrero-Higueras, Á. M., & Conde, M. Á. (2021). Measuring students acceptance and usability of a cloud virtual desktop solution for a programming course. *Applied Sciences (Basel, Switzerland)*, *11*(15), 7157. DOI: 10.3390/app11157157

Malik, A., & Om, H. (2017). Cloud computing and Internet of Things Integration: architecture, applications, issues, and challenges. In *Springer eBooks*. 1–24. https://doi.org/DOI: 10.1007/978-3-319-62238-5_1

Mehta, P. S., Shedge, P. P., Vyavahare, S., Soundarrajan, S., & Chillarge, G. (2024). Next-Gen Learning cloud integration for academic and industrial synergy. In Mahalle, P. N. (Ed.), *Advances in educational technologies and instructional design book series* (pp. 73–108)., DOI: 10.4018/979-8-3693-6705-6.ch004

Mondal, S., & Goswami, S. S. (2024). A narrative literature review on the economic impact of cloud computing: Opportunities and challenges. *Computing and AI., 1934*, 1934. Advance online publication. DOI: 10.59400/cai1934

Naveed, Q. N., Alam, M. M., Qahmash, A. I., & Quadri, K. M. (2021). Exploring the determinants of service quality of cloud E-Learning system for active system usage. *Applied Sciences (Basel, Switzerland), 11*(9), 4176. DOI: 10.3390/app11094176

Odhiambo, M. O. (2024). Availability of academic resources presents opportunities and challenges. In *Advances in linguistics and communication studies.* 435–458. https://doi.org/DOI: 10.4018/979-8-3693-3579-6.ch017

Ok, E., Grace, J., & John, M. (2024). *Collaborative Learning Tools.* ResearchGate.

Padhy, R. P., Patra, M. R., & Satapathy, S. C. (2011). Cloud computing: security issues and research challenges. In *IRACST - International Journal of Computer Science and Information Technology & Security.* IJCSITS.

Padmanaban, H. (2024). *Quantum computing and AI in the cloud.* Journal of Computational Intelligence and Robotics.

Paul, S., & Das, D. (2025). Future Trends and Innovations in Cloud-Driven Education. In *Advances in educational technologies and instructional design book series.* 321–346. https://doi.org/DOI: 10.4018/979-8-3693-6705-6.ch013

Ponde, P. (2024). Security and privacy considerations in Cloud-Based education. In *Advances in educational technologies and instructional design book series.* 175–194. https://doi.org/DOI: 10.4018/979-8-3693-6705-6.ch007

Qasem, Y. M., Abdullah, R., Yaha, Y., & Atana, R. (2020). Continuance Use of Cloud computing in Higher Education Institutions: A Conceptual model. *Applied Sciences (Basel, Switzerland), 10*(19), 6628. DOI: 10.3390/app10196628

Qashou, A. M. A., Bahar, N., & Mohamed, H. (2025). Qualitative exploration of data security risks in mobile cloud computing for higher education. *Security and Privacy, 8*(2), e70001. Advance online publication. DOI: 10.1002/spy2.70001

Raza, S. A., & Khan, K. A. (2022). Knowledge and innovative factors: How cloud computing improves students' academic performance. *Interactive Technology and Smart Education, 19*(2), 161–183. DOI: 10.1108/ITSE-04-2020-0047

Rimal, B. P., Jukan, A., Katsaros, D., & Goeleven, Y. (2011). Architectural Requirements for cloud computing Systems: An Enterprise Cloud approach. *Journal of Grid Computing*, *9*(1), 3–26. DOI: 10.1007/s10723-010-9171-y

Runiasih, M., Rugaiyah, R., & Ika, L. (2025). Systematic Literature Review (SLR): Implementation of Flip Classroom using Learning Management System (LMS) to improve students' learning independence. *International Education Trend Issues*, *3*(1), 1–8. DOI: 10.56442/ieti.v3i1.963

Sabi, H. M., Uzoka, F. E., Langmia, K., & Njeh, F. N. (2016). Conceptualizing a model for adoption of cloud computing in education. *International Journal of Information Management*, *36*(2), 183–191. DOI: 10.1016/j.ijinfomgt.2015.11.010

Scalera, M., Gentile, E., Plantamura, P., & Dimauro, G. (2020). A Systematic mapping study in cloud for Educational innovation. *Applied Sciences (Basel, Switzerland)*, *10*(13), 4531. DOI: 10.3390/app10134531

Sein-Echaluce, M. L., Fidalgo-Blanco, A., García-Peñalvo, F. J., & Fonseca, D. (2021). Impact of Transparency in the Teamwork Development through Cloud Computing. *Applied Sciences (Basel, Switzerland)*, *11*(9), 3887. DOI: 10.3390/app11093887

Sharma, C., & Sharma, C. (2024). Cloud Computing Security: Threats and Mitigation Strategies. In *2024 International Conference on Signal Processing and Advance Research in Computing (SPARC)*. 1–6. IEEE. https://doi.org/DOI: 10.1109/SPARC61891.2024.10829335

Sharma, M., Vadalkar, S., Singh, A., Chavan, G., & Tsagarakis, K. P. (2025). How does cloud computing adoption accelerate education? – Enhancing academic results by integrating knowledge management pre and post pandemic. *International Journal of Human-Computer Interaction*, •••, 1–21. DOI: 10.1080/10447318.2025.2558030

Sharma, S. U. (2025). Using Cloud Technology to Maximise Resources for Education. *Journal of Scientific research and Technology*, *3*(2), 33–36.

Tari, M., Ghobaei-Arani, M., Pouramini, J., & Ghorbian, M. (2024). Auto-scaling mechanisms in serverless computing: A comprehensive review. *Computer Science Review*, *53*, 100650. DOI: 10.1016/j.cosrev.2024.100650

VOSviewer - Visualizing scientific landscapes. VOSviewer. Retrieved October, 2025, from https://www.vosviewer.com/

Yasser, I., El-Khalek, A. A., Khalid, H. E., Salama, A. A., & Essa, A. K. (2025). *Cloud Computing and neutrosophic Logic: Navigating uncertainty in Educational administration*. UNM Digital Repository.

ADDITIONAL READINGS

Mehta, P. S., Doshi, S. T., Butle, S. N., Inamdar, S. C., & Shinde, S. D. (2025). The Fusion of Academia and Industry via Cloud Integration in Education: Customized Learning and Adaptive Systems. In Bridging Academia and Industry Through Cloud Integration in Education (pp. 109-144). IGI Global Scientific Publishing.

Ramadugu, G. (2025). Leveraging AI for Continuous Integration and Delivery Enhancing Developer Productivity in Smart Education and Sustainable Learning. In Smart Education and Sustainable Learning Environments in Smart Cities (pp. 287-300). IGI Global Scientific Publishing.

Soni, R., Bhatia, K., & Rajput, N. (2025). A thorough analysis of cloud computing technology: Present, past, and future. In Recent Advances in Sciences, Engineering, Information Technology & Management (pp. 137-145). CRC Press. Bhosle, A. (2025). Real-Time Analytics and AI Transforming Data Engineering for Sustainable Learning in Cloud Environments. In Smart Education and Sustainable Learning Environments in Smart Cities (pp. 347-364). IGI Global Scientific Publishing.

Zhang, X. (2025). Internet of Things and Cloud Computing-Powered Virtual Laboratories for Enhanced STEM Education. Journal of The Institution of Engineers (India): Series B, 1-10. Andrews, M., Owen, A., & Oye, E. (2025). Integrating MuleSoft with Learning Management Systems (LMS): Enhancing Educational Experiences through Seamless Data Connectivity.

KEY TERMS AND DEFINITIONS

Flexibility and Availability: The way that instructors and students access educational information has been completely transformed by cloud computing. Learning resources are accessible around-the-clock from any location with internet access thanks to cloud software. Cloud-powered virtual classrooms remove geographical obstacles to education by facilitating smooth remote learning experiences.

Enhanced Cooperation: The capacity of cloud computing to enable real-time collaboration is one of its main benefits. No matter where they are physically located, teachers and students may collaborate on projects, share documents, and take part in group discussions thanks to cloud-based applications. This cooperative setting prepares students for the demands of the workplace of the future by simulating real-world professional settings.

Cost-Efficiency: Cloud computing in education has significant financial advantages. By doing away with the requirement for large amounts of on-premise hardware, educational institutions can drastically lower their infrastructure expenses. Additionally, compared to traditional licensing arrangements, cloud-based software licensing models frequently turn out to be more cost-effective. Because of its cost-effectiveness, educational institutions can shift their resources from maintaining IT systems to enhancing the quality of instruction.

Security of Data: Since educational institutions manage sensitive data, including student records, financial information, and proprietary research data, data security is a critical component. For this reason, educational establishments like colleges and schools are increasingly becoming the focus of cyberattacks.

Personalized Learning: Through the use of data analytics and machine learning platforms, cloud technology makes personalized learning possible. In order to customize content and provide targeted interventions, these platforms collect information about students' strengths and limitations. With personalized learning paths, students can focus on areas that need more help while making rapid progress in their strong points. Teachers can adjust their teaching strategies to meet the needs of each individual student by monitoring student performance and learning trends. This data-driven method enables the development of personalized learning routes and assists in identifying areas in which students want more assistance.

Chapter 8
Blending Computational Thinking and Learner-Paced Segment:
Content Expert Evaluation of Digital Video Courseware Development

Wan Nor Ashiqin Wan Ali
http://orcid.org/0000-0003-2489-2843
Universiti Malaysia Perlis, Malaysia

Wan Ahmad Jaafar Wan Yahaya
http://orcid.org/0000-0002-8605-0062
Universiti Sains Malaysia, Malaysia

Syed Zulkarnain Syed Idrus
Universiti Malaysia Perlis, Malaysia

Mohd Noorul Fakhri Yaacob
http://orcid.org/0000-0002-0938-7833
Universiti Malaysia Perlis, Malaysia

ABSTRACT

This study explores the development and implementation of digital video courseware (DVC) combining computational thinking (CT) and segmentation techniques to enhance educational outcomes. Segmentation, particularly self-explanation, improves engagement and achievement, while system-paced segmentation reduces cognitive load for novices, boosting retention. The courseware fosters CT skills like

DOI: 10.4018/979-8-3373-1190-6.ch008

algorithmic thinking and supports learner-paced flexibility for better information processing. Content experts evaluated its design, pedagogical relevance, and CT alignment. Results highlight improved understanding, retention, and performance, offering insights for advancing learner-centered education through CT integration and adaptive pacing.

1. INTRODUCTION

The collaboration between government, industries, and educational institutions has become increasingly essential in addressing the skills gap among university graduates to support Malaysia's vision of achieving high-income status by 2020. The 11th Malaysia Plan forecasts an increase in job vacancies requiring Technical and Vocational Education and Training (TVET)-related skills from 2016 to 2020. However, many institutions have struggled to produce graduates with TVET-related skills that align with industry demands (Adnan et al., 2021; Affero & Hassan, 2013; Economic Planning Unit, 2015; Ministry of Higher Education, 2017; UNESCO, 2021). Juškevičiene and Dagiene (2018), along with Rodgers and LaBoskey (2018), have emphasized that digital competence and computational thinking (CT) are key 21st-century skills, particularly crucial for university students in technical fields.

The concept of CT was first introduced by Papert in 1980 as part of computer science. However, Wing (2006) later popularized CT as a fundamental skill for all students, sparking a growing body of research on its integration into education. According to Lockwood and Mooney (2017), many countries are still in the process of adopting CT within their curricula at all educational levels. Educators are encouraged to apply diverse tools and materials to teach CT across subjects. When considering teaching and learning approaches, designing educational technology that effectively presents learning materials has become a key strategy. Research by Stiller et al. (2011) and Wang (2017) indicates that appropriate pacing in learning materials can significantly enhance motivation, student engagement, and achievement. Additionally, the Cognitive Theory of Multimedia Learning (CTML) principles, as discussed by Zavgorodniaia et al. (2021), highlight the importance of modality effects, segmentation, and system-paced learning in the design of educational materials. Their study also identified the need for further research to optimize learner control in managing learning materials.

Numerous studies have demonstrated the benefits of learner-paced instruction, including reduced cognitive load Hasler, Kersten, & Sweller (2007) and Moreno and Mayer (2007), improved knowledge achievement (Hasler, Kersten, & Sweller; 2007; Lusk et al., 2009; Mayer & Chandler, 2001), and more efficient skill acquisition. However, Stiller et al. (2009) pointed out that previous studies providing empirical evidence of the advantages of learner-paced instruction often lack experimental manipulation in real-world settings. Furthermore, as the use of videos in education, such as for learning materials, assessments, and content delivery, continues to rise, emphasizing learner control over the pace of learning becomes increasingly important (Biard et al., 2018). Several studies have investigated how learner control features (such as play, pause, stop, next module, and speed control) impact cognitive load (Biard et al., 2018; Hasler, Kersten, & Sweller, 2007). Hasler et al. (2007) argued that whether through pre-defined segments or allowing students to control their learning materials, learner-paced design should be an integral part of educational technology to improve instructional strategies. They also called for further research to explore these aspects in greater depth. Given the growing importance of digital competence and CT as essential 21st-century skills, and their potential to be emphasized among university students, this study aims to integrate digital video learning and CT into the design of educational materials.

2. LITERATURE REVIEW

The global COVID-19 pandemic has revealed varying levels of educators' willingness to adopt technology to support student learning (Pérez et al., 2020). In line with this, Malaysia aims to equally value and nurture both academic and Technical and Vocational Education and Training (TVET) pathways, leveraging technology as outlined in the Ten Shifts of the Malaysia Education Blueprint 2015–2025 (Higher Education). The country aims to produce high-quality TVET graduates and transform higher education by promoting globalized online learning through technology integration. The widespread availability of digital technology among teenagers and young adults has made technology a central aspect of their lives, with this group often referred to as "digital natives" (Prensky, 2009). These students interact with technology on a daily basis and should possess at least a moderate level of digital competence to manage their activities effectively. Rodgers and LaBoskey (2018) argue that even digital natives, as 21st-century students, need to be trained and exposed to the effective use of digital technology.

According to Juškevičiene and Dagiene (2018), digital competence encompasses the development of digital content such as images, audio, and video, and involves the self-efficacy and engagement with digital technology for learning and work. McCaslin and Young (2015) highlight the significance of using digital video in technical subjects, particularly as an assessment tool for university students. Studies by Huang et al.(2020) and McCaslin and Young (2015) suggest that digital video can encourage student engagement and focus on their studies, offering an efficient way to present learning material. Digital video has become a crucial element in undergraduate courses, not only in engineering but across various disciplines (McCaslin & Young, 2015). Similarly, Mazin et al. (2020) regard digital video as an effective tool for assessing TVET students' competencies, aligning with technological advancements in education. Snelson et al. (2021) emphasize the role of digital video production in helping students learn content-based knowledge while fostering technological skills.

A wealth of research supports the positive effects of digital video in education, including enhancing knowledge and achievement (Bell & Bull, 2010; McCaslin & Young, 2015; Means et al., 2009; Rodgers & LaBoskey, 2018; Shin, 2018), boosting interest and motivation (Bell & Bull, 2010; Ebrahimzadeh & Alavi, 2017; Forte & Guzdial, 2004; Juškevičiene & Dagiene, 2018; Liao, Chen, & Shih, 2019; McCaslin & Young, 2015; Shin, 2018), and improving self-efficacy (Juškevičiene & Dagiene, 2018; Shin, 2018) among young adult students. Ismail et al. (2017) further highlight the impact of multimedia educational videos in boosting knowledge and motivation among TVET students. Shin (2018) suggests that self-efficacy is a subset of motivation and can increase alongside it. However, McCaslin and Young (2015) and Rodgers and LaBoskey (2018) contend that students should receive more exposure and training in video production skills, including camera practice, editing, and video production techniques, to enhance the learning process. They point out that few studies focus on using educational technology to promote video production knowledge among students. Snelson et al. (2021) also note the lack of research into the use of video production as a method for learning content-based knowledge and as part of coursework assessment.

Juškevičiene and Dagiene (2018) argue that digital competence and computational thinking (CT) are closely related, both being critical 21st-century skills for the younger generation. They also suggest that more research is needed to explore the relationship between CT and digital competence. Despite the growing interest in promoting CT among students, challenges remain in integrating it into educational curricula (2018). The expanding volume of media, tools, and technology on the internet requires students to acquire a broader set of skills and competencies. Voogt et al. (2015) highlight the potential benefits of integrating CT into curricula and call for further research to advance the CT agenda in education. They argue that CT can be applied beyond programming subjects, emphasizing its relevance in fields outside of computer science, and that understanding CT concepts does not necessarily require a predefined set of conditions. Denning (2009) and Lu and Fletcher (2009) emphasize that algorithmic thinking, a core component of CT, is fundamental to students' ability to understand and address problems.

Researchers often define CT by its key components, including algorithms (Juškevičiene & Dagiene, 2018). The concept of algorithmic thinking, explored by Catlin and Wollard (2014), investigates the relationship between CT, student activities, and curriculum subjects. Wing (2008) and Voogt et al. (2015) stress that computational thinking is essential for conceptual understanding across disciplines, with algorithmic thinking playing a central role. Aho (2012) argues that CT skills involve computational steps and algorithms that form part of the students' problem-solving process. They also highlighted that, integrating CT across disciplines such as social sciences can enhance students' use of computational terminology. Voogt et al. (2015) conclude that CT is widely regarded as an essential 21st-century skill in educational technology studies.

Mayer et al. (2019) discuss the widespread use of educational technology in university courses, including courseware and online learning platforms and highlighted call for more research into how to design effective online lessons and educational technology. They emphasize the need for studies that examine how segmenting principles, such as those used in course-based online lessons or educational technologies, impact learning outcomes. They also caution that educators should be mindful of the risk of cognitive overload caused by extraneous information and the simultaneous presentation of text, images, and narration in instructional design. With the increasing use of digital video as an assessment tool during the pandemic and the rise of educational technology, this study aims to promote basic video production skills through a courseware platform. Bell and Bull [26] assert that learning with digital video requires effective teaching and instructional strategies. Recent research by Castro-Alonso et al. (2021) explores the best practices for instructional strategies, particularly regarding cognitive load theory and the Cognitive Theory of Multimedia Learning (CTML). They argue that pacing techniques, such as segmenting learning materials and incorporating interactive features like scrollbars (Hatsidimitris & Kalyuga, 2013) or next buttons (Mayer & Chandler, 2001; Stiller, Petzold, & Zinnbauer, 2011), allow students to manage information more effectively, enhancing learning outcomes.

Castro-Alonso et al. (2021) also highlight the positive impact of learner-paced instruction on student achievement. Studies by Hatsidimitris and Kalyuga (2013) and Stiller et al. (2011) show that students using learner-paced learning materials outperform those using instructor-paced (system-paced) materials. However, Castro-Alonso et al. (2021) argue that students' expertise and prior knowledge influence their performance, with expert students benefiting from learner-paced materials while novices may perform better with instructor-paced materials. They recommend further research into the effects of learner-paced and instructor-paced learning, considering moderating factors such as gender and modality, as well as their influence on students' psychological and academic outcomes. Given the importance of digital competence and CT as complementary skills for the 21st century, this study seeks to bridge gaps in the promotion of digital video learning, the role of algorithmic thinking in supporting digital competence, and the significance of learner-paced segmentation in educational design.

The current landscape of education for TVET (Technical and Vocational Education and Training) students emphasizes the need to equip them with 21st-century skills, particularly considering the advancements in Industry 4.0 (Azid et al. 2019; Mashelkar, 2018; Rodgers & LaBoskey, 2018). Among these essential skills are digital competence and computational thinking (CT), both of which are crucial for technical students (Juškevičiene & Dagiene, 2018; Rodgers & LaBoskey, 2018). Digital competence encompasses the self-efficacy and engagement with digital technologies like photographs, audio, and video, for both study and work (Juškevičiene & Dagiene, 2018). The application of digital video in technical subjects, particularly as an assessment tool, has been emphasized by McCaslin and Young (2015), who suggest that it helps increase student focus and motivation while enabling more effective communication of ideas (McCaslin & Young, 2015; Huang et al., 2020). Furthermore, digital video production offers an indirect way to nurture students' digital competence skills (Snelson, Yang, & Temple, 2021).

This study seeks to bridge the gap by developing a digital video courseware that incorporates algorithmic thinking and two presentation modes: learner-paced and system-paced. The research aims to evaluate the effects of these modes on students' achievement and CT self-efficacy. By combining digital video production with algorithmic thinking, this study aims to equip students with both CT skills and digital competence, preparing them for success in the 21st century.

3. METHODS

3.1 Preliminary Investigation (PI)

The researcher conducted a preliminary investigation (PI) to validate the dependent variables and research problem. The study examined whether students' learning challenges were tied to student-centered learning and explored the feasibility of integrating algorithmic thinking, learner-control, and learner-paced approaches into Malaysia's TVET curriculum.

Semi-structured interviews were used to assess the awareness and knowledge of Computational Thinking (CT) among TVET students, as well as issues related to current teaching methods, online learning, and the technologies employed. This method, known for its flexibility and effectiveness in qualitative research (Bradford & Cullen, 2013), encouraged participants to share insights into their experiences and challenges in digital video learning.

The PI included interviews with a CT expert, three lecturers specializing in digital video, and a focus group of six TVET students. These interviews provided in-depth insights, while a survey of first-year students gathered additional data on their learning experiences.

A senior lecturer with 18 years of experience teaching TVET students and serving as a Master Trainer for MDEC's Computational Thinking Computer Science (CTCS) program highlighted low awareness of CT in Malaysian higher education institutions (HEIs). She noted that while TVET students excel in practical tasks, they lack problem-solving and CT skills.

The expert emphasized the importance of algorithmic thinking, which students found relatable and applicable in their studies and daily lives. However, she expressed concern that CT education in Malaysia is fragmented, with training efforts focused only on primary and secondary schools. She advocated for integrating CT concepts at all education levels, particularly in universities, to prepare students for a competitive workforce.

3.1.1 Interviews with Digital Video Lecturers

Three lecturers shared their experiences teaching digital video production:

- Lecturer A: Five years teaching animation and instructional design.
- Lecturer B: Twenty years in the video production industry and seven years teaching digital video.
- Lecturer C: Twelve years in photography and seven years teaching the subject.

Thematic analysis of their interviews identified key themes: students, lecturers, classroom environment, and teaching approaches.

- Traditional Teaching Challenges

Lecturers noted that conventional teaching methods, such as lectures and step-by-step tutorials, often led to students quickly forgetting newly learned material. Challenges were amplified in large classes, particularly for technical subjects like animation and video production. Reports and final-year projects revealed that many students struggled with documenting workflows, particularly in creating accurate flowcharts and applying algorithmic thinking.

- Online Learning Challenges

The shift to online learning during the COVID-19 pandemic introduced communication barriers and reduced student engagement. Large class sizes and heavy teaching loads further hindered effective teaching and interaction.

- Proposed Technological Interventions

The lecturers agreed that integrating technology into teaching could help students learn at their own pace, improving knowledge retention and problem-solving skills. They stressed the need for CT concepts, particularly algorithmic thinking, to support better learning outcomes in digital video education.

- Implications for Curriculum Development

The findings highlight gaps in digital video education, including limited CT integration and reliance on traditional methods that are insufficient for the digital age. Embedding CT skills into the curriculum can enhance students' digital competence, problem-solving abilities, and readiness for the evolving job market. This study emphasizes the importance of modernizing teaching methods to support student-centered learning and improve digital video education outcomes.

3.2 Design and Development of Digital Video Courseware

The development of digital video courseware followed the integrated Waterfall-ADDIE model, with particular focus on producing, evaluating, and refining multimedia educational materials. Content creation for the courseware was based on a standardized syllabus from a Malaysian university, ensuring alignment with similar global courses. Key resources included videos, animations, and scripts derived from books, expert inputs, and online sources. Content delivery emphasized algorithmic thinking, leveraging computational thinking frameworks like those in Computational Thinking Education. Supporting visuals and animations enhanced comprehension and engagement, with some adapted directly from platforms like YouTube.

To support multimedia delivery, the development team utilized tools like Adobe Premiere Pro, Microsoft PowerPoint, and Powtoon for creating and editing animations, graphics, and videos. These elements were integrated into a main platform developed using Articulate 360 software. The process, which took five months, ensured technical and aesthetic cohesion, with iterative testing to refine outputs. Features like breadcrumb trails, sitemaps, and highlighted text were incorporated to simplify navigation and emphasize key learning points, making the platform user-friendly for students.

The courseware also provided structured guides for both students and facilitators. For students, modules included step-by-step instructions, a welcome video, navigation tips, and summaries to reinforce learning. Facilitators received detailed procedural guides outlining their roles, tasks, and best practices during courseware use, as well as access to all necessary materials and tools for monitoring and supporting student progress. This dual guidance system ensured smooth implementation and consistent student experiences.

Rigorous formative and summative evaluations ensured the quality of the courseware. Testing involved system evaluations, user acceptance testing (alpha and beta tests), and pilot studies. These tests identified bugs, optimized functionality, and refined content based on feedback from experts and students. For instance, adjustments included revising video pacing, improving menu structures, and addressing usability issues. Comments from experts also shaped the design, confirming that features like machine-generated voice narration were suitable for the target audience. This iterative process ensured the courseware met both educational objectives and technical requirements.

4. EXPERT EVALUATION: SYSTEM TESTING

During the development phase, unit and integration testing were conducted to ensure functionality and compatibility of the digital video courseware. Unit testing assessed the performance of individual components, while integration testing verified that these components worked seamlessly when combined into modules. This process was essential for detecting and resolving potential errors or bugs. Once integration testing was successfully completed, the system testing phase began.

System testing, also known as black-box testing, evaluated the entire application as a unified system to confirm that all integrated components operated effectively. The researcher performed functional system testing to ensure the prototype application met expected operational standards. This testing included assessments such as functional, load, performance, and usability tests. Following functional system testing, usability testing was conducted through expert evaluations before advancing to user acceptance testing.

Content expert evaluations involved six experts; three content experts and three instructional design experts; from five public universities. Their evaluations were analyzed using descriptive statistics, which indicated consistent results across experts. The mean score for the content evaluation was 4.36, as shown in Tables 1. This high score demonstrated that the courseware met quality standards and was deemed suitable for user acceptance testing.

Table 1. Content Expert Evaluation Summary

Item	N	Min	Max	Sum	Mean	Std. Deviation
1. Instructional approach is appropriate for student level.	6	3	5	26	4.33	0.82
2. Teaching strategy is appropriate for subject content.	6	4	5	26	4.33	0.52
3. Language used is appropriate for student level.	6	4	5	25	4.17	0.41
4. Graphics facilitate better student learning.	6	4	5	29	4.83	0.41
5. Interactivity levels are sufficient.	6	3	5	25	4.17	0.75
6. Feedback provided is adequate.	6	3	5	25	4.17	0.75
7. Content delivery is well-organized.	6	3	5	24	4.00	0.89
8. No grammatical errors.	6	4	5	25	4.17	0.41
9. Content is accurate.	6	4	5	28	4.67	0.52
10. No gender biases.	6	3	5	25	4.17	0.98
11. Content is appropriate for learning objectives.	6	4	5	27	4.50	0.55
12. Learning objectives reflect the expected outcomes.	6	3	5	25	4.17	0.75
13. User can control the learning process.	6	3	5	24	4.00	0.89
14. Interfaces are interesting.	6	4	5	26	4.33	0.52
15. Menu is appropriate.	6	4	5	28	4.67	0.52
16. No technical issues (works on computer/tablet/phone).	6	4	5	29	4.83	0.41
17. No programming errors affecting functionality.	6	4	5	29	4.83	0.41
18. Media elements (text, graphic, audio, video, animation) are appropriate.	6	4	5	26	4.33	0.52

Explanation:

1. **N (Number of Evaluators)**: All evaluations were conducted by 6 content experts.
2. **Min/Max**: These values show the range of scores given for each item, with Min being the lowest score and Max the highest score. The range shows the diversity in expert opinions.
3. **Sum**: This is the total score given for each item by all evaluators combined.
4. **Mean**: The mean is the average score for each item. It is a measure of the central tendency of the evaluations. Most items had means ranging from 4.00 to 4.83, indicating a generally positive evaluation.
5. **Std. Deviation**: This represents how spread out the scores are from the mean. A low standard deviation (close to 0.41 for most items) means the experts were in agreement, while a higher standard deviation (e.g., 0.98 for "No gender biases") indicates more variability in the scores.

Overall, the evaluation scores were high, with the mean ratings for most items falling between 4.00 and 4.83, showing that the content experts found the digital video courseware effective and suitable. The lowest rating of 4.00 was for items such as "Content delivery is well-organized" and "User can control the learning process," but these were still considered acceptable. The highest ratings (4.83) were for items related to technical functioning, such as "No technical issues" and "No programming errors affecting functionality," indicating that the courseware is stable and works well across different devices. The content accuracy, learning objectives, and media elements were also positively evaluated, with scores mostly in the range of 4.17 to 4.67.

This evaluation suggests that the courseware is well-designed and acceptable for use, with no major issues identified.

4.1 User Acceptance Testing

User Acceptance Testing (UAT) is a crucial phase in the software development lifecycle where end users verify and validate the developed system or application. Conducted after unit testing, integration testing, and system testing, UAT serves as the final stage of functional testing. It is considered a form of black-box testing and typically involves two or more end users. UAT can be further classified into alpha and beta testing(Nadiyah & Faaizah, 2015).

4.1.1 Alpha Testing

Alpha testing is a software evaluation method performed to identify bugs and issues before the product is released to end users. In contrast, beta testing occurs in a real-world environment with actual users. For this study, alpha testing was conducted using a usability heuristic evaluation checklist, which was administered to six multimedia experts. The checklist was adapted from Erenler's (2018) translation of Nielsen's 10 Usability Heuristics. The evaluation was analyzed using descriptive statistics, which revealed that there were no significant differences between the experts' evaluations. The mean score for the heuristic evaluation checklist was 4.37, as shown in Table 2. Based on this score, the researcher concluded that the digital video courseware was acceptable and ready for pilot testing. However, revisions were made to the courseware based on the experts' feedback to improve the application.

Table 2. Summary of Descriptive Statistics for Heuristic Evaluation Checklist

Type of Analysis	Mean	Min	Max	Range	Variance	N of Items
Heuristic Evaluation Checklist	4.37	3.50	4.83	1.33	0.09	36

The results from the expert evaluations indicated that, with some modifications based on their feedback, the digital video courseware could proceed to pilot testing. Most experts' comments focused on improving the structure of the menus and making slight adjustments to the interface design. All expert suggestions were carefully considered by the researcher to enhance the courseware's functionality. Following these revisions, the beta test was conducted with the target users: university students.

4.1.2 Beta Testing

Beta testing encompasses a comprehensive evaluation of all features of a system from the perspective of end-user interaction. Various measurement systems are employed to assess user acceptance. This study focuses on usability measurements related to user satisfaction, categorized into task-level and test-level satisfaction. Both forms involve simple questionnaires, with task-level satisfaction being evaluated immediately after users complete a task. Seventeen university students participated in this beta test, all of whom had no prior knowledge of the digital video subject matter. According to Nielsen (2012), five to twenty users are typically sufficient to obtain statistically significant results for usability testing.

Post-test and post-task questionnaires were employed during the beta test. The post-test aimed to assess users' overall perception of the digital video courseware's usability, while the post-task questionnaire sought to capture users' impressions of individual tasks. The post-test was conducted using the System Usability Scale (SUS), and the post-task evaluation was based on the NASA Task Load Index (NASA-TLX). SUS has been widely validated as a reliable instrument for measuring usability (Sauro, 2011), and its use is further supported by Laubheimer (2018) in the Nielsen Norman Group.

The SUS scores were analyzed based on the SUS Score Calculation method, with scores below 50 considered unacceptable, between 50 and 70 as marginal, and above 70 as acceptable. The NASA-TLX scores were analyzed using the NASA-TLX Scoring Worksheet, developed by the TestScience Team, with interpretations based on previous studies (Hart & Staveland, 1988).

The results from the SUS and NASA-TLX indicate that the digital video courseware is considered appropriate by users, showing high user satisfaction and a manageable mental workload. With a SUS score of 71.76, the courseware is rated as a "good" application. Although the NASA-TLX score of 54.28 suggests a high workload, it is still considered acceptable, especially since the students were not fully familiar with the NASA-TLX process. Based on these results, the researcher moved forward with the pilot testing to assess the impact of the courseware on university students' knowledge and CT self-efficacy.

5. PILOT TESTING

To prepare for the main study, a pilot test was conducted to assess logistics and gather preliminary data. Pilot testing helps identify potential weaknesses in the research design or procedures, allowing for improvements in the overall quality and efficiency of the study. The pilot test followed the same procedures as the main study to mitigate risks and minimize the likelihood of issues in the larger-scale project. Thirty-two participants from the "Bachelor of Computer Science (Interactive Media)" degree program at the same university participated in the pilot test. The participants were divided into two groups: Group A (16 participants) was exposed to the Digital Video Courseware-Learner Segment (DVC-LS), and Group B (16 participants) was exposed to the Digital Video Courseware-System Segment (DVC-SS). According to Nielsen (2012), a sample size of five participants is typically sufficient for obtaining generalizable results. However, to ensure the robustness and generalizability of the findings, the pilot test was conducted with 32 students.

The results were analyzed using descriptive statistics to examine significant differences between pre-test and post-test scores for the dependent variables. Table 3 shows significant improvements in knowledge scores for both treatment groups, DVC-LS and DVC-SS, with an overall increase in students' scores after the intervention.

Table 3. Descriptive Statistics for Knowledge Score in Pilot Testing

	DVC-LS	DVC-SS	Overall
Pre-test	21.63	22.19	21.91
Post-test	28.06	31.25	29.66
Sample size	16	16	32

Similarly, the results for Computational Thinking (CT) self-efficacy perception, shown in Table 4, indicate an increase in mean scores from pre-test to post-test. The significant difference in mean scores for DVC-LS was greater than that for DVC-SS.

Table 4. Descriptive Statistics for CT Self-Efficacy Perception Score in Pilot Testing

	DVC-LS	DVC-SS	Overall
Pre-test	3.86	3.67	3.76
Post-test	4.27	4.17	4.22
Sample size	16	16	32

Overall, the results from the pilot testing, as shown in Tables 3 and 4, demonstrate significant improvements in knowledge, and CT self-efficacy after the intervention. The mean scores indicate increased knowledge and self-efficacy among the students who used the digital video courseware, whether learner-paced (DVC-LS) or system-paced (DVC-SS). Based on these findings, the researcher concludes that the concept of Computational Thinking (CT) can be successfully integrated with the segmenting principle, regardless of the type of segmentation used. Both learner-paced and system-paced segments show improvements in the pre-test and post-test results.

These pilot testing results are considered preliminary, and the researcher plans to conduct the main experiment following the same procedures with a larger and different sample. However, it is important to note that the results from the pilot testing cannot be generalized to the full experiment, as a larger number of participants may yield different outcomes. Therefore, the findings from the pilot study should be viewed as an initial indication, with the real experiment providing more definitive results.

6. ENSURING VALIDITY IN DIGITAL VIDEO COURSEWARE EVALUATION

The process of developing and evaluating digital video courseware (DVC) demands a systematic and methodologically rigorous approach to ensure the credibility and reliability of its outcomes. This rigor is especially vital when incorporating computational thinking and learner-paced segments, as these elements must be seamlessly integrated to create meaningful and effective learning experiences. Ensuring validity in this context involves employing robust measures to address potential sources of bias and errors that could undermine the study's findings. The ultimate goal is to provide educators and learners with a dependable tool that supports educational objectives while maintaining high standards of research integrity.

A significant aspect of this evaluation process involves addressing internal validity, which pertains to the accuracy and trustworthiness of the study's findings within the research setting. This includes identifying and mitigating factors such as participant selection biases, testing inconsistencies, and the influence of external events on the results. For instance, by selecting participants who had no prior exposure to digital video production, the study ensured a uniform starting point for assessing the courseware's effectiveness. Additionally, implementing rigorous testing protocols minimized threats such as participant fatigue, while validated instruments like the Digital Video Learning Tool (DVLt) further enhanced the reliability of collected data. These measures ensured that observed outcomes were genuinely attributable to the intervention and not extraneous factors.

Equally important is external validity, which ensures that the findings of the study are generalizable to broader educational settings. By carefully designing the study to reflect real-world classroom environments and selecting a sample representative of the target population—first-year students in Malaysian MTUN universities—the research successfully mirrored authentic teaching and learning conditions. This approach not only enhances the applicability of the findings but also underscores the potential of the courseware to be adopted across diverse educational contexts. Furthermore, the involvement of field experts in validating the instruments and assessing the courseware underscores the reliability of the tools and the quality of the content, ensuring that the DVC meets established educational standards. The research employed comprehensive internal and external validity measures to uphold the accuracy and generalizability of its results.

6.1 Internal Validity Measures

Internal validity was a key priority in this study, with specific measures implemented to mitigate potential threats that could undermine the credibility and accuracy of the findings. The internal validity of this research are as follows:

- History and Maturation: The study was carefully scheduled to minimize external factors that might influence participants' responses. Testing was conducted immediately after interventions and at strategic intervals to assess both immediate and long-term knowledge retention, reducing the risks of external distractions or participant fatigue impacting the results.
- Selection and Mortality: Participants were systematically selected based on their lack of prior knowledge in digital video production, ensuring a baseline level of experience. A well-defined timeline minimized participant dropouts, maintaining the integrity of the sample group.
- Diffusion and Rivalry Threats: To prevent bias, participants were not informed of their group assignments or the purpose of the study. This approach reduced the likelihood of inter-group diffusion of treatment effects or rivalry among participants.
- Testing and Instrumentation: Validated instruments such as the Digital Video Learning Tool (DVLt) and Computational Thinking Self-Efficacy (CTSe) were used. These tools were designed to mitigate potential testing effects, ensuring consistent and reliable data collection throughout the study.

6.2 External Validity Measures

Ensuring external validity was a crucial component of the study, as it determined the extent to which the findings could be applied to broader educational contexts beyond the specific research setting. To achieve this, the research focused on creating a representative sample and replicating authentic learning environments. The sample comprised first-year students from Malaysian MTUN universities, selected using stratified sampling to reflect the diversity of the target population and ensure that the results could be generalized to similar demographics. Moreover, the study was conducted in real classroom settings, mirroring typical teaching and learning scenarios to enhance ecological validity. This approach ensured that the evaluation of the digital video courseware (DVC) aligned with practical, everyday educational conditions. Additionally, the study prioritized the reliability of its instruments by engaging experts in computational thinking and video production to validate their effectiveness. This validation ensured that the tools used for data collection were both accurate and relevant, further supporting the generalizability of the results. By meticulously addressing these aspects, the research laid a solid foundation for the broader adoption and implementation of the DVC in diverse educational settings, underscoring its potential to facilitate computational thinking and personalized learning across various learner groups. The external validity of this research are as follows:

- Population Validity: The participant sample was drawn from first-year students in Malaysian MTUN universities, representing the target demographic. Stratified sampling methods ensured diversity and inclusivity within the study population, enhancing the generalizability of the findings.
- Ecological Validity: The research was conducted in natural classroom settings, closely mimicking real-world educational environments. This approach strengthened the applicability of the results to similar educational contexts.
- Instrument Reliability: All instruments underwent rigorous validation by experts in computational thinking and digital video production. This process ensured that the tools were both reliable and suitable for assessing the intended constructs.

6.3 Outcomes and Content Expert Evaluation

To further validate the courseware, a content expert evaluation was conducted involving six specialists in relevant fields. These experts assessed the quality, relevance, and suitability of the developed DVC using descriptive statistical methods. The results revealed a high mean score of 4.36, indicating strong effectiveness and readiness for broader user testing. This evaluation underscored the successful integration of computational thinking elements and learner-paced segments, affirming the courseware's potential as an innovative educational tool.

In conclusion, the study's rigorous approach to ensuring both internal and external validity highlights the robustness of its findings. By addressing key validity threats and engaging experts in evaluation, the research establishes a strong foundation for future investigations and the broader application of DVC in enhancing digital education.

7. DISCUSSION

The title of this study highlights two central themes: the integration of computational thinking (CT) and learner-paced learning in the development of digital video courseware. The study's aim was to assess how blending these elements could impact university students' knowledge acquisition, and computational thinking self-efficacy. Through content expert evaluations and pilot testing, the findings suggest that the approach of combining CT with learner-paced segments in digital video courseware can be a powerful educational tool that enhances students' learning experiences.

7.1 Blending Computational Thinking with Digital Video Courseware

Computational thinking (CT) is a critical skill in modern education, particularly in disciplines like computer science. In this study, the integration of CT concepts into the digital video courseware was a central aspect of the intervention. Content expert evaluations ensured that the digital video courseware was aligned with key CT principles, thereby ensuring that the courseware was both accurate and effective in promoting CT skills. By blending CT with digital video, the courseware not only delivered content but also encouraged students to engage in problem-solving, algorithmic thinking, and logical reasoning—core components of computational thinking.

The positive results from the pilot testing, such as the improvement in knowledge scores and the increased CT self-efficacy among students, confirm the effectiveness of this integration. The results suggest that students exposed to digital video courseware designed with computational thinking principles gained a deeper understanding of the content and felt more confident in their ability to apply these principles, particularly when using the learner-paced segment (DVC-LS).

7.2 Learner-Paced Segment and Its Role in Enhancing Learning

The study also emphasizes the role of learner-paced segments in the development of digital video courseware. Learner-paced learning, where students have control over the speed and timing of their learning, allows for personalized and more flexible educational experiences. In this study, the learner-paced segment (DVC-LS) was shown to be particularly effective in boosting students' self-efficacy. This finding highlights the importance of pacing in online and technology-mediated learning environments, suggesting that when students can control their learning pace, they are more likely to stay engaged, feel competent, and experience greater intrinsic motivation.

The learner-paced format's effectiveness, particularly in the context of computational thinking, aligns with the study's hypothesis that such an approach would foster deeper engagement and better learning outcomes. The evaluation of the courseware by content experts ensured that the learner-paced segments were appropriately structured, balancing the need for flexibility with the instructional objectives of the courseware. This careful design was crucial in maximizing both knowledge gain and the development of CT self-efficacy.

7.3 Content Expert Evaluation

Content expert evaluation played a vital role in this study, ensuring the validity and relevance of the digital video courseware. Experts in the fields of computational thinking and multimedia education evaluated the content for accuracy, relevance, and effectiveness. This evaluation process helped refine the courseware and ensured that the integration of CT concepts was both pedagogically sound and aligned with educational standards. The expert feedback also guided the design of the learner-paced segments, ensuring that they provided sufficient guidance while allowing for flexibility in the learning process.

7.4 Implications for Digital Video Courseware Development

The findings from this study offer valuable insights for the development of digital video courseware that blends computational thinking and learner-paced learning. By integrating CT principles into the content and offering learner-paced options, educators can create more engaging, personalized, and effective learning experiences for students. The content expert evaluations not only ensured the quality and relevance of the courseware but also highlighted the importance of careful design and thoughtful integration of instructional principles.

In conclusion, the study demonstrates that blending computational thinking with learner-paced digital video courseware has the potential to significantly enhance students' knowledge, and computational thinking self-efficacy. The positive outcomes observed in this study, particularly in the DVC-LS group, underscore the importance of incorporating both CT principles and flexible learning modalities to support students' cognitive and motivational development.

8.0 CONCLUSION

This study's findings indicate that the learner-paced predefined segment format of the digital video courseware can significantly enhance students' knowledge and computational thinking (CT) self-efficacy. Although the courseware demonstrated high usability, the moderate cognitive load reflected in the NASA-TLX results suggests the need for further refinement to optimize the balance between educational content and user-friendly design. Data from pilot testing provided critical insights to guide the next phase of the study, which will involve a larger participant sample. The study's rigorous consideration of internal and external validity ensures that the findings are both reliable and generalizable to similar educational contexts.

Additionally, the content expert evaluation confirmed the courseware's robustness and acceptability, with a mean evaluation score of 4.36 reflecting strong alignment with pedagogical standards. This feedback validated the courseware's design and content quality, affirming its readiness for broader application. Incorporating expert recommendations during development ensured that key instructional objectives were met, creating a solid foundation for scalability and future testing of its effectiveness in diverse learning environments.

REFERENCES

Adnan, A. H. M., Rahmat, A. M., Mohtar, N. M., & Anuar, N. (2021). Industry 4.0 critical skills and career readiness of ASEAN TVET tertiary students in Malaysia, Indonesia and Brunei. *Journal of Physics: Conference Series, 1793*(1), 0–11. DOI: 10.1088/1742-6596/1793/1/012004

Affero, I., & Hassan, R. (2013). Issues and challenges of technical and vocational education & training in Malaysia for knowledge worker driven. *National Conference on Engineering Technology,* February 2015. https://doi.org/DOI: 10.13140/2.1.4555.2961

Aho, A. V. (2012). Computation and computational thinking. *The Computer Journal, 55*(7), 833–835. DOI: 10.1093/comjnl/bxs074

Azid, N. W., Rawian, R., Sarimah, S.-A., & Kiong, T. T. (2019). The development of interactive case-based smart thinking and industrial problem-solving stimulator to enhance TVET students' thinking skills. *Journal of Engineering Science and Technology*, *14*(5), 2643–2656.

Bell, L., & Bull, G. (2010). Digital video and teaching. *Contemporary Issues in Technology & Teacher Education, 10,* 1–6. http://www.editlib.org/p/34120

Biard, N., Cojean, S., & Jamet, E. (2018). Effects of segmentation and pacing on procedural learning by video. *Computers in Human Behavior*, *89*, 411–417. DOI: 10.1016/j.chb.2017.12.002

Bradford, S., & Cullen, F. (2013). Research and research methods for youth practitioners. In *Research and Research Methods for Youth Practitioners*. DOI: 10.4324/9780203802571

Castro-Alonso, J. C., Koning, B. B. De, Fiorella, L., & Paas, F. (2021). *Five strategies for optimizing instructional materials: Instructor- and learner-managed cognitive load*. 1379–1407.

Catlin, D., & Wollard, J. (2014). Educational robots and computational thinking. *Proceedings of 4th International Workshop Teaching Robotics, Teaching with Robotics & 5th International Conference Robotics in Education*, 144–151.

Denning, P. J. (2009). The profession of IT: Beyond computational thinking. *Communications of the ACM*, *52*(6), 28–30. DOI: 10.1145/1516046.1516054

Ebrahimzadeh, M., & Alavi, S. (2017). The effect of digital video games on efl students' language learning motivation. *Teaching English with Technology*, *17*(2), 87–112.

Economic Planning Unit. (2015). Strategy Paper 9: Transforming technical and vocational education and training to meet industry demand. In *Eleventh Malaysia Plan,2016-2020:Way Forward.* rmk11.epu.gov.my/pdf/strategy-paper/Strategy Paper 09.pdf

Erenler, H. H. T. (2018). Heuristic evaluation of e-elearning. *International Journal of Organizational Leadership*, *7*(2), 195–210. DOI: 10.33844/ijol.2018.60235

Forte, A., & Guzdial, M. (2004). Computers for communication, not calculation: Media as a motivation and context for learning. *Proceedings of the Hawaii International Conference on System Sciences, 37*(January), 1513–1522. DOI: 10.1109/HICSS.2004.1265259

Hart, S. G., & Staveland, L. E. (1988). Development of NASA-TLX (Task Load Index): Results of empirical and theoretical research. In Hancock, P. A., & Meshkati, N. (Eds.), *Human Mental Workload* (pp. 139–183). North Holland Press., DOI: 10.1007/s10749-010-0111-6

Hasler, B. S., Kersten, B., & Sweller, J. (2007, September). Learner control, cognitive load and instructional animation. *Applied Cognitive Psychology*, *21*(6), 713–729. DOI: 10.1002/acp.1345

Hatsidimitris, G., & Kalyuga, S. (2013). Guided self-management of transient information in animations through pacing and sequencing strategies. *Educational Technology Research and Development*, *61*(1), 91–105. DOI: 10.1007/s11423-012-9276-z

Huang, R., Tlili, A., Yang, J., & Chang, T.-W. (2020). The Chinese experience in maintaining undisrupted learning in COVID-19 outbreak. In *Handbook on Facilitating Flexible Learning During Educational Disruption* (Issue March). Beijing: Smart Learning Institute of Beijing Normal University. https://www.researchgate.net/publication/339939064

Ismail, M., Ismail, I., & Razali, N. (2017). Video animation as teaching aid for engineering drawing course in Malaysia vocational college. *International Journal of Trend in Research and Development*, *4*(5), 475–479.

Juškevičiene, A., & Dagiene, V. (2018). Computational thinking relationship with digital competence. *Informatics in Education*, *17*(2), 265–284. DOI: 10.15388/infedu.2018.14

Laubheimer, P. (2018). *Beyond the NPS: Measuring perceived usability with the SUS, NASA-TLX, and the single ease question after tasks and usability tests experience*. Nielsen Norman Group: World Leaders in Research-Based User. https://www.nngroup.com/articles/measuring-perceived-usability/

Liao, C. W., Chen, C. H., & Shih, S. J. (2019). The interactivity of video and collaboration for learning achievement, intrinsic motivation, cognitive load, and behavior patterns in a digital game-based learning environment. *Computers & Education, 133*, 43–55. DOI: 10.1016/j.compedu.2019.01.013

Lockwood, J., & Mooney, A. (2017). *Computational thinking in education: Where does it fit? A systematic literary review. March.*

Lu, J. J., & Fletcher, G. H. L. (2009). Thinking about computational thinking categories and subject descriptors. *Sigcse*, 260–264.

Lusk, D. L., Evans, A. D., Jeffrey, T. R., Palmer, K. R., Wikstrom, C. S., & Doolittle, P. E. (2009). Multimedia learning and individual differences: Mediating the effects of working memory capacity with segmentation. *British Journal of Educational Technology, 40*(4), 636–651. DOI: 10.1111/j.1467-8535.2008.00848.x

Mashelkar, R. A. (2018). Exponential technology, industry 4.0 and future of jobs in India. *Review of Market Integration, 10*(2), 1–20. DOI: 10.1177/0974929218774408

Mayer, R. E., & Chandler, P. (2001). When learning is just a click away: Does simple user interaction foster deeper understanding of multimedia messages? *Journal of Educational Psychology, 93*(2), 390–397. DOI: 10.1037/0022-0663.93.2.390

Mayer, R. E., Wells, A., Parong, J., & Howarth, J. T. (2019). Learner control of the pacing of an online slideshow lesson: Does segmenting help? *Applied Cognitive Psychology, 33*(5), 930–935. DOI: 10.1002/acp.3560

Mazin, K. A., Norman, H., Nordin, N., & Ibrahim, R. (2020). Student self-recording videos for TVET competency in MOOCs. *Journal of Physics: Conference Series, 1529*(4), 042061. Advance online publication. DOI: 10.1088/1742-6596/1529/4/042061

McCaslin, S. E., & Young, M. (2015). Increasing student motivation and knowledge in mechanical engineering by using action cameras and video productions. *Advances in Production Engineering & Management, 10*(2), 87–96. DOI: 10.14743/apem2015.2.194

Means, B., Toyama, Y., Murphy, R., Bakia, M., & Jones, K. (2009). Evaluation of evidence-based practices in online learning: A meta-analysis and review of online learning studies. In *U.S. Department of Education*. www.ed.gov/about/offices/list/opepd/ppss/reports.html

Ministry of Higher Education. (2017). TVET for human resource development. In *Technical Vocational Education & Training (Tvet) in Malaysia : Selected Works*. http://mycc.my/document/files/PDF Dokumen/Technical Vocational Education %26 Training (TVET) in Malaysia Selected Works.pdf

Moreno, R., & Mayer, R. (2007). Interactive multimodal learning environments: Special issue on interactive learning environments: Contemporary issues and trends. *Educational Psychology Review*, *19*(3), 309–326. DOI: 10.1007/s10648-007-9047-2

Nadiyah, R. S., & Faaizah, S. (2015). The development of online project based collaborative learning using ADDIE model. *Procedia: Social and Behavioral Sciences*, *195*, 1803–1812. DOI: 10.1016/j.sbspro.2015.06.392

Nielsen, J. (2012). *How many test users in a usability study*. https://www.nngroup.com/articles/how-many-test-users/

Papert, S. (1980). MINDSTORMS: Children, Computers, and Powerful Ideas. In *Basic Books* (Vol. 66).

Pérez, M. P., Pesek, I., Zmazek, B., & Lipovec, A. (2020). Video explanations as a useful digital source of education in the COVID 19 situation. *Journal of Elementary Education*, *13*(4), 395–412. DOI: 10.18690/rei.13.4.395-412.2020

Prensky, M. (2009). H . Sapiens digital: From digital immigrants and digital natives to digital wisdom. *Innovate: Journal of Online Education.*, *5*(3).

Rodgers, C., & LaBoskey, V. K. (2018). Digital video creation in the LSP classroom. *The EUROCALL Review*, *26*(1), 71–104. DOI: 10.4995/eurocall.2018.9666

Sauro, J. (2011). *Measuring usability with the system usability scale (SUS)*. https://measuringu.com/sus/

Shin, M.-H. (2018). Effects of project-based learning on students' motivation and self-efficacy. *English Teaching*, *73*(1), 95–115. DOI: 10.15858/engtea.73.1.201803.95

Snelson, C., Yang, D., & Temple, T. (2021). Addressing the challenges of online video analysis in qualitative studies: A worked example from computational thinking research. *The Qualitative Report*, *26*(6), 1974–1988. DOI: 10.46743/2160-3715/2021.4734

Stiller, K. D., Freitag, A., Zinnbauer, P., & Freitag, C. (2009). How pacing of multimedia instructions can influence modality effects: A case of superiority of visual texts. *Australasian Journal of Educational Technology*, *25*(2), 184–203. DOI: 10.14742/ajet.1149

Stiller, K. D., Petzold, K., & Zinnbauer, P. (2011). Presentation time concerning system-paced multimedia instructions and the superiority of learner pacing. *Australasian Journal of Educational Technology*, *27*(4), 693–708. DOI: 10.14742/ajet.945

UNESCO. (2021). Technical Vocational and Education Training (TVET) 1. *Sub-Education Policy Review Report: Technical and Vocational and Training (TVET)*, 74. https://en.unesco.org/sites/default/files/tvet_final_-_january_2021.pdf

Voogt, J., Fisser, P., Good, J., Mishra, P., & Yadav, A. (2015). Computational thinking in compulsory education: Towards an agenda for research and practice. *Education and Information Technologies*, *20*(4), 715–728. DOI: 10.1007/s10639-015-9412-6

Wang, Y.-H. (2017). Integrating self-paced mobile learning into language instruction: Impact on reading comprehension and learner satisfaction. *Interactive Learning Environments*, *25*(3), 397–411. DOI: 10.1080/10494820.2015.1131170

Wing, J. (2006). Computational thinking. *Communications of the ACM*, *49*(3), 22–24. DOI: 10.1145/1118178.1118215

Wing, J. M. (2008). Computational thinking and thinking about computing. *Philosophical Transactions of the Royal Society A: Mathematical, Physical and Engineering Sciences, 366*(1881), 3717–3725. DOI: 10.1098/rsta.2008.0118

Zavgorodniaia, A., Tilanterä, A., Korhonen, A., Seppälä, O., Hellas, A., & Sorva, J. (2021). Algorithm visualization and the elusive modality effect. *ICER 2021 - Proceedings of the 17th*

Chapter 9
The Use of ICTs Tools for Students With ADD/ADHD in a Spanish EFL Classroom

Isabel María García Conesa
http://orcid.org/0000-0001-7005-2509
Centro Universitario de la Defensa de San Javier, Spain

ABSTRACT

This chapter explores the implementation of Information and Communication Technologies (ICT) as teaching support for students with Attention Deficit Disorder (ADD) and Attention Deficit Hyperactivity Disorder (ADHD) in the context of a Spanish EFL classroom in Compulsory Secondary Education. The analysis focuses on a review of studies that use technological tools to enhance attention, concentration, and participation among students with ADD/ADHD in English language teaching. Overall, the chapter highlights the effectiveness of ICT in interventions with the learners studied. A comprehensive view is provided on these tools as a viable and effective solution for addressing the educational needs of students with ADD/ADHD in English language learning. The purpose of this chapter is to conduct a literature review on the use of ICT in the English as a Foreign Language (EFL) classroom for learners with ADD/ADHD.

INTRODUCTION

One of the challenges faced by any secondary education teacher is how to help and support students with neurobiological attention deficit disorders (ADD) not

DOI: 10.4018/979-8-3373-1190-6.ch009

to hinder their learning and participation in the classroom, given that during this educational stage students are developing and acquiring fundamental knowledge for their social and educational evolution. For this reason, it is important to provide teachers with skills and strategies to help all students to learn in the best possible way.

Attention Deficit Disorder (ADD) is a disorder characterised by inattention, impulsivity and hyperactivity. It is a neurobiological disorder, caused by an alteration in the functioning of two brain neurotransmitters: dopamine and noradrenaline. These substances are released more slowly in the prefrontal cortex, an area involved in the control of attention, concentration and impulsivity. The development of the disorder may be due to environmental factors such as prematurity, genetic factors and may be aggravated using drugs, alcohol or tobacco during pregnancy (Alanazi & Al Turki, 2021).

Attention Deficit Hyperactivity Disorder (ADHD) is an important issue in the field of conduct disorders in children and adolescents, not only because it is one of the most common diagnostic categories in school-age clinical populations, but also because the school-age period has the greatest impact on children and adolescents in terms of morbidity and functional impairment.

As an educational strategy for students with attention disorders, this chapter proposes the use of ICT in the English classroom within the framework of Compulsory Secondary Education, to improve their attention span, organisation, planning and impulse control. Different guidelines to follow and solutions will help not only students in the classroom, but also teachers who are constantly challenged to adapt their teaching to meet the individual needs of each student. As Pilamunga Valla et al. (2021) mention, to educate for the future, it is important to break away from the rigidity of educational patterns and opt for a flexible school that adapts to the environment and the needs of every student.

According to some estimates, in Spain it affects between 5% and 8% of schoolchildren and almost 50% of them have other associated pathologies. According to the Flores Nieves et al. (2024), 20% of children with ADHD experience specific learning disorders in reading and writing, resulting in poor school performance. However, it is not only school performance that is affected, but also behavioural problems in the classroom, grade repetition, school dropout or even expulsion of some students.

According to Estrella Quintana et al. (2024), it is one of the most important disorders within Child and Adolescent Psychiatry, constituting nearly 50% of its clinical population. ADHD is chronic and begins to reveal before the age of 7. It is estimated that more than 80% of children will continue to present problems into adolescence, and between 30-65% into adulthood, and it tends to be an under-diagnosed and under-treated disorder.

The use of ICT in an ADHD case can be applied along three different lines (Fernández Gacho et al., 2020). First, they can be used to assess students at all stages of the process, not only in the diagnostic assessment. Second, they also serve as a secondary reinforcer that speeds up the resolution of activities and helps to minimise student impulsivity, as well as not spending money on printed materials. Third, ICT can also be seen as a tool that reinforces learning and helps to regulate attention, which can have a positive impact on the student's behavioural control and improve their ability to learn.

In this chapter, we address the issue of ADHD and its impact on students' learning in secondary education, especially in the context of learning English as a foreign language. The chapter begins with a clear definition of ADHD and a description of its main characteristics. It then explores how this disorder may affect students' academic performance in secondary education and looks specifically at its influence on English language learning. It then examines the role of ICT as a classroom support tool for students with ADHD, highlighting its advantages in terms of adaptability, interactivity and motivation. However, it also addresses the limitations and challenges that may arise when using ICT, such as the need for supervision and possible distraction. Finally, previous studies that have investigated the use of ICT in the English classroom for students with ADHD are reviewed, summarising the findings and highlighting conclusions relevant to the topic at hand.

OBJECTIVES

The purpose of this chapter is to conduct a literature review on the use of ICT in the classroom of English as a foreign language (EFL) for learners with ADHD. To develop it, the following general objectives are presented: to explore new current methodologies in the English classroom, and to analyse the results and experiences of previous studies related to ADHD learners in the English classroom with the support of ICT.

Among the specific objectives, we can include the following: to analyse the situations and challenges faced by students with ADHD in the English classroom, considering factors such as attention, concentration and organisation; to identify existing pedagogical strategies and adaptations to support learners with ADHD in the English classroom; to evaluate the effectiveness of the pedagogical strategies and adaptations used in the English classroom for students with ADHD; and to propose recommendations and suggestions to improve the support and inclusion of students with ADHD.

METHOD

Once the general and specific objectives have been established, we proceed to define the methodology used. This chapter will focus on a literature review to search for and analyse all existing publications on the support of technological tools with students with ADHD in the English classroom. The main purpose is to conduct an exhaustive search for information with the aim of making a critical reflection based on evidence. The aim is to collect all the essential data that will support and enrich the analysis, while avoiding the inclusion of outdated information that may detract the analysis. The goal is to ensure that the reflection is supported by more relevant information, ensuring that each piece of data makes a meaningful contribution to the critical and evidence-based analysis.

This chapter is a journey through the historical and theoretical frameworks to which ICT and ADHD in the English language classroom are related. An analysis will be made to clarify the main strengths and weaknesses of this methodology to foresee scenarios and guarantee universal accessibility for all students, allowing them to enjoy a comprehensive education.

At the same time, the chapter will make an exposition and an itinerary explaining the definition of the disorder and its characteristics, the students' role in secondary education and in the learning of English as a foreign language, the use of ICT as a support tool in the classroom and previous studies on its uses in the English classroom for students with ADHD with the aim of exploring new current methodologies in the English classroom and analysing the results and experiences of previous studies.

First, Google Scholar has been used, as it is an easily accessible search engine which allows finding relevant academic sources, accessing scientific articles and obtaining reliable citations to support research arguments. In addition, it has many articles that have been highly useful for this work.

Second, Ellibro.net has been used as it is based on an online library which has proved beneficial providing a wide selection of e-books specialised in the field of education with up-to-date sources. The bibliographic portal Dialnet has also been used as a source of scientific readings and quality information in Spanish, from which the following have been obtained: journal articles, bibliographic reviews and doctoral theses by other authors.

Finally, we also used articles from ProQuest, a multidisciplinary database specialising in social sciences, technology, history, language and literature, from which we obtained books in electronic format and doctoral theses. Secondary sources used in the primary sources that met the search criteria were included in this review. These secondary sources are duly cited in the body of the text.

In addition, the websites of official bodies such as the American Psychiatric Association, the Spanish Federation of Attention Deficit and Hyperactivity Disorder

Associations, the Ministry of Education, Culture and Sport, and the Official State Gazette have been consulted. As a result, inclusion criteria have been described for the results obtained from the searches:

- Documents published in the 21st century, articles prior to the year 2000 are discarded.
- Authors, articles and data related to the academic and psychological field are included.
- The search is limited to languages such as Spanish and English.
- References to the Spanish educational system are included and articles referring to other educational systems are discarded.

From all the sources consulted, applying the inclusion criteria and after reading the titles and abstracts of the papers, 37 papers were finally included and consulted for this literature review:

- Google Scholar (13)
- Ellibro.net (8)
- ProQuest (5)
- Dialnet (1)

DISCUSSION

Definition of ADHD and its Characteristics

ADHD is considered a neurodevelopmental disorder whose aetiology is mainly hereditary and affects the development and function of different regions of the brain, especially the prefrontal region (Quintero Gutiérrez del Álamo, 2019). As mentioned earlier, ADHD is divided into types according to its predominant symptoms which are: impulsivity, inattention and hyperactivity. Hyperactivity is not found in all cases of attention deficit disorder.

In Spain, the prevalence of ADHD among children and adolescents is estimated at 6.8% (Quintero Gutiérrez del Álamo, 2019). Therefore, students with this disorder may require support or adaptations in the educational environment, including different teaching strategies, changes in the curriculum or the implementation of intervention plans.

Because of the term 'hyperactivity', students with this disorder are often mistaken for children or adolescents who are unruly, misbehaving and disruptive. The problems of ADHD go much further; the main problems are difficulty in self-control and inattention, and the resulting problems in academic performance, but it is not a problem of intelligence or reasoning. Gardner (2021) explains that this disorder is not easy to diagnose, as those responsible for diagnosing it will be people related to specialised healthcare and with experience in ADHD.

The subtypes of ADHD that can be found and their criteria for diagnosis are inattention, hyperactivity, impulsivity. Knowledge of all these criteria and types will be important for the development of the work, as it will provide a better understanding of the type of pupils referred to. According to Latorre Cosculluela (2018), this information is relevant because it allows us to understand the specific challenges faced by learners with ADHD in the English classroom, which can help us design strategies and use ICT effectively to address these challenges.

According to the inattention criterion, the following behaviours can be observed: the student does not show attention to detail or has problems with carelessness in school tasks, has difficulty sustaining attention on tasks, does not seem to be listening when being spoken to directly, does not follow task instructions or does not complete tasks, avoids engaging in tasks that require mental effort. In addition, he/she loses objects needed for tasks, is distracted by irrelevant stimuli or is careless in daily activities.

Regarding hyperactivity, the following principles can be noted: the person often moves from the seat or moves the limbs frequently, leaves his seat when he is expected to remain seated, runs, jumps or has feelings of restlessness in situations where it is inappropriate to do so, has problems playing or working quietly at leisure activities and often acts as if he has a motor and often talks excessively.

On impulsivity, the following criteria can be perceived: the person frequently rushes answers without having fully formulated questions, has difficulties in respecting speaking time and interrupts or intrudes into other people's conversations.

Furthermore, we must clarify the existence of the 4 subtypes of attention deficit disorder according to the criteria they meet: ADD with hyperactivity (meeting criteria of hyperactivity, inattention and impulsivity), ADD without hyperactivity (with a predominance of inattention), ADD with hyperactivity (with a predominance of impulsive hyperactivity) and ADD with unspecified hyperactivity.

It is also worth noting that most scientific studies have focused on men while women have been little studied, identified or addressed (Kaldonek, 2020). This author describes that the voices of women with ADHD have not been heard strongly enough, reflecting the social position of women in society. Furthermore, he described that women with ADHD feel that they are inadequate, that they are complaining about nonsense and feel like complete losers.

Women and girls with ADHD have been socially silenced with their problems as is currently the case in society. Girls with hyperactivity learn to inhibit their behaviour to an unnatural degree or, on the contrary, risk the consequences of being labelled as 'unladylike girls'. Those women and girls without hyperactivity simply go unnoticed into the background (Maryanti et al., 2021).

Since the diagnosis of ADHD is made based on the medical clinic, health professionals collect all information about the child's situation through questionnaires and interviews with parents. In addition, a series of clinical tests are performed to rule out other problems such as an organic disease, psychological tests to show their abilities and limitations and psycho-pedagogical tests to analyse the risk of academic failure (Moro, 2021).

Pupils With ADHD in Secondary Education Learning English

Any student (with or without educational support needs) faces changes during secondary education. These years are based on constant challenges: change of teachers, classmates, subjects and even making decisions about their academic future. Moreover, not only are these changes important, but they also face psychological and physical changes in their development (Riga et al., 2019).

However, when we talk about a child with ADHD, we must consider that in secondary education we will have some specific needs. In addition, we will have to look for a safe, structured place with specific support for this type of student. Teachers and families will need strategies to help this type of students with all the challenges they face. Without such strategies these students can rarely be fully supported.

According to a review by Mphebbi (2023), a large percentage of parents of students with ADHD report a lack of curricular adaptation, school support, knowledge of the disorder, individualised attention and human and material resources in secondary schools. Consequently, there is a certain impact on the academic performance of this type of students: 50% require tutoring, 30% repeat school years, between 30-40% attend support classes, between 10-35% do not successfully complete secondary education and 50% of students have specific learning difficulties (SLD). This information clearly shows that there are deficiencies in the education system in terms of care and support for pupils with ADHD. Measures need to be taken to improve their curricular adaptation, provide resources and individualised support, and increase awareness of the disorder among both educational professionals and parents. In this way, it will be possible to reduce academic difficulties and improve the chances of success.

Table 1. Impact on students' academic performance

Reason	Impact
require academic tutoring	55
repeat school years	30
attend support classes	40
do not complete secondary education	35
have specific learning difficulties	50

On the other hand, students with ADHD report feeling forgetful, bored, lost and overwhelmed. The pupils interviewed say that they do not feel understood by the teaching staff, as the teachers do not understand their behaviour, they send them out of class without explaining the reason or they constantly reproach them for their attitude, thus feeling little understood. It is worrying that students report feeling misunderstood by teachers, as this can negatively affect their self-esteem and motivation to learn. There is a need to promote greater empathy and understanding on the part of teachers, as well as to implement inclusive educational strategies that are adapted to the individual needs of these students.

Table 2. ADHD students' feelings

Feeling	Number of students
comprehensive	5
imaginative	3
funny	6
friendly	9
forgetful	11
frustrated	8
important	6
anxious	4
overwhelmed	12
understood	4
lost	9
uncomfortable	6
integrated	7
loved	5
overwhelmed	9

In turn, this study also analyses teachers' experience with this type of students in the secondary school classroom. In this study, none of the participants in the study consider that their teacher training is extremely good for dealing with this group of students. These teachers state that to improve the performance of these pupils in question, they need more training in the subject, sufficient resources, more individualised attention by reducing the pupil ratio and a greater understanding of the needs of these pupils (De la Cruz et al., 2020).

Furthermore, Cortés Pascual et al. (2019) also indicate that teachers' knowledge of the disorder is currently still low or moderate. Moreover, the level of teachers' knowledge about this disorder is related to the type of education received previously and the self-efficacy that each teacher possesses to be able to meet the demands of these students. These authors suggest that there is a need for further teacher training on this disorder.

In their review, 50 secondary school teachers from 7 schools were asked about the existence of a specific working methodology at the school for pupils with ADHD. Of all of them, 54% did not know if a different methodology existed, 31% confirmed that it did not and 15% said that it did (Figure 1). These results reflect a lack of knowledge and awareness of the importance of adapting teaching methodology for students with ADHD.

Figure 1.

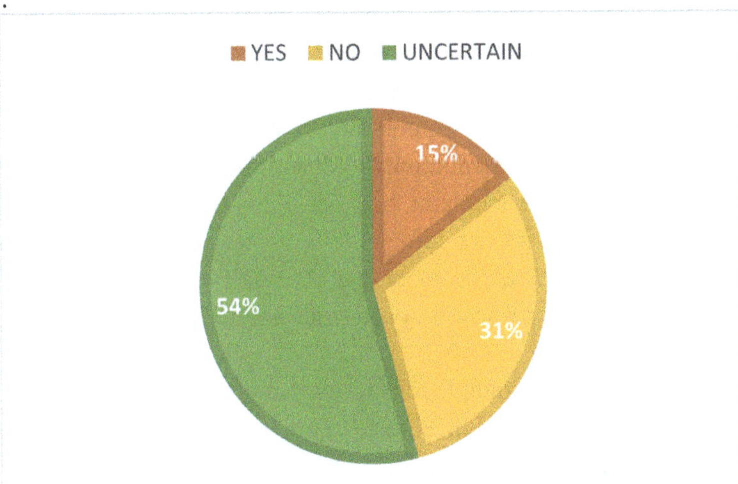

Moreover, it should be borne in mind that, in secondary education, the figure of the tutor is not as present as in primary education. The tutor spends fewer teaching hours with the students, making it more difficult to monitor their learning than in the previous stage and, in addition, there are more teachers per subject, so that each methodology is different.

This information is relevant as it highlights the urgency of considering the characteristics of the educational environment in secondary education when designing strategies for the use of ICT for students with ADHD in the English classroom. It is important to consider how to ensure effective coordination between different teachers and how to adapt strategies to the specific characteristics of each subject.

Understanding the disorder and its ramifications is the first step for teachers to identify students with ADHD and provide them with the appropriate tools to help them optimally develop their academic skills. In other words, fully understanding ADHD and its management information will help teachers and parents to understand this condition in an objective and balanced way, and to face the condition of students with ADHD with the right attitude, thus contributing to improve their prognosis and evolution. On the other hand, it is very important to provide students with tools that allow them to reduce or even suppress some inappropriate behaviours in the classroom, and for this motivation is essential, as these are not easily self-motivated, so these students should receive positive reinforcement to complete tasks.

Some of the guidelines and recommendations to combat learning problems in secondary school students are as follows (García et al., 2021):

- According to recommendations to teachers on how to teach, it is important to offer a high degree of motivation, avoid an abundance of verbal explanation, explain in detail the process, teach study techniques, divide the work into small amounts, encourage cooperative work and empathise with the student through attitudes that foster support and trust.
- According to the recommendations to ensure that students learn to learn, it is essential to act as a mediator between the young person and the strategies to assimilate and master content, to apply teaching for all, knowing their strengths and weaknesses, to encourage the analysis of information, to motivate the development of practical activities and to have fluid communication with parents.
- According to the recommendations to motivate students to learn, it is necessary to offer some kind of incentive, to encourage insistence to obtain success, to help them to achieve independence, to achieve a sense of belonging and to encourage stimulation, through finding pleasure in what they do.

When talking about ADHD, it should be borne in mind that hyperactivity and impulsivity are two characteristics that can be aggravated in the classroom, so finding guidelines to combat them will be of vital importance with these pupils. Some of these guidelines may include creating an environment that fosters communication, establishing rules in a consistent manner, explaining clearly and precisely the expected behaviour pattern to deal with situations of anger, frustration and even aggression, adopting a firm attitude when a rule is enforced, avoiding permissive attitudes, exchanging punishments for consequences or using negotiation for conflict resolution (Perales & Ballespí, 2023).

Table 3. Proposed adaptation of assessment processes for students with ADD(H).

Schedule	Before the exam	During the exam
Coordinate the schedule of exams.	Advise the use of time markers or stopwatches.	Give written exams to the student to avoid slow procedures.
Schedule tests for at least one week in advance.	Make sure you have all the material you need to carry out the tests.	Allow them to take the exam in one or two sessions if they are tired.
Conduct tests in the first hours of school.	Check that you have repeated material, so that if it falls on the floor, don't waste time looking for it.	Make sure that the student has understood the questions asked.
Communicate the tests to parents before the deadline.		Allow the student to go to the teacher's desk.
		If the learner loses focus, the teacher can guide him/her.
		Check that all questions have been answered and remember to check all answers before submitting them.

The aim is to assess students affected by ADHD to help and overcome academic failure while respecting the teaching principles of quality, equality and attention to student diversity inherent to our education system. To guarantee this, the Spanish Autonomous Communities include in their legislation a Diversity Attention Plan (PAD in Spanish) which includes a variety of curricular adjustments, organisational measures and support and improvements that schools must consider to adequately attend to the diversity of their students. In this way, each school must adapt to the characteristics of its environment and its students. In this sense, teachers must select and implement the strategies indicated in the PAD to provide all children in the centre with attention adapted to their general and special educational needs, to reduce and/or prevent possible occurrences (Cocha Millingalle et al., 2021).

Encouraging support for students with tools (educational, social and psychological) could be a key to their development and avoid frustrating or negative feelings. Furthermore, fostering empathy and understanding among classmates can contribute to a more inclusive and supportive environment for students with ADHD. In addition, equipping teachers with the necessary resources will ensure that all students with

ADHD can successfully acquire knowledge. In other words, preparing both the learner and the teacher is important to achieve a perfect development in all their activities.

Technology as a Support Tool in the Classroom

It is undeniable that technology has revolutionised the way we receive information. The importance of ICT in all fields of knowledge has shown an important dynamic as a means of communication, especially in the field of education (Pilamunga et al, 2021). Classrooms are enriched with laptops, tablets, projectors, or digital whiteboards. Many classroom tasks, such as taking notes, doing pencil and paper exercises, writing with chalk on the blackboard, consulting information in printed books, have been replaced using digital books and the creation of multimedia content, video games, educational apps, etc.

According to Latorre Cosculluela et al. (2018), there are many technological applications that have helped to make learning clearer. A clear example is the model of learning to read, in which subjects trace a line of coloured dots with their fingers to draw a letter, and the application keeps repeating the sounds of the letters until the subject learns them completely. The use of ICT has many advantages for pupils as it is quite attractive for them, encouraging attention, motivation and interest in the classroom. Moreover, it is considered a tool to reinforce the learning process in the classroom and to promote autonomy in their own learning process.

On the other hand, numerous studies reflect the change in the role of the teacher. Uyaguari Paredes et al. (2024) state that the consideration of the teacher as the sole source of acquisition and dissemination of knowledge, has been replaced by passive studies. The study argues that the teacher moves from a role of transmitter of knowledge to a guide and facilitator of student learning. This has been the most relevant change with the incorporation of ICT in classroom learning processes and where the role of the teacher has completely changed, as students can access any data they need through the Internet.

The teacher must be in control of ICT and use them at an advanced level, as well as generating content adapted to these tools and establish methodologies adapted to them. The aim should not be to introduce the most modern technologies, but to assess their didactic possibilities. To do so, we must assess their didactic possibilities and how they can benefit our students.

With the help of ICT, teachers become a fundamental part of an educational triangle, without which the teaching and learning process would be difficult or impossible. Therefore, according to Fernández Menor (2021), the teacher is a support centre as an essential reference for students, allowing active work dynamics to help students make the most of the content. Furthermore, the effectiveness of the use of ICT in the classroom will depend solely on the organisation of the teaching staff and their adaptive capacity to include them in their teaching methodologies.

However, the use of ICT in the classroom can also pose difficulties. Research by Robles Villalba and Reyes Moreno (2024) shows that one of the main obstacles to developing the educational potential of ICT is related to the organisation and traditional culture of schools. Most schools have computers, but they are used by a small number of teachers and, furthermore, teachers who use computers have problems modifying their work in the classroom and the expectations of their students.

It is interesting to consider the challenges and obstacles that may arise when using ICT in the classroom and the need to overcome them to take advantage of their educational potential. Also, it is important to consider how relevant it is to advance the integration of ICT in education, especially in the current context, and how educational legislation supports this incorporation.

Benefits and Limitations of Technology for Students With ADHD

Teachers should consider how to deal with all the different learner profiles, in this case, learners with ADHD. To do so, it is important that they study the characteristics of these students and gather all relevant information on this topic. In other words, knowing how to manage these students to offer them the best possible attention so that they can achieve their academic goals. Students with ADHD must cope with the emotions of their disorder, such as inattention, drowsiness, restlessness, disorientation and boredom.

Furthermore, the review by Rubiano Caballero and Lozano Zárate (2024)) states that there are numerous positive aspects regarding the use of ICT, for example, that they allow the development of motivating and attractive activities for pupils, favour independent learning, promote cooperative work and the inclusion of pupils with Specific Educational Support Needs. However, this study corroborates that there are also some disadvantages, such as the problem of generating spontaneous situations, as well as the lack of teacher training and the negative view of the use of ICT resources.

There is a clear inclination to use ICT with students with ADHD to improve attention, improve control of impulsivity, improve literacy, mathematics and social skills. In addition, this review notes that the use of ICT resources reduces time wasted in completing tasks, favours social support through groups of people with ADHD and reduces school failure.

Kojayan et al. (2021) highlight the positive factors of the use of ICT in classrooms with students with ADHD. Firstly, it is noted that when pupils switch from writing with pencil and paper to using an ICT resource, pupils avoid wasting time searching for their material or having the material ready for the activity, such as sharpening the pencil. By using ICT resources with word processors, pupils would avoid these actions. Nevertheless, care should be taken to ensure that during computer activities there are no other distractions that attract their attention more than the task at hand.

Pupils with ADHD have difficulties paying attention to written language. However, they have a high visual memory, which is why visual support through different tools favours the process. The use of image-based ICT is enriching for these students and reinforces learning and attention, at the same time as they manage to reduce their impulses, favouring self-control (Latorre Cosculluela & Quintas Hijós, 2022). Students with ADHD are motivated to use the computer in educational activities as it allows them to work at their own pace and the computer does not rely on writing skills allowing them to manipulate their own work and erase or use grammar checkers.

Another tool that is beneficial for students with ADHD is educational software. According to Del Cisne Saldarriaga and Briones Palacios (2022), these tools should be novel and have attractive graphics so that the student focuses their attention. In addition, they should not include too many illustrations to avoid any distractions, they should be short and of gradual difficulty. They should also incorporate mechanisms to avoid frustration (such as sending messages to encourage them to continue with the activity).

In the same study, it is revealed that most teachers surveyed in this research work (over 87%) are aware of the educational possibilities of ICT in relation to attention to diversity, but just over half of the sample (53%) confirm that they have not used software as a measure for attention to diversity and more than (70%) recognise that they have never used this software specifically for students with ADHD in the classroom. Approximately 43% of teachers use the same activities or assessment tools for students with ADHD as for other students.

Another study by Campos Soto (2022) analyses parents' opinions on whether they consider ICT to be useful for reinforcing the learning of their children with attention problems. Of those surveyed, 73.97% answered 'yes', while 23.29% responded that they are not useful for this purpose and finally, 2.74% abstained from answering the survey (Figure 2). This study considers the positive view that parents have of the use of technologies in the educational setting while there is a considerable group of parents who do not see clear benefits in the use of technologies for the learning of their children with attention problems. There may be concerns about possible distractions or lack of personal interaction in the learning process. Furthermore, on the other hand, the fact that a small percentage, 2.74%, abstained from answering the survey is also relevant.

Figure 2.

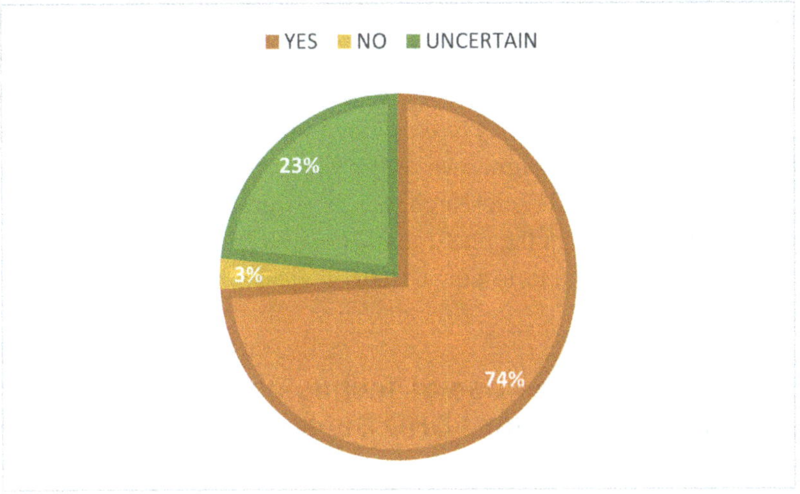

Furthermore, although the use of ICT in teaching practice is more integrated by teachers at state schools, it is the teachers at public schools who use educational software to a greater extent as a measure of attention to diversity and especially with pupils with ADHD. This difference invites us to delve deeper into the causes of this difference and promote an equitable integration of ICT in all educational centres.

Educational software allows students to encourage participation, to have more dynamic and motivating classes, to encourage socialisation and inclusion of students with ADHD with the rest of their classmates and to consult, review and self-evaluate the contents of the subject whenever they consider it necessary. Furthermore, ICT have a positive effect on the training of these students who are complemented by other kinds of psychological and medical treatment, to improve concentration and motivation.

However, activities created in multimedia materials are not necessarily better for students with ADHD. In fact, they can sometimes include some extra load, a multimedia material can be harmful if it is not well designed, if it includes irrelevant stimuli - too much movement or too bright colours - or is not well adapted to the needs of ADHD. Caution is needed to ensure that they are well designed and adapted to the specific needs of these learners. Inappropriate multimedia material can be detrimental and distracting rather than facilitating learning.

Furthermore, after analysing the effect of ICT in enhancing the learning of students with ADHD, several weaknesses and threats were found. The weaknesses found are based on the shortcomings of ICT management or excessive use of ICT, exposure to digital screens and electromagnetic fields as they produce stress, anxiety and eyestrain. In conclusion, this study shows that while ICTs can be useful to reinforce the learning of students with ADHD, there are also weaknesses and threats associated with their use. It is essential to address these issues and promote responsible, balanced and safe use of technologies in the educational environment.

Although studies relating ICT to students with ADHD are scarce, the studies developed are encouraging, as they have led to improvements in the learning of this type of students in the classroom, such as the work of Figueroa Pozo et al. (2023) or De Araujo et al. (2024). This research indicates significant improvements in the learning of these students in the classroom. However, more research is needed to explore this topic further and to adapt technological strategies to the individual needs of each student.

Previous Studies on the Use of Technology in the English Classroom for ADHD Students

The current generation is immersed in a technological world and therefore their environment and daily life must respond to what their reality demands. In terms of education, it is vital that teachers implement technological tools and strategies that go hand in hand with today's new advances, which are of course of great interest to students. According to Al-Mutairi (2024), learning a foreign language can become a rich exercise depending on the means and tools that are used for it.

Considering the premise that children with ADHD are intelligent, capable, imaginative and everything may interest them, many researchers consider that this disorder is closely related to a divergent cognitive style. The main characteristic is the ability to generate a wide variety of imaginative responses to certain problems and from different points of view. It is therefore possible to capture the attention of these learners if they are taught with the appropriate styles (Ewe, 2019).

To help learners with ADHD to learn English, some authors have conducted literature reviews on the subject. One of them by Gkora (2024) in which he presents a didactic technological proposal (Scratch) that facilitates and favours the development of concentration, reflexivity, creativity and self-esteem. The Scratch tool is a digital tool that is presented as an effective means of application and creation of knowledge dedicated to adolescents so that they can manage their own learning, be active creators of content and applications in any area of knowledge. This tool can be used in the English classroom thanks to its visual language and its multiple content possibilities, creating themes based on the use and creation of different situations, games, stories, dialogues, presentations, animations, projects or activities using multimedia resources, as well as sharing their own projects through the Scratch website.

Students diagnosed with ADHD generally have difficulty adapting to the normal dynamics of a class, as they need more time to perform the same activity, paying attention or following directions. Therefore, a tool that allows the student to manage time, ease of use and a motivation that allows them to overcome external distractions can be an extremely important tool in the methodology aimed at this type of student. The programming language allows to merge the use of creativity, imagination, academic and playful motivations leading to the use of personalised and meaningful education. The use of Scratch for students with ADHD offers a very simple graphical environment in structure that is an essential element when working with these students, as it is one of the difficulties associated with this disorder.

The aim of the proposal is to use a programming environment based on the Scratch language through which students design and create their own tasks with the contents of the English subject. The content addressed is, on the one hand, to encourage the use of ICT as a resource to promote learning; on the other hand, to contribute to the development of English language proficiency with the aim of stimulating learning according to the needs of students based on dynamics that favour autonomy, creativity and provide sufficient conditions. The environment where English is learned is a favourable space due to its environmental conditions, and it provides spaces for autonomous and creative learning. This learning should be related to the interests of each student to evaluate their knowledge of the subject and where the teacher in the use of the Scratch tool is always a guide of the process, seeking maximum autonomy on the part of the student.

Some of the sessions proposed with the Scratch tool for the English class include the use of the creation of questions on the contents of daily routines where students must use verbs (get up, have breakfast, go to school...), adverbs (never, sometimes, often), and nouns (bag, bus, notebook...) that their classmates will have to answer.

Another session proposed with this tool is the computational creation of stories by designing collaborative narratives together with a classmate and including an oral format within the programme. In this session, characters are created, as well as scenes of movement between them, using the discursive elements of English.

Different from the Scratch model, Rodrigo Yanguas et al. (2022) proposed the video game 'Wordland' aimed at students with special educational needs (SEN). ADHD is found to improve the processes of vocabulary acquisition in English, the cognitive skills of this type of students and strengthen lexical competence in English from the interaction of the different activities proposed in the video game.

Through the educational software, 'Wordland', different interactive games are proposed for the student to relate in a different way English words contextualised in various scenarios. When students enter the resource, they find different phases called Big Body (activity that relates to English words related to the human body), Fear House (referring to the parts of the house), Family Merry Go Round (relating to words of the members and roles of the family), Ferry Wheel Jobs (referring to professions) and the last phase Roller Coaster Test (which can only be played once all the activities have been passed and corresponds to the evaluation stage of the words acquired in the previous stages). In each phase there is a different activity per competence of the area of English to be strengthened, so all phases will work the communicative competence.

The contributions of the resource lay in the increase of words related to the house, the body, the family and professions, and in turn, his lexical competence improved as the students develop the activities. At this stage of implementation (Roller Coaster Test), the students manage to learn a series of words that are remembered at first thanks to the repetition of the games, but mainly thanks to the audios of the words and their context. Cognitive processes such as the association between audio, images and context are considerably favourable to their learning. In turn, another advantage of this resource is related to the strengthening of autonomy, knowledge of tools and the functioning of the computer.

However, factors such as the number of students, school times and curricula not adapted to their needs interfere with the progress of this type of students. But initiatives such as 'Wordland' can help to improve this situation and provide an opportunity to demonstrate the successes that these students can achieve when proposals are designed for them. These tailored proposals designed specifically for them can make a significant difference to their learning and allow them to demonstrate their potential.

Furthermore, considering the characteristics of the ADHD population and their difficulties in acquiring vocabulary in English in the specific case of this research, the motivation that the teacher offers during the performance of each activity is key in building self-confidence and achieving their goals. They need constant approval and encouragement from their teacher, and it is at that moment when a change of attitude towards knowledge and what they can achieve is observed.

Aspects such as time and the lack of training on ADHD can affect the implementation of the use of ICT, since they require time and observation of their processes. Besides, the hourly intensity of the English subject is three hours a week, an aspect that does not fully allow to demonstrate a process and is maximised when at home there is no feedback of what has been taught. However, although there are challenges, these initiatives can make a significant difference to students with ADHD in their learning and allow them to demonstrate their potential.

In another review by Villafuerte Holguín and Alonzo Rezabala (2020), the use of ICT in the English classroom is evaluated for students with ADHD) in Primary Education. The aim of this work was to study the effectiveness of the use of touch screens in curricular adaptations to teach English through play to children with this disorder. This study is based on the assertion that the use of touch screens has spread to various educational settings and has led to the emergence of technology for learning and knowledge.

In this study, three children between the ages of six and ten, diagnosed with ADHD and who were behind in English, were exposed to curricular adaptations for the teaching of English vocabulary. During this process, digital screens, recreational games and a significant motivational load were used for the participants. The proposed process comprised seven 40-minute sessions and a pre-test and post-test were used to show progress in the variables of listening comprehension, number of new words and pronunciation. The suggestions made by experts in this study seek to add aspects to be improved, attention span, motivation and desire to learn, as well as others to be considered (Villafuerte Holguín & Alonzo Rezabala, 2020).

The English teacher's report on these pupils shows that there are specific characteristics of these three pupils during English lessons: the first student participates in few activities and projects in class, needs external help to do his homework and delays progress in English classes; the second student also needs permanent help and his pronunciation in English is below the class average; and the third student is notoriously behind the rest of the students in the English class, his participation in class is very sporadic, he fails to do his homework and needs help to take notes during class (Villafuerte Holguín and Alonzo Rezabala, 2020).

During the seven sessions, the introduced vocabulary was repeated to improve students' knowledge retention and exercise short- and long-term memory. Each session began with a review of the content of the previous session to reinforce and connect with the new content to be taught. The sessions included a brief assessment to verify each child's learning and ICT resources such as touch screens were used during the sessions (Villafuerte Holguín and Alonzo Rezabala, 2020).

During these sessions recorded in the study by Villafuerte Holguín and Alonzo Rezabala (2020), the result achieved was to maintain the attention of the participants in learning from 40 seconds at the beginning to 135 seconds in the last session, which allows improving the process of teaching English with ADHD children. In addition, the use of touch screens in curricular adaptations provides dynamism to each session, thus serving to encourage and stimulate the learning of these students who have low motivation towards the foreign language.

Furthermore, in the results of the tests administered at the beginning (pre-tests) and the tests applied after the curriculum adaptation sessions (post-tests), progressive improvement can be observed in the variables of pronunciation of vocabulary, number of new words learned in English and listening comprehension of vocabulary. In conclusion, all the results of the study show optimisation in all the students with respect to the variables indicated.

In summary, although the results presented are encouraging and suggest optimisation in the students in the variables mentioned, it is important to analyse them critically and consider other factors that could have influenced the results. In addition, it is necessary to consider the limitations of the study, such as the sample size and the validity of the tests used.

Table 4. English language learning achieved with the curricular adaptations with digital screens.

Informants	Pronunciation		Comprehension		Vocabulary	
	Pretest	Post-test	Pretest	Post-test	Pretest	Post-test
Student 1	4/10	8/10	3/10	7/10	11	24
Student 2	3/10	7/10	4/10	8/10	14	25
Student 3	4/10	7/10	3/10	7/10	9	21

As a conclusion of this study, Villafuerte Holguín and Alonzo Rezabala (2020) state that the use of touch screens in curricular adaptations to work on English has allowed them to increase the real knowledge of concentration of children with special educational needs such as ADHD. However, the pupils in this study also expressed the need for rest, which is why the researchers consider periods of between three and five minutes to be important in language practice. Furthermore, this study corroborates the need for teachers with the availability and knowledge to make curricular adaptations that offer meaningful learning experiences.

CONCLUSIONS

To fulfil the objectives of this literature review, to explore new methodologies in the English classroom adapted to students with ADHD and to compare the results and experiences of these students in the English classroom with the support of ICT, it has been necessary to analyse a series of concepts, definitions and information related to this case, as well as to establish its importance in the Spanish educational system to place it in context. Consequently, we have studied the benefits and limitations of ICT, the characteristics of ADHD, the students with this disorder in secondary education and in foreign language learning, as well as a review of studies on the use of ICT in the English classroom for students with ADHD to compare their results and experiences.

This chapter has fulfilled the objectives that have been set. The work has fulfilled its objective of analysing the situations and challenges faced by students with ADHD in the English classroom, considering factors such as attention, concentration and organisation. However, further research is needed on the possible challenges specifically in English. Although some authors explain that many of their proposals are based on the objective of improving the processes of vocabulary acquisition in

English, improving cognitive skills and strengthening lexical competence in English, they do not delve into the real challenges and difficulties that students have when learning this foreign language.

In addition, these studies have also provided an overview of the strengths and weaknesses of previous research. Thus, it is important to continue research on ADHD, considering women with ADHD in these same studies. It is also necessary to continue working to raise awareness in the educational community to ensure attention adapted to the educational needs of students affected by ADHD. Therefore, it is essential to equip all educational institutions with ICT resources given all the benefits they bring, as well as for teachers to continue training to keep their teaching tools up to date and inclusive.

It has become necessary to establish as accurate a definition of the disorder as possible. This issue has proved complex given the variety of criteria that are established within it (hyperactivity, impulsivity and inattention). While the criteria have been described and the behaviours of students with these characteristics have been analysed, there has been a lack of research or scientific studies on women with ADHD. Most scientific studies have focused on men while women have been little studied, identified or treated. Therefore, this chapter proposes to advance and integrate women in all studies or treatments related to ADHD to stop silencing their voices, to be heard and to be treated correctly.

This work reveals the need to implement support measures and adapt educational methodologies to meet the individual needs of students, more specifically the use of ICT in the English classroom with students with ADHD. In addition, it is essential to address specific learning difficulties to ensure a successful educational process.

Furthermore, the data highlight the need to improve initial teacher training, provide adequate resources, promote more individualised attention and foster a greater understanding of the needs of students with special needs. These findings point to the importance of implementing changes in teacher training and education policies to ensure inclusive and quality education.

This work considers the need to overcome difficulties and make progress in the implementation of technology in education. At the same time, it highlights the importance of teachers' professional development and access to technological resources to generalise the use of ICT in English classes. This implies that teachers must be willing to adapt and acquire new technological skills to improve teaching and learning.

Within the existing pedagogical strategies and adaptations to support students with ADHD in the English classroom, it has been observed that the use of ICT has been beneficial for these students as it offers them strengthened autonomy, greater concentration, attention and motivation. Despite all these advantages found in these adaptations, these pupils also encounter certain difficulties.

The difficulties and limitations that have been found in reviewing the use of these strategies are the rest time that these pupils need after the language activities developed and the teachers' knowledge when it comes to implementing ICT in the English classroom. On the other hand, it can be intuited from the information reviewed that the implementation of ICT in the classroom depends largely on the willingness of the teachers. Many teachers fail to implement these tools, in some cases due to a lack of knowledge and in others due to a lack of will.

In addition to the positive assessments of teachers, it is important to highlight the limitations found during the literature review. A paucity of specific studies on the teaching of English using ICT in the secondary classroom with students with ADHD has been observed. This lack of research on the topic is a significant limitation that needs to be addressed. It would be beneficial to conduct case studies that would allow for a more detailed analysis of the effects and benefits of ICT in teaching English with students with ADHD in secondary education.

In terms of future work, by promoting further research, it will be possible to obtain more solid data and evidence to support the effective use of ICT in English language teaching with these learners. This will enable teachers and education professionals to make informed decisions and develop more effective strategies to support the learning of pupils with ADHD in English language teaching.

In conclusion, the use of ICT with students with ADHD in the secondary English classroom presents both opportunities and challenges. While it is important to overcome constraints such as lack of access to technological resources and insufficient teacher training, we cannot ignore the potential benefits that ICT can bring. However, it is essential to strike the right balance and encourage their responsible use, ensuring that they are effectively integrated into the curriculum and promote an enriching and inclusive learning environment. Ultimately, the use of ICT in the secondary English classroom can be a powerful tool to enhance the learning of these students and prepare them for an increasingly digitalised world.

REFERENCES

Al-Mutairi, M. A. (2024). Unlocking Potential: Harnessing Language Learning Apps to Support Young Learners with ADHD. *International Journal of Scientific Research and Management, 12*(4), 3313–3319. DOI: 10.18535/ijsrm/v12i04.el02

Alanazi, F., & Al Turki, Y. (2021). Knowledge and attitude of Attention-deficit and Hyperactivity Disorder (ADHD) among male primary school teachers. *Journal of Family Medicine and Primary Care, 10*(3), 1218–1226. DOI: 10.4103/jfmpc.jfmpc_2194_20 PMID: 34041155

Campos Soto, M. N. (2022). *Implementación y mejora de procesos lectoescritores en alumnos con trastorno por déficit de atención e hiperactividad (TDAH) a través de las Tecnologías de la Información y la Comunicación (TIC) basadas en la animación 3D*. Editorial Universidad de Granada.

Cocha Millingalle, M.C., Naveda Bonilla, D.J., & Tenelanda López, D.V. (2021). Estrategias metodológicas docentes y su asociación en el proceso de enseñanza-aprendizaje del idioma inglés de escolares con TDAH. *Journal of Science and Research, 6*(2), 132.145. https://doi.org/DOI: 10.5281/zenodo.5514319

Cortés Pascual, A., Moyano Muñoz, N., & Quilez Robres, A. (2019). The relationship between executive functions and academic performance: Review and meta-analysis. *Frontiers in Psychology, 10*(1582), 1–18. DOI: 10.3389/fpsyg.2019.01582 PMID: 31354585

De Araujo, C. S., Aragão, A. O., Boechat, G. P., de Carvalho, I. E., Costa, J. W., Lopes, J. M., Castro, T. S., & Honório, V. A. (2024). Technology as a support in teaching students with ADHD: A view from educators. *Contribuciones a aas. Ciencia y Sociedad, 17*(2), 1–15. https://doi.org/https://doi.org/10.55905/revconv.17n.2-144

De la Cruz, G., Ullauri Moreno, M. I., & Freire, J. (2020). Strategies of teaching English as a Foreign Language towards Attention Deficit Hyperactivity Disorder Students (ADHD). *Colombian Applied Linguistics Journal, 22*(2), 169–182. DOI: 10.14483/22487085.16118

Del Cisne Saldarriaga, A., & Briones Palacios, Y. M. (2022). El aprendizaje de estudiantes con trastorno por déficit de atención e hiperactividad desde las TIC. *Episteme Koinonia, 5*(1), 405–416. DOI: 10.35381/e.k.v5i1.1821

Estrella Quintana, P. A., Guerrero Paredes, K. A., & Chiluisa Chiluisa, M. J. (2024). Attention deficit hyperactivity disorder and English Language teaching and learning. *Revista Científica de Innovación Educativa y Sociedad Actual, 4*(1), 76–87. DOI: 10.62305/alcon.v4i1.66

Ewe, L. P. (2019). ADHD symptoms and the teacher–student relationship: A systematic literature review. *Emotional & Behavioural Difficulties*, *24*(2), 1–20. DOI: 10.1080/13632752.2019.1597562

Fernández Gacho, L., Arias Gonzáles, V., Rodríguez Navarro, H., & Manzano Soto, N. (2020). Estudio e intervención en niños con Trastorno por Déficit de Atención e Hiperactividad de educación primaria. *Revista interuniversitaria de formación del profesorado, 34*(2), 247-274. https://doi.org/DOI: 10.47553/rifop.v34i2.77535

Fernández Menor, I. (2021). Evaluación de un estudiante con TDAH tras una intervención con recursos tecnológicos. *Páginas de Educación*, *14*(2), 121–131. DOI: 10.22235/pe.v14i2.2565

Figueroa Pozo, D. C., Avilés Castellanos, N. F., Narváez Carvajal, L. I., Recalde Drouet, E. M., & Crespín Crespín, E. E. (2023). Recursos digitales para fortalecer el aprendizaje de niños con TDAH. *Ciencia Latina Revista Científica Multidisciplinar, 7*(2), 7648-7662. https://doi.org/https://doi.org/10.37811/cl_rcm.v7i2.5902

Flores Nieves, C. H., Jiménez Guerrero, A. M., Vargas Galarza, M. V., Delgado Mora, L. E., Changoluisa Chicaiza, K. S., & Tapia Lincango, M. C. (2024). Relevancia del Uso de Recursos Tecnológicos en el Proceso de Aprendizaje de los Estudiantes con TDAH. *Ciencia Latina Revista Científica Multidisciplinar*, *8*(4), 9958–9972. DOI: 10.37811/cl_rcm.v8i4.13143

García, J., Pérez, M., & López, A. (2021). Diagnóstico y manejo del TDAH en el entorno educativo. *Revista de Neurociencia Clínica*, *34*(2), 123–134.

Gardner, H. (2021). *Disciplined mind: what all students should understand*. Simon & Schuster.

Gkora, V. (2024). Advancing ADHD Education: Autonomy, technology, and inclusive strategies. *GSC Advanced Research and Reviews*, *18*(3), 101–121. DOI: 10.30574/gscarr.2024.18.3.0084

Kojayan, A., Statti, A. L., & Torres, K. (2021). The Effects of Technology Integration in the Classroom for Students With ADHD. *International Journal of Curriculum Development and Learning Measurement*, *11*(1), 1–10. DOI: 10.4018/IJCDLM.2021010101

Latorre Cosculluela, C., Liesa-Orús, M., & Vázquez-Toledo, S. (2018). Escuelas inclusivas: Aprendizaje cooperativo y TAC con alumnado con TDAH. *MAGIS Revista Internacional de Investigación en Educación*, *10*(21), 137–152. DOI: 10.11144/Javeriana.m10-21.eatt

Latorre Cosculluela, C., & Quintas Hijós, A. (2022). *Inclusión educativa y tecnologías para el aprendizaje*. Ediciones Octaedro. DOI: 10.36006/16298-4

Maryanti, R., Nandiyanto, A. B., Hufad, A., & Sunardi, S. (2021). Science education for students with special needs in Indonesia: From definition, systematic review, education system, to curriculum. *Indonesian Journal of Community and Special Needs Education, 1*(1), 1–8. DOI: 10.17509/ijcsne.v1i1.32653

Moro, S. (2021). Educational Intervention Strategies in the English Language Classroom for Students With Attention-Deficit/Hyperactivity Disorder in Primary Education. *Revista Electrónica Educare, 25*(3), 1–20. DOI: 10.15359/ree.25-3.20

Perales, F., & Ballespí, S. (2023). Prevalencia y manifestaciones clínicas del TDAH en población infantil. *Journal of Child Psychology and Psychiatry, and Allied Disciplines, 45*(1), 12–24.

Pilamunga Valla, E. V., Jiménez Pacheco, N. J., Mendiburu Rojas, A. F., & Medina Gonzáles, R. H. (2021). Las TICS y su influencia en el proceso de aprendizaje en los estudiantes con TDAH. *Journal of Science and Research, 7*, 1163–1180.

Quintero Gutiérrez del Álamo, F. J. (2019). *Actualización en el manejo del TDAH*. Lúa Ediciones.

Riga, A., Ifanti, T., & Papayiannis, N. (2019). Recommended educational approaches for teaching children with Attention Deficit Hyperactivity Disorder (ADHD) coordination and organization skills. *IJRDO-Journal of Educational Research, 2*(12), 1–14.

Robles Villalba, F. J., & Reyes Moreno, E. R. (2024). Evaluando el impacto de las TIC's en estudiantes con TDAH. *Una Ciencia Revista de Estudios e Investigaciones, 17*(32), 68–83. DOI: 10.35997/unaciencia.v17i32.771

Rodrigo Yanguas, M., González Tardón, C., Bella Fernández, M., & Blasco Fontecilla, H. (2022). Serious Video Games: Angels or Demons in Patients with Attention-Deficit Hyperactivity Disorder. *Frontiers in Psychiatry, 13*, 1–19. DOI: 10.3389/fpsyt.2022.798480 PMID: 35573357

Rubiano Caballero, M. L., & Lozano Zárate, Y. Y. (2024). Educación inclusiva y TIC: Prácticas pedagógicas para el rendimiento escolar de estudiantes con TDAH en instituciones públicas. *Ciencia Latina Revista Científica Multidisciplinar, 8*(5), 9857–9878. DOI: 10.37811/cl_rcm.v8i5.14364

Uyaguari Paredes, D.R., Uyaguari Paredes, M.J., Ríos Chichande, D.A. & Vásquez Villacis, M.I. (2024). Desarrollo de un entorno educativo inclusivo para niños con trastorno por déficit de atención e hiperactividad mediante el uso de tecnologías de la información y comunicación. *South Florida Journal of Development, 5*(3), 1-17. https://doi.org/ 5n3-017DOI: 10.46932/sfjdv

Villafuerte Holguín, J. S., & Alonzo Rezabala, M. E. (2020). Touchscreens and teaching English to children with attention deficit disorder: language practices and recreational games. *Apertura, 12*(2), 52.73. https://doi.org/DOI: 10.32870/ Ap.v12n2.1908

Villafuerte Holguín, J. S., & Alonzo Rezabala, M. E. (2020). Pantallas táctiles y enseñanza del inglés a niños con trastorno por déficit de atención: Prácticas idiomáticas y juegos recreativos. *Apertura (Guadalajara, Jal.), 12*(2), 52–73. DOI: 10.32870/Ap.v12n2.1908

Chapter 10
Leveraging ICTs for Transforming English Language Teaching:
Facilitating Educational Development and Reform

Vivit Rosmayanti
http://orcid.org/0000-0002-3648-9878
Universitas Negeri Makassar, Indonesia

Muliaty Ibrahim
Universitas Megarezky, Indonesia

Amra Ariyani
Universitas Negeri Makassar, Indonesia

ABSTRACT

The integration of Information and Communication Technologies (ICTs) in English Language Teaching (ELT) has significantly transformed teaching methods and improved the accessibility of education resources. ICTs, including web-enabled resources, mobile devices, Artificial Intelligence, and computer software, have made ELT more engaging and interactive for learners of all levels. They can bridge gaps in quality education, especially in remote and marginalized regions, and promote self-paced learning, instant response, and active participation. ICTs also help overcome the digital divide by providing learners with knowledge about different cultures and peoples, enhancing international cooperation. They contribute to education

DOI: 10.4018/979-8-3373-1190-6.ch010

reform processes in line with global agendas like Sustainable Development Goal 4 (Equitable Quality Education). However, concerns like lack of technology, teacher training, and culture can limit the efficient use of ICT in ELT.

INTRODUCTION: THE ROLE OF ICT IN EDUCATIONAL DEVELOPMENT AND REFORM

The integration of Information and Communication Technologies (ICTs) into education as a whole has modified the teaching approach and the level of learning achieved in different contexts. ICTs are a range of technologies that enable information to be sent and received such as the internet, cellular phones, and other computer applications. These technologies are in support of normal teaching in learning, but they also have the ability to induce higher levels of creativity during the learning process (Anastasopoulou *et al.*, 2024). The use of ICTs in education is greatly emphasized due to their capacity to help bridge geographical divides that inhibit the availability of resources and opportunities to many learners especially in developing nations (Yang *et al.*, 2018; Anastasopoulou *et al.*, 2024; Islam, 2015).

The incorporation of ICTs into the educational systems of different nations is indispensable, particularly regarding the improvement of education's quality and effectiveness in the context of international competition. They aid the educational systems in changing policies towards increasing access to education while improving its quality. The movements like 'Education for All' have utilized the ICT tools to reach the objectives of inclusiveness and social equity by incorporating more marginalized groups into the education (Zubković et al., 2022; Zeng, 2022) Furthermore, ICTs assist in fostering a collaborative learning environment where learners are able to engage and communicate with other learners and educators beyond their immediate environment, which in turn expands their educational landscape (Yang *et al.*, 2018; Anastasopoulou *et al.*, 2024; Avisteva, 2020). The adoption of ICTs in educational reform can be best exemplified during the COVID–19 pandemic where the need to shift to online learning was instant and emphasized the need for having sufficient digital readiness and competent teachers (Mahdum *et al.*, 2019; Avisteva, 2020).

The intersection of ICTs and English Language Teaching (ELT) is especially captivating considering the impact technology has had on the language teaching and learning process. The incorporation of ICT in ELT caters for learners with varying styles and preferences because of the interactivity and immersion that is provided. For instance, mobile language learning applications, websites, and online classrooms help learners to use the target language in real life context which enhances their communicative skills (Hidayatulloh, 2024; Mustofa & Sari, 2020). Moreover, the incorporation of multimedia materials such as videos and podcasts facilitate effective language learning by exposing students to various accents, dialects, and cultures that they can use in the language (Anastasopoulou *et al.*, 2024; Liang *et al.*, 2022). The application of ICT in ELT enhances the motivation of students towards lessons more readily while simultaneously enabling the teachers to be more flexible, active, and responsive to the learners of the twenty-first century (Anastasopoulou *et al.*, 2024; Buttar, 2016).

In this section, I explore the myriad impacts that information and communication technologies (ICTs) have on the teaching of English language, particularly its use and evolution in English Language Teaching (ELT) throughout the globe. It is also intended to chapters focus on the strengths that the use of technology brings toward achieving educational goals and success and the subsequent relief of gaps at all levels of society. Moreover, it will analyze the barriers to the integration of ICT into language teaching due to factors like infrastructure, lack of pedagogical preparation, and resources (Mustofa & Sari, 2020; Ofosu-Asare, 2024). This chapter attempts to review existing literature and case studies and analyze how these practices within ELT can be enhanced with the use of ICTs to contribute to the discourse on the use of ICT in the teaching and learning processes in this time of increasing educational reforms and globalization (Zeng, 2022; Buttar, 2016).

The ICT impact on education development and reform is very deep and complex. As educational institutions across the globe grapple with technology's integration into their practices, it remains critical to understand how ICTs can improve learn-er"s outcomes and facilitate equity and inclusion in education. The relationship between ICTs and English Language Teaching serves as a pointer to this phenomenon, because, as it has been demonstrated, it impacts the ther\'e are many new opportunities to interact, partner, and learn that were nunca antes disponibles en la clas. This chapter illustrates and expands critical aspects of these themes ICTs use in English language education.

UNDERSTANDING THE VALUE OF ICT IN ENGLISH LANGUAGE EDUCATION

What are ICTs?

The incorporation of Information and Communication Technologies (ICTs) into English Language Teaching (ELT) is slowly gaining recognition as a critical factor in improving educational practices and outcomes. ICTs denote a wide variety of tools and resources used in communication, learning, and exchange of information. In ELT, these tools include computers, mobile phones, the Internet, multimedia, educational software, and apps designed for language learning. These tools help in devising effective methods of language teaching by adding different levels of interactivity to the lessons. For instance, computers and mobile devices allow students to access a wide range of online materials, enabling learners to use authentic resources and practice their language skills in real-life contexts (Poudel, 2022; Dianti & Atmanegara, 2019). In addition, Internet communication facilitates language learning through collaboration as it allows students to interact with other learners and teachers around the world (Dianti & Atmanegara, 2019; Kjellsdotter, 2020).

The application of multimedia resources such as videos, podcasts, and interactive simulations enhances learning processes due to their powerful capabilities of meeting different learning styles needs. These resources motivate learners while exposing them to various accents, dialects, and cultures and make the process learning easier which is a significant contributing factor to communicative competence in a second language (Dianti & Atmanegara, 2019; Khadka, 2021). Other educational software and applications such as language platforms and gamified learning tools facilitate language acquisition through targeted learning designs that provide immediate feedback which is critical for effective language learning (Peng *et al.*, 2024; Fitria, 2023). Teachers can utilize these information and communication technology (ICT) tools to improve flexibility and interactivity in ELT classrooms so they can better meet the needs of the learners, which enhances the students' learning results (Dianti & Atmanegara, 2019; Bhandari, 2020).

The modifications brought English Language Teaching with the incorporation of devices enabled the shift from the classical pedagogy of teaching to ELT. Previously, the approach was centered around rote learning and passive systems were the norm where learners were to receive information without any activity on their part. On the contrary, students are brought into participation where they take interest in collaboration and critical thinking in Technology Enhanced Language Learning (Peng et al., 2024; Khadka, 2021). This is evident in the use and practice of communicative language teaching which considers language as an instrument for interaction and communication within the classroom setting (Poudel, 2022; Dianti & Atmanegara, 2019). It is also possible through the application of ICT education to be learner centered to increase motivation among the learners and learning based on the interests and the needs of the students (Kjellsdotter, 2020; Bhandari, 2020).

Additionally, the incorporation of ICTs within the scope of English Language Teaching (ELT) usage is consistent with constructivist learning theories that promote self-exploratory learning (Fitria, 2023; Bohara, 2024). With the aid of technology, students can participate in collaborative projects, webinars, and self-access learning resources which help them to learn the language. This helps improve their language proficiency and also sparks the desire of learners being self-directed (Bohara, 2024; Kjellsdotter, 2020). Therefore, the incorporation of ICTs transforms the previously existing teacher dominated view that regards pedagogical effect in ELT, which is ICT inspired, as self directed learning environment (Kjellsdotter, 2020; Fitria, 2023).

How ICTs Enhance English Language Teaching

Information and Communication Technologies (ICT) use in English Language Teaching (ELT) is perhaps one of the most controversial aspects of educational discourse today. The implementation of ICT in the learning system definitely fosters interaction in greatly increases the quality of learning a foreign language. Presentations, games, and virtual classes are taken as examples of materials that make learners participants of the learning process. These media include videos, audio, and interactive simulations that provide teachers with different ways of teaching various language skills to students with different levels of understanding and learning so that students can remember technique easily and effectively (Saud & Laudari, 2023). Adoption of gamification, the use of game elements to enhance motivation in blended learning, enables learners to work on language tasks thinking that they are actually doing something meaningful. Students are not the passive consumers of these technologies. In Zoom or Google Classroom students and learners can communicate with each other without restrictions (Hersi, 2024).

Language learners' engagement, preparation, and learning outcomes improves using digital tools. Some studies show that students who use ICT in their language learning register higher levels of interest and attention in their work. Learners are encouraged to participate actively in their education because of the interactive nature of ICT tools which promotes self-directed learning. For example, applications and websites to learn foreign languages give instant feedback and have features that cultivate the learner's confidence and motivation (Fitri *et al.*, 2024; Habók & Nguyen, 2024). Also, students can now access ICT for authentic materials such as podcasts, videos, and articles which allow them to use the language in a real context, which increases their interest and willingness to learn. This transformation toward more stimulating and participative forms of the learning environment is one way of dealing with undesirable attitudes and behaviors towards language learning because students will be willing to continue their studies if they find the process of learning simple and applicable to their day-to-day life (Alismaiel *et al.*, 2022; Fitri *et al.*, 2024).

A further ICT's importance is seen in improving communication, collaboration, and access to resources around the globe with respect to ELT. Digital tools allow the students to participate in collaborative projects and interact across borders. This enables students to carry out language tasks with their peers irrespective of their geographical locations (Ragi, 2024). This collective effort improves their language performance and develops important soft skills such as collaboration and communication that are needed in the modern world. Besides, the learners have access to the vast of information on the Internet resources including language learning sites, language practice forums, on-line dictionaries, etc. (Saud & Laudari, 2023; Habók & Nguyen, 2024). With this access, students can take control of their learning and encourage them to investigate topics from different angles, thereby improving their language education (Fitri *et al.*, 2024; Habók & Nguyen, 2024).

ICT AND EDUCATIONAL REFORM: ALIGNING WITH GLOBAL EDUCATION GOALS

The Role of ICTs in Advancing SDG 4: Quality Education

The adoption of Information Communication Technologies (ICTs) is regarded as essential in engaging education at the global level to meet targets under the SDG 4 and especially ensuring inclusive equitable quality education and promoting lifelong learning. ICTs are key to supporting inclusive education because they offer digital devices that increase access to learning opportunities and resources among various learners including those with disabilities and from marginalized groups. Through technology, teachers are able to design more flexible and customized education systems that meet the requirements of every student, hence, encouraging equity in education results (Zhou *et al.*, 2020; Zeng, 2022). For example, assistive technologies can prompt the active participation of learners with disabilities in the education process as well as high education learners in remote places and low education regions can be served through online educational resources.

The use of ICT can help to overcome educational gaps which continue to be a persistent issue in almost all parts of the world. Notably, developing nations struggle with inadequate infrastructure and trained personnel as well as profound socio-economic disparities that act as barriers to education. These gaps can be fulfilled using ICT through marginalized groups and regions surpassing traditional barriers to education. For example, different rural popups have mobile learning projects to provide educational resources where schools are not feasible/remotely located. These activities not only enhance learning scope but also promote the communication and interaction amongst learners and instructors.

The same authors clearly show in their work that the importance of ICT is increasing and is central to fulfilling SDG 4. It is acknowledged by almost all policymakers and leaders in education that the introduction of ICT into the education system can cause positive systemic change to the teaching and learning processes. For example, Zeng (2022) points out that China and other countries focus government policy on the use of ICT in basic education because it serves educational equity and quality. These reforms aim to achieve more equal education by investing in educational technology infrastructure and teachers' ICT training to have positive effects irrespective of the socio-economic factors of the learners.

ICTs as a Driver of Pedagogical Reform

The use of Information and Communication Technologies (ICTs) in education is perhaps the most important development in teaching and learning because it catalyzes a pedagogical change. One of the most notable changes is the departure from conventional, teacher dominated schooling methods to more engaging, participatory, and cooperative frameworks. In a standard classroom scenario, the educator is considered as the dominant knowledge distributor and learned in a pedagogical monologue. It is true that students may remain passive which can lead to minimal engagement in critical thinking and self solving exercises. Nevertheless, with the incorporation of ICTs into pedagogical practice, there is a marked shift towards instruction which promotes the use of student agency and collaboration (Umar *et al.,* 2024). As an example, students can use Google Docs and discussion boards to collaborate in class projects, share ideas, and give feedback. These tools help to increase interaction among students, which improves the overall learning atmosphere (Tapalova & Zhiyenbayeva, 2022; Murtaza *et al.*, 2022). This modification goes beyond improving student motivation and engagement, but rather equips students with skills to work in teams in the contemporary business world (Abbas *et al.,* 2023; Umar *et al.,* 2024).

Self-paced learning and differentiated e-instruction is made possible through AI's self-learning technologies. These technologies support teachers with catering lessons to the various skill levels and student paces in class. Adaptive learning systems, for example, use algorithms to compute how students are performing and dynamically adjust the difficulty and content of learning materials (Guo et al., 2024;, Pendy, 2023). This strategy enables learners to assume responsibility for their own learning because they can identify areas they need to focus on. Lastly, some artificial intelligence tools offer instant feedback. Such feedback enables learners to self-assess their strengths and weaknesses and select appropriate learning strategies (Domènech, 2023). In second language acquisition, such individualization is critical because learners generally come from diverse linguistic backgrounds and have varying degrees of proficiency (Konyrova, 2024).

The development of teacher professional development is important because it sustains pedagodical improvement facilitated with ICT. Technologies as well as the resources that educators enable their students to work with fuels a change in their teaching practices fostering TPACK development (Umar *et al.*, 2024; Kiwonde, 2023). TPACK is a term used to describe what teachers are constrained to possess in order to use ICT for pedagogy. Adequately designed professional development courses centered on TPACK will allow teachers to safely tackle the challenge of integrating technology as part of their students' learning which will help improve the effectiveness and innovativeness of teaching (Tapalova & Zhiyenbayeva, 2022). In addition, teachers' increasing competence in the use of ICT can encourage students to develop general ICT skills for use in other subjects leading to improved digital competence and self-directed learning in the classroom (Thomas, 2024).

Supporting Lifelong Learning Through ICTs

As stated above, the use of Information and Communication Technologies (ICTs) in China has increased in the past few years, especially to foster English language acquisition. ICTs provide English language learners with a tremendous amount of resources to the such as online courses, educational videos, podcasts, language learning applications, as well as websites that aim at providing practice and skill improvement in the target language. For instance, students can learn even when they are out of school because there are some language courses that are sequenced pedagogically on Phones and Tablets (Islamov et al., 2020). This form of language learning is characterized by a disregard for strict time schedules. Such materials are useful for adult learners who have busy schedules because they do not have to attend a classroom. The flexibility that ICTs afford learners enables them to organize their work in a way which is most advantageous to them, thus making it possible for the learners to achieve their objectives (Fadillah et al., 2024).

Apart from adults, Information Communication Technologies also provides a mean for young learners to acquire learning materials which promotes equal opportunities within the education system. A learner from a rural or economically underdeveloped area now has the possibility of accessing educational opportunities, irrespective of any geographical constraints, due to the availability of the internet (Rodríguez-Jiménez *et al.,* 2023). Teachers and students can now excel with the help of free online libraries and other OER which provide access to textbooks, videos, and research articles for students learning a new language (Rashid et al., 2019). The expansion of such materials is vital for learners of all socio-economic classes to actively participate in continuous education (Iskrenović-Momčilović & Momčilović, 2021; Pratama *et al.,* 2023). Moreover, special education students can also be supported with appropriate tools to meet their educational needs through information and communication technologies (Basham *et al.,* 2020).

The importance of ICTs in lifelong learning is more than accessing resources; they also allow learners to take charge of their learning journeys. Lifelong learning is dependent on E-learning tools available these days that permit monitoring and goal setting at a personal level, where individuals can reflect on their accomplishments (Porozovs *et al.,* 2019; Omboto *et al.,* 2022). For example, many learning management systems (LMS) and educational applications have built in tracking tools that provide feedback to guide learning activity modifications by learners. Autonomy of this nature increase motivation and provides a sense of responsibility for the learning actively integrated, which is crucial for achieving lifelong learning habits (Fitri *et al.,* 2024).

Additionally, the use of social networks and other online communities in the learning process can promote collaboration and help among learners. These networks allow people to meet others with the same objectives, share ideas, and work together on projects to enhance the entire learning process (Gümüş *et al.,* 2021). That is to say, one motivation to nurture social interactions in education is for the purpose of mastering a language as it involves productive and receptive skills which have to be twined with interactions. Using, ICTs to create environments for collaborative learning makes it possible for some learners to establish a learning community that continues learning together.

ICT TOOLS AND STRATEGIES IN ENGLISH LANGUAGE TEACHING

Digital Platforms and Tools for Language Learning

The introduction of Information and Communication Technologies (ICTs) in English Language Teaching (ELT) has significantly altered language acquisition techniques, resulting in the development of many digital platforms, particularly Duolingo and Babbel. These platforms are recognised for their creative methods in promoting language acquisition through interactive activities, including games and exercises customised to the specific requirements of individual learners (Fadillah *et al.*, 2024). Studies demonstrate that these technologies can improve student outcomes when accompanied by timely feedback and regular practice in a flexible learning setting (Raj *et al.*, 2019). These applications facilitate a self-paced learning approach, enabling users to include language usage into their regular activities, thereby ensuring continual enhancement (Atmowardoyo *et al.,* 2021).

Moreover, the benefits of these platforms beyond vocabulary development; they cultivate a collaborative community among learners. Users can communicate with peers and participate in discussions, which are crucial for acquiring proficiency in a foreign language (Kurniasari *et al.,* 2022; Zhou *et al.,* 2020). Such interactions promote genuine conversation and practice, motivating users to exchange experiences and improving their overall language acquisition process (Zhou *et al.*, 2020). Furthermore, data analytics derived from user involvement on these platforms can guide pedagogical techniques, enabling educators to customise their methods to better respond to learners' interactions (Fadillah *et al.*, 2024).

Supplementary digital resources, like podcasts, YouTube, and blogs, play a crucial part in the English language learning environment. Podcasts provide learners with exposure to actual spoken language and a variety of subjects, enhancing listening and comprehension abilities while accommodating personal interests (Nst *et al.*, 2023). YouTube functions as a valuable repository of captivating information that offers visual and aural stimuli, hence augmenting comprehension and retention of the language (Michael & Shah, 2020). YouTube channels often provide instruction on diverse linguistic subjects and cultural subtleties, rendering language acquisition relevant and practical for learners (Mustafa *et al.*, 2020; Yaacob & Saad, 2020).

Blogs function as valuable resources for English learners, offering insights into practical language application, grammatical guidance, and cultural context. They promote active engagement with the text through comments and conversations, hence enhancing language development (Saud, 2023). The amalgamation of numerous digital platforms establishes an all-encompassing learning environment that surpasses conventional classroom limitations, offering stimulating and pertinent experiences tailored to diverse learners (Fitri *et al.*, 2024).

Virtual and Augmented Reality (VR/AR) in ELT

The incorporation of Virtual Reality (VR) and Augmented Reality (AR) technology in English Language Teaching (ELT) is transforming the methods by which language abilities are cultivated through immersive learning experiences. These technologies allow learners to replicate real-world scenarios in which they can practise contextual language application, such as ordering meals in a restaurant or navigating an unfamiliar city (Wu, 2024). Studies demonstrate that VR and AR create immersive learning environments that improve language retention and reduce the anxiety typically linked to language acquisition, fostering a more supportive setting for practice and risk-taking (Bahari *et al.*, 2024).

Cokay et al. (2023) demonstrate that VR immersion in English as a Foreign Language (EFL) classrooms markedly improves authentic communicative language use and competency in students. Their research emphasised that interaction with VR environments enables learners to practise the language in authentic circumstances, a feature often absent in conventional classroom techniques (Bahari et al., 2024). Peixoto *et al.* (2019) conducted experiments demonstrating that students who received VR-enhanced lessons showed significant improvements in their speaking and listening skills compared to those taught through traditional methods, highlighting the effectiveness of VR in providing contextual language education.

Augmented reality technologies have acquired prominence in language acquisition, notably in improving vocabulary and comprehension abilities. A thorough review by Parmaxi and Demetriou (2020) demonstrated that augmented reality tools enhance learning activities by providing fascinating and engaging experiences, enabling learners to contextualise terminology and interact with digital representations superimposed on real-world situations. This interactive model promotes collaborative learning, crucial for language development as it accommodates many learning styles (Peixoto *et al.*, 2019; Parmaxi & Demetriou, 2020). Through interaction with AR applications, students can engage in contextualised learning experiences that improve retention and comprehension while actively participating in their classes (Huang et al., 2021).

Furthermore, teamwork is a crucial element of successful language acquisition facilitated by VR and AR technologies. Collaborative engagement in AR/VR projects enhances students' communication abilities and fosters confidence via realistic speaking experiences (Batool, 2022). Engagement in these immersive activities may involve interaction with native speakers or fellow learners, hence enhancing the language acquisition process (Chen, 2020). Collaborative frameworks are vital as they replicate authentic conversational practices necessary for fluency enhancement.

In conclusion, VR and AR technologies signify a dramatic advancement in English Language Teaching, providing immersive, contextually enriched settings that improve learning results. By enabling learners to connect with real-world scenarios and participate in collaborative activities, these technologies enhance the enjoyment of learning while markedly advancing language proficiency through contextual practice and genuine communication experiences. The expansion of these innovations offers a hopeful future for language instruction by enhancing student engagement and learning efficacy.

AI and Machine Learning in ELT

The use of Artificial Intelligence (AI) and Machine Learning (ML) in the English Language Teaching (ELT) sector is efficiently changing the instructional delivery of languages by providing personalized learning and self-paced instruction. One of the most notable improvements in the country is the introduction of adaptive learning systems and AI language tutors. These technologies implement algorithms that examine learner data. Hence, instruction can be customized according to each learner's proficiency level, learning model, and pace Wu (2024). For example, Rosetta Stone and Busuu utilize AI to develop individualized learning paths, which ensure that learners are given practice where they are actually lacking. This type of personalization helps learners to be more focused and motivated, and more importantly, help them achieve effective language learning because learners are no longer restrained to a certain pace of speed without feeling inundated (Bahari *et al.*, 2024).

Also, AI-powered language tutors provide tremendous assistance in learning a foreign language through instant feedback and support. These systems have the ability to evaluate a learner's performance in real time, enabling them to provide corrective measures and suggestions that assist in sharpening a student's language proficiency (Bahari et al., 2024; Peixoto *et al.*, 2019). For instance, AI applications can offer analysis of pronunciation and the use of grammar and vocabulary, thereby providing learners with detailed analytic information with regards to their achievements and deficiencies (Cokay *et al.*, 2023). Language learners find this system crucial because it provides a means to enhance and substitute approaches to learning almost instantly. In addition, AI can improve classroom experiences for students who require language tutoring because these students can speak and write without the usual classroom fear of being ridiculed (Peixoto *et al.,* 2019).

The application of AI does not stop at instructional diabetes; it influences language evaluation and feedback processes as well. Standardized testing has been the mainstay for assessing learners, but these traditional tests seldom capture the real abilities of a learner. Unlike these methods, AI assessment tools can analyse multiple dimensions of a student's performance on a given task, such as fluency, coherence, and contextually appropriate usage, and provide a more nuanced picture of their language proficiency (Parmaxi & Demetriou, 2020). For example, AI can grade students' written works and oral presentations, providing comprehensive comments on each task with pinpoint feedback on what needs to be improved (Peixoto *et al.,* 2019;, Parmaxi & Demetriou, 2020). These assessments, which are based on AI, do not only increase the precision of grade estimation, but also enable teachers to meaningfully adjust their pedagogical procedures and remedial actions (Huang *et al.,* 2021).

Moreover, AI expands the possibilities of the incorporation of Formative Assessment which is more advanced towards the learning processes. AI networks can help teachers continuously monitor learners' achievements and, through feedback, diagnose issues and correct them at the same time (Batool, 2022). In order for assessment to be facilitated, students have to appreciate the fact that learning is a continuous process and view challenges as opportunities to learn – not obstacles to achievement (Chen, 2020). Besides, AI can also be used to develop adaptive tests that change in accordance with the learner's actions so as to attain optimum challenge level and modification for the learner.

OVERCOMING CHALLENGES IN LEVERAGING ICTs FOR ELT

Technological Barriers

The challenges posed by the use of ICTs in English Language Teaching (ELT) are complex, including possible technological problems, but their greatest impact will be improvement on aspects of educational processes. One of the major problems is the digital divide, which highlights the gaps distinguished by technology usage, region of the world, geography, or even socioeconomic class. Most students, particularly those living in the underserved lcfzjezflcq regions where the level of living is substandard will suffer from this divide, thus severely affecting how effective ICT will be in language (Paudel, 2022; Alokozay *et al.*, 2022). For instance, students from the rural areas, as well as those from the poorer sections of society, have limited access to online language learning programs because of poor infrastructure in terms of internet connectivity, inadequate devices, adequate technical support (Abraham *et al.,* 2022).

To achieve equity in access to language education for all, there is an increased need to eliminate the digital divide phenomenon. An effective solution would entail the formulation and implementation of mobile learning tools intended for users with limited access to computing resources (Paudel, 2020). Mobile learning applications can be created and designed to run efficiently on smartphones and tablets, meaning that students are able to retrieve the language teaching and learning materials from almost any location, even those that have very weak internet connections. In addition, these mobile learning applications can have offline components so learners can download instructional materials for later use to help address some of the connectivity problems (Shrestha *et al.,* 2021).

Initiatives from the community are just as important in closing the digital gap and improving the accessibility of the ICT resources to the learners of languages. School and educational institutions may partner with local government, non-government organizations, and civic groups to set up programs that supply students with proper equipment and skills needed to harness technology for learning languages (Adipat, 2021; Saud, 2023). For example, those focusing on providing computers and Internet connectivity in educational community centers or conducting training programs on instructional technologies need to be designed for both learners and teachers to give them a voice in the educational process (Basri & Paramma, 2019). In addition, students could be supported through networks which enable them to exchange information and resources thereby making it possible to foster a supportive learning environment which, promotes participation and motivation (Khadka, 2021; Ryn & Sandaran, 2020).

Teacher Training and Professional Development

The incorporation of Information and Communication Technologies (ICTs) in English Language Teaching practices is successful by how effective the teacher training is conducted. As teachers are expected to increasingly integrate technology tools and resources in their instructions, they must also be trained in a way that prepares them to effectively utilize these technologies. Teachers who are not sufficiently trained may find it difficult to integrate ICTs in their instruction which results in missed chances to improve students' engagement and learning outcomes Kaur & (Ahuja, 2023). In addition, the speed of change in technology requires continuous professional training to make sure that teachers are familiar with the most recent tools and teaching methods (Gündüzalp, 2021).

Explaining the scope of teacher induction programs should not be restricted to the application of ICT skills, but should include new pedagogy that enhance the language acquisition processes. For instance, the training should also involve mobile learning applications, online assessment tools, and other Web 2.0 social media applications in the curriculum (Hidayat, 2019). Besides, teacher professional development programs need to account for and address teachers' levels of technologic sophistication and their experiences with digital content. If training that is specific to a particular situation or requirement is provided, concerning the teachers' context, schools will facilitate the incorporation of ICT in ELT. Without a doubt, instructors must modify their structures on how to employ ICTs in practice, otherwise they will be ineffective. For this goal, many initiatives and resources have been developed. As an example, educators may improve their digital literacy skills by attending online courses, webinars and workshops (Li *et al.*, 2024)

Additionally, professional learning communities (PLCs) constitute an important place where teachers collaborate on common problems encountered in the implementation of ICTs in teaching (Hafifah & Sulistyo, 2020). Such cooperative structures contribute not only to teachers' technological proficiency, but also serve as a means of continuous improvement and innovation in language teaching (Hussain *et al.*, 2021).

In addition, the role of educational institutions in the improvement of pedagogy cannot be disputed. Schools and universities must earmark resources for training teachers in digital skills and the use of ICTs in (Sintema & Jita, 2022). This approach is about giving technological support and encouraging professional development, as well as partnership with teacher training institutions (Kithungu *et al.*, 2021). Such support for professional development of teachers will enable these institutions train teachers to apply ICTs in language instruction, thereby improving the students' learning benefits.

Cultural and Linguistic Considerations

The application of Information and Communication technologies (ICT) within English for the Teaching of English as a Foreign Language (TEFL) has to take into account the cultural and linguistic aspects because these learners are from different backgrounds. The relevance of ICT tools to learners' culture and language is fundamental in the promotion of effective inclusivity. Relevance can be attained from the choice of the materials in the digital form in which they are available. Language learning mobile applications and online materials, for example, should not only be appealing to the interests of the learners but also be adequate to their sociocultural realities (Paudel, 2020). Students' motivation comes from seeing their culture represented within the resources they use for learning, in framing, they deepen the understanding of the language in relevant cultural contexts.

In addition, the integration of global tools in local contexts in language teaching is crucial for the success of ELT. Although a number of ICT tools are created and developed for a global audience, teachers need to tailor these resources to the specific language and culture of their students. This might mean amending the lesson plans to include local language varieties and culturally appropriate illustrations that depict the local lives of the learners. For instance, teachers can focus on vernacular styles of speech by augmenting general English platforms like Duolingo with locally designed materials that deal with region-specific details and inter-cultural aspects. This technique not only enhances learner participation, but also enables students to use the language more skilfully within different and more complex contexts.

Apart from adopting tools from outside, educators can use local materials and knowledge for enhancing language teaching. Community members may participate in learning as guest speakers, or experts which would enable the learners to practice the language and culture in context. Educators are then able to anchor their lessons to the community enabling more effectiveness and giving students an authentic context where they can step out of the classroom.

The deep cultural and linguistic aspects within ICT integration are radically important since they affect learners' motivation and achievement. Studies indicate that students are more likely to engage with and persist in their studies when they deem the materials and methods used in their language education to be relevant to their culture and linguistically. Thus, educators constantly need to think about and change their teaching approaches so that the multicultural needs of the constituents in an ever globalized world are addressed.

CONCLUSION

The implementation of Information and Communication Technologies (ICTs) in English Language Teaching (ELT) has changed educational practice in recent years. Moreover, It has triggered more changes. In this regard, this chapter analyzed the use of ICTs in Educational practices, its influences on digitalization, the application of AI and machine learning, the problems technology creates and its prospective benefits for society. As the case studies show, ICTs can be successfully utilized in various situations, with language learning being one of them, which in turn promotes integration and allows disabled children as well as children who do not speak English natively to be taught in an inclusive setting.

Last but not least, the most critical point is towards ICTs making ELT more inclusive, available, and flexible. Technology-enhanced teaching gives a broader scope to teachers for innovative instruction which captures learner interests and motivates them to learn. Students' resources and teaching materials can be modified according to the learner's location as well as the conditions in which ICTs are applied. Thus, these steps not only make learning more enjoyable and active, but also prepare students to live and operate in a highly interrelated and interdependent multilingual world.

Most importantly, the role of policymakers, educators, and software developers in promoting the use of ICTs for educational purposes is vital. Investment is necessary in the use of training tools that assist educators apply technology to teaching, and in digital development. On their part, teachers should actively seek new instructional improvement workshops aimed at collaboration among peers to share best practices and sophisticated tools. At the same time, the creators of the tools must try to design contextually relevant and culturally appropriate tools for all types of learners. This combined effort can enhance the teaching learning environment and eliminate barriers to equity and effectiveness in education through technology and English language teaching and learning.

REFERENCES

Abbas, N., Ali, I., Manzoor, R., Hussain, T., & Hussain, M. (2023). Role of artificial intelligence tools in enhancing students' educational performance at higher levels. *Journal of Artificial Intelligence Machine Learning and Neural Network*, (35), 36–49. DOI: 10.55529/jaimlnn.35.36.49

Abraham, M., Arficho, Z., & Habtemariam, T. (2022). Effects of training in ict-assisted english language teaching on secondary school english language teachers' knowledge, skills, and practice of using ict tools for teaching english. *Education Research International*, *2022*, 1–10. DOI: 10.1155/2022/6233407

Adipat, S. (2021). Developing technological pedagogical content knowledge (tpack) through technology-enhanced content and language-integrated learning (t-clil) instruction. *Education and Information Technologies*, *26*(5), 6461–6477. DOI: 10.1007/s10639-021-10648-3 PMID: 34220281

Alismaiel, O., Cifuentes-Faura, J., & Al-Rahmi, W. (2022). Social media technologies used for education: An empirical study on tam model during the covid-19 pandemic. *Frontiers in Education*, *7*, 882831. Advance online publication. DOI: 10.3389/feduc.2022.882831

Alokozay, W., Shadab, N., & Dauodzai, R. (2022). Utlilizing ict in english language teaching and learning from the perspective of afghan efl tearchers. *Randwick International of Education and Linguistics Science Journal*, *3*(2), 256–271. DOI: 10.47175/rielsj.v3i2.451

Anastasopoulou, E., Tsagri, A., Avramidi, E., Lourida, K., Mitroyanni, E., Tsogka, D., & Katsikis, I. (2024). The impact of ict on education. *Technium Social Sciences Journal*, *58*, 48–55. DOI: 10.47577/tssj.v58i1.11144

Atmowardoyo, H., Weda, S., & Sakkir, G. (2021). Learning strategies in english skills used by good language learners in millennial era: A positive case study in universitas negeri makassar. *Elt Worldwide Journal of English Language Teaching*, *8*(1), 28. DOI: 10.26858/eltww.v8i1.20339

Avisteva, R. (2020). Teachers' perspectives on the implementation of information and communication technology in language teaching.. DOI: 10.2991/assehr.k.200325.046

Bahari, A., Smith, M., & Scott, H. (2024). Examining the impact of chatbot-based language learning support, adaptive learning algorithms, and virtual reality language immersion on efl learners' language learning proficiency and self-regulated learning skills. DOI: 10.20944/preprints202403.1715.v1

Basham, J., Gardner, J., & Smith, S. (2020). Measuring the implementation of udl in classrooms and schools: Initial field test results. *Remedial and Special Education*, *41*(4), 231–243. DOI: 10.1177/0741932520908015

Basri, M., & Paramma, M. (2019). Efl students' perspective on the usefulness of ict based learning in indonesian higher education. *Elt Worldwide Journal of English Language Teaching*, *6*(2), 105. DOI: 10.26858/eltww.v6i2.10515

Batool, H. (2022). Augmented reality applications as a digital learning innovation in response to the pandemic. *Frontiers in Education*, *7*, 937074. Advance online publication. DOI: 10.3389/feduc.2022.937074

Bhandari, B. (2020). English teachers' perceptions and practices of integrating ict in elt. *Butwal Campus Journal*, *3*(1), 87–102. DOI: 10.3126/bcj.v3i1.36509

Bohara, D. (2024). Teachers' perceived value of integrating ict in transposing english language teaching in nepal. *KMC Journal*, *6*(1), 106–121. DOI: 10.3126/kmcj.v6i1.62332

Buttar, S. (2016). Ict in higher education. *People International Journal of Social Sciences*, *2*(1), 1686–1696. DOI: 10.20319/pijss.2016.s21.16861696

Chen, C. (2020). Ar videos as scaffolding to foster students' learning achievements and motivation in efl learning. *British Journal of Educational Technology*, *51*(3), 657–672. DOI: 10.1111/bjet.12902

Dianti, R. and Atmanegara, Y. (2019). Ict impact on efl secondary students' english achievements.. DOI: 10.2991/conaplin-18.2019.305

Domènech, J. (2023). Chatgpt in the classroom: friend or foe?., 339-347. DOI: 10.4995/HEAd23.2023.16179

Fadillah, E., Widiastuti, S., Sulaeman, D., Sugiartini, P., & Setiawan, D. (2024). Efl learners' attitudes on the use of ict-based learning as efforts in improving english language achievement. *Journal of Languages and Language Teaching*, *12*(3), 1573. DOI: 10.33394/jollt.v12i3.11388

Fitri, N., Afrida, N., & Meutia, C. (2024). English foreign language teachers' perception on the use of information and communication technology in teaching english at smkn 3 langsa. *Jolcc*, *2*(1), 1–17. DOI: 10.61320/jolcc.v2i1.1-17

Fitria, T. (2023). Using information and communication technology (ict) in the education sector: Opportunities and challenges in english language teaching (elt). *Teaching English as Foreign Language Literature and Linguistics*, *3*(1), 9–30. DOI: 10.33752/teflics.v3i1.4038

Gümüş, M., Çakır, R., Korkmaz, Ö., & Erdoğmuş, F. (2021). Analysis of it acceptance levels, ict attitudes, and individual innovation levels of special education teachers and their opinions. *International Journal of Technology in Education, 4*(3), 553–588. DOI: 10.46328/ijte.87

Gündüzalp, S. (2021). 21st century skills for sustainable education: Prediction level of teachers' information literacy skills on their digital literacy skills. *Discourse and Communication for Sustainable Education, 12*(1), 85–101. DOI: 10.2478/dcse-2021-0007

Guo, B., Mondol, E., Karim, A., & Musallami, N. (2024). Impact of artificial intelligence on the enhancement of quality of teaching in the private sector tertiary education: International perspective. *International Journal of Academic Research in Progressive Education and Development, 13*(4). Advance online publication. DOI: 10.6007/IJARPED/v13-i4/23546

Habók, A., & Nguyen, L. (2024). A comparative study of language learners' ict attitudes and behavior pre- and post-covid-19. *Scientific Reports, 14*(1), 600. Advance online publication. DOI: 10.1038/s41598-023-50872-0 PMID: 38182683

Hafifah, G., & Sulistyo, G. (2020). Teachers' ict literacy and ict integration in elt in the indonesian higher education setting. *Turkish Online Journal of Distance Education*, •••, 186–198. DOI: 10.17718/tojde.762050

Hersi, M. (2024). Gamification-cum-motivational strategies in english language learning. *International Journal of Linguistics, 16*(3), 1. DOI: 10.5296/ijl.v16i3.21797

Hidayat, M. (2019). The analysis of teacher's belief on digital literacy in forign language teaching. [Eeal Journal]. *English Education and Applied Linguistics Journal, 2*(1), 36–41. DOI: 10.31980/eealjournal.v2i1.1080

Hidayatulloh, S. (2024). Book review: Integrating information and communication technologies in english for specific purposes. *LLT Journal a Journal on Language and Language Teaching, 27*(1), 557–560. DOI: 10.24071/llt.v27i1.6750

Huang, X., Zou, D., Cheng, G., & Xie, H. (2021). A systematic review of ar and vr enhanced language learning. *Sustainability (Basel), 13*(9), 4639. DOI: 10.3390/su13094639

Hussain, I., Angaiz, D., & Jan, S. (2021). Relationship between prospective teachers' attitude towards information and communication technology (ict) and its use for educational purposes in gilgit-baltistan. *Pakistan Journal of Social Research, 03*(04), 398–411. DOI: 10.52567/pjsr.v3i4.302

Iskrenović-Momčilović, O., & Momčilović, A. (2021). Ict in teaching for children with intellectual disabilities. *Journal of Intellectual Disabilities*, *26*(2), 365–374. DOI: 10.1177/1744629521995376 PMID: 33827310

Islam, M. (2015). Impact of ict on women empowerment in south asia. *Journal of Economic & Financial Studies*, *3*(03), 80. DOI: 10.18533/jefs.v3i03.166

Islamov, R., Greenwald, O., & Tunyova, N. (2020). Information and communication technologies in foreign language teaching in multinational higher education environment of mining region. E3s Web of Conferences, 174, 04042. DOI: 10.1051/e3sconf/202017404042

Khadka, K. (2021). Prospects and challenges faced by secondary teachers while integrating ict in nepalese esl classrooms. *Dristikon a Multidisciplinary Journal*, *11*(1), 89–99. DOI: 10.3126/dristikon.v11i1.39141

Kithungu, R., Gakunga, D., Nungu, M., & Nungu, M. (2021). A teaching strategy based upon a model of agentic learning. *Journal of Studies in Education*, *11*(1), 58. DOI: 10.5296/jse.v11i1.17285

Kiwonde, F. (2023). E-learning training as a strategy for teachers' professional development in tanzania: A study of selected primary schools in musoma district, tanzania. *Asian Research Journal of Arts & Social Sciences*, *21*(4), 43–53. DOI: 10.9734/arjass/2023/v21i4491

Kjellsdotter, A. (2020). What matter(s)? a didactical analysis of primary school teachers' ict integration. *Journal of Curriculum Studies*, *52*(6), 823–839. DOI: 10.1080/00220272.2020.1759144

Konyrova, L. (2024). The evolution of language learning: Exploring ai's impact on teaching english as a second language. *Eurasian Science Review*, *2*(2), 133–138. DOI: 10.63034/esr-42

Kurniasari, N., Widodo, J., & Yappi, S. (2022). The effectiveness of youtube on elementary school students' reading skills during the covid-19 pandemic. *Jurnal Basicedu*, *6*(3), 3840–3846. DOI: 10.31004/basicedu.v6i3.2756

Li, W., Yu, S., Wang, M., Li, X., Ma, G., Ju, X., & Ling, C. (2024). How do life course events affect the accumulation of digital literacy?: Based on qualitative research of 16 medical university teachers of china. *Medicine*, *103*(27), e38755. DOI: 10.1097/MD.0000000000038755 PMID: 38968540

Liang, M., Lim, C., Park, J., & Mendoza, N. (2022). A review of ict-enabled learning for schoolgirls in asia and its impacts on education equity. *Educational Technology Research and Development, 71*(2), 267–293. DOI: 10.1007/s11423-022-10178-w PMID: 36533220

Mahdum, M., Hadriana, H., & Safriyanti, M. (2019). Exploring teacher perceptions and motivations to ict use in learning activities in indonesia. *Journal of Information Technology Education, 18*, 293–317. DOI: 10.28945/4366

Michael, E., & Shah, P. (2020). Students' perception on youtube usage in rural esl classroom. *International Journal of Academic Research in Business & Social Sciences, 10*(3). Advance online publication. DOI: 10.6007/IJARBSS/v10-i3/7059

Murtaza, M., Ahmed, Y., Shamsi, J., Sherwani, F., & Usman, M. (2022). Ai-based personalized e-learning systems: Issues, challenges, and solutions. *IEEE Access : Practical Innovations, Open Solutions, 10*, 81323–81342. DOI: 10.1109/ACCESS.2022.3193938

Mustafa, A., Taha, N., Al-Shboul, O., Alsalem, M., & Malki, M. (2020). Using youtube to learn anatomy: perspectives of jordanian medical students. Biomed Research International, 2020(1). DOI: 10.1155/2020/6861416

Mustofa, M., & Sari, A. (2020). Video subtitle to teach listening skill of junior high school students. [Journal of English Educators Society]. *Jees, 5*(2), 149–153. DOI: 10.21070/jees.v5i2.874

Nst, A., Daulay, S., & Dewi, U. (2023). Using rosetta stone application as a mobile assited language learning (mall) in efl: documentary study. Scope Journal of English Language Teaching, 8(1), 08. DOI: 10.30998/scope.v8i1.17653

Ofosu-Asare, Y. (2024). Developing classroom ict teaching techniques, principles and practice for teachers in rural ghana without access to computers or internet: A framework based on literature review. *International Journal of Information and Learning Technology, 41*(3), 262–279. DOI: 10.1108/IJILT-04-2023-0045

Omboto, C., Kanga, A., & Njageh, A. (2022). Myth or reality: Digital literacy programme implementation in primary special schools in nairobi, kenya. *European Journal of Education, 5*(2), 51–66. DOI: 10.2478/ejed-2022-0014

Parmaxi, A., & Demetriou, A. (2020). Augmented reality in language learning: A state-of-the-art review of 2014–2019. *Journal of Computer Assisted Learning, 36*(6), 861–875. DOI: 10.1111/jcal.12486

Paudel, P. (2020). Information and communication technology in foreign language classes in english: Roles and practices. *International Journal of Technology in Education and Science*, *5*(1), 37–55. DOI: 10.46328/ijtes.179

Paudel, P. (2022). Challenges of using information and communication technology in teaching english in nepal. *KMC Journal*, *4*(2), 1–18. DOI: 10.3126/kmcj.v4i2.47725

Peixoto, B., Pinto, D., Krassmann, A., Melo, M., Cabral, L., & Bessa, M. (2019). Using virtual reality tools for teaching foreign languages., 581-588. DOI: 10.1007/978-3-030-16187-3_56

Pendy, B. (2023). Artificial intelligence: The future of education. *Jurnal Indonesia Sosial Sains*, *2*(11). Advance online publication. DOI: 10.59141/jiss.v2i11.801

Peng, R., Razak, R., & Halili, S. (2024). Integration of information and communication technology in teaching english among in-service teachers. *International Journal of Evaluation and Research in Education (Ijere)*, *13*(2), 1235. DOI: 10.11591/ijere.v13i2.26347

Porozovs, J., Dudkina, A., & Valdemiers, A. (2019). The use of information and communication technologies of pedagogical specialities students in the study process., 104-110. DOI: 10.22616/REEP.2019.013

Poudel, A. (2022). Information and communication technology in english language teaching: Some opportunities and challenges. *Journal of Comparative & International Higher Education*, *14*(4). Advance online publication. DOI: 10.32674/jcihe.v14i4.3874

Pratama, A., Faroqi, A., & Mandyartha, E. (2023). Design of integrated school learning information system using codeigniter framework.. DOI: 10.11594/nstp.2023.3326

Ragi, T. (2024). Unveiling influence: Assessing the effect of ict tools on student activities through knowledge management processes. *CRR*, *1*(2), 50–57. DOI: 10.21844/crr.v1i02.1112

Raj, S., Ann, W., Subramaniam, P., & Yunus, M. (2019). Using youtube as a platform to learn social expression. *Creative Education*, *10*(02), 288–296. DOI: 10.4236/ce.2019.102023

Rashid, S., Yasin, M., & Ashaari, N. (2019). Undergraduate students of special education's readiness towards the use of information and technology (ict) in teaching and learning the sign language. *Creative Education*, *10*(11), 2374–2385. DOI: 10.4236/ce.2019.1011169

Rodríguez-Jiménez, M., Pérez-Jorge, D., Contreras, A., & Larumbe-Zabala, E. (2023). Development and validation of edu-cov-nee-neae questionnaire to measure educational response, ict usage and impact of covid-19 pandemic on students with special educational needs. *Heliyon*, *9*(9), e19231. DOI: 10.1016/j.heliyon.2023.e19231 PMID: 37674819

Ryn, A., & Sandaran, S. (2020). Teachers' practices and perceptions of the use of ict in elt classrooms in the pre-covid 19 pandemic era and suggestions for the 'new normal'. *LSP International Journal*, *7*(1). Advance online publication. DOI: 10.11113/lspi.v7n1.100

Saud, D. (2023). Teachers and students' perceptions towards using ict in elt in model schools. Journal of Tikapur Multiple Campus, 78-93. DOI: 10.3126/jotmc.v6i01.56347

Saud, D. (2023). Teachers and students' perceptions towards using ict in elt in model schools. Journal of Tikapur Multiple Campus, 78-93. DOI: 10.3126/jotmc.v6i01.56347

Shrestha, S., Haque, S., Dawadi, S., & Giri, R. (2021). Preparations for and practices of online education during the covid-19 pandemic: A study of bangladesh and nepal. *Education and Information Technologies*, *27*(1), 243–265. DOI: 10.1007/s10639-021-10659-0 PMID: 34341654

Sintema, E., & Jita, T. (2022). Pre-service teachers' self-concept and views toward using ict for teaching science. *Eurasia Journal of Mathematics, Science and Technology Education*, *18*(9), em2154. Advance online publication. DOI: 10.29333/ejmste/12396

Tapalova, O., & Zhiyenbayeva, N. (2022). Artificial intelligence in education: Aied for personalised learning pathways. *Electronic Journal of e-Learning*, *20*(5), 639–653. DOI: 10.34190/ejel.20.5.2597

Thomas, T. (2024). The role of artificial intelligence in formal and informal education for students. *International Journal for Research in Applied Science and Engineering Technology*, *12*(3), 69–71. DOI: 10.22214/ijraset.2024.58738

Umar, I., Isah, U., A, I., Shuaibu, B., & M, M. (2024). Advancing teachers' professional development: exploring models and impact assessment on ict integration in colleges of education. IJIR, 2(1), 45-56. DOI: 10.59890/ijir.v2i1.792

Wu, T. (2024). Virtual research laboratory in foreign language teaching: Strategies and role. Scientific Bulletin of Mukachevo State University Series. *Pedagogy and Psychology*, *10*(1), 107–115. DOI: 10.52534/msu-pp1.2024.107

Yaacob, Z., & Saad, N. (2020). Acceptance of youtube as a learning platform during the covid-19 pandemic: the moderating effect of subscription status. Tem Journal, 1732-1739. DOI: 10.18421/tem94-54

Yang, H., Zhu, S., & MacLeod, J. (2018). Promoting education equity in rural and underdeveloped areas: Cases on computer-supported collaborative teaching in china. *Eurasia Journal of Mathematics, Science and Technology Education, 14*(6). Advance online publication. DOI: 10.29333/ejmste/89841

Zeng, W. (2022). An empirical research on china's policy for ict integration in basic education from 1988 to 2021. *Educational Technology Research and Development, 70*(3), 1059–1082. DOI: 10.1007/s11423-022-10079-y PMID: 35221627

Zhou, Q., Lee, C., Sin, S., Lin, S., Hu, H., & Ismail, M. (2020). Understanding the use of youtube as a learning resource: A social cognitive perspective. *Aslib Journal of Information Management, 72*(3), 339–359. DOI: 10.1108/AJIM-10-2019-0290

Zubković, B., Pahljina-Reinić, R., & Kolić-Vehovec, S. (2022). Predictors of ict use in teaching in different educational domains. *Humanities Today Proceedings, 1*(1), 75–91. DOI: 10.2478/htpr-2022-0006

Chapter 11
Designing an Interactive Website Based on "Life Mathematics" to Develop Mathematical Problem-Solving Skills and Reduce the Level of Anxiety

Abdelrahman Mohammad Abu Sarah
http://orcid.org/0009-0003-1438-077X
Ministry of Education and Higher Education, Palestine

Yasmeen Majed Shkokani
Universiti Sains Islam Malaysia, Malaysia

Husam Tewfeeq Herzallah
Al-Quds Open University, Palestine

Riyad Abdalrazzaq Shkokani
Universiti Sains Islam Malaysia, Malaysia

Abedallah Mohammad Abu Sarah
Ministry of Education, Palestine

ABSTRACT

The aim of the research is to design an "interactive website based on life mathematics" to develop mathematical problem-solving skills and reduce the level of

DOI: 10.4018/979-8-3373-1190-6.ch011

anxiety among secondary students in Palestine. Researchers used the experimental curriculum with a semi-experimental design. The sample consisted of (40) students, from eleventh grade scientific secondary students, who were distributed equally to two groups: Experimental, examined the content of the module (equations and discrepancies) using the interactive website based on life mathematics, and control group who studied the same content in the usual way. The data was collected using the mathematical problem-solving skills test and mathematical anxiety scale. The results showed a statistically different between the two groups' students (experimental and control), in the testing of mathematical problem-solving skills for the students of the experimental group, and the absence of a difference between the students of the two groups in the mathematical anxiety scale.

INTRODUCTION

This search aims to designing an interactive website based on "life Mathematics" to develop problem-solving skill and reduce the anxiety among secondary students, in scientific stream in Palestine. To achieve the research objectives, an interactive website has been designed specifically for the units of equations and inequalities using the (Google site) application, and it has been connected to a collection of virtual techniques that mimic some actual circumstances involving students utilizing the cloud computing service displayed by the (Geogebra) program. To ensure that students can interact and respond to the activities presented on the website, a set of electronic tests has been designed through (Google Forms). Additionally, a unique group was created for the pupils using the (WhatsApp), to create a space where students can talk about and inquire about unclear things. And in order to collect the required data, the following research tools were designed: an equivalence test in the unit of equation and inequalities, a test of mathematical problem solving skills, and a scale of mathematics anxiety. The search sample consisted of 40 students from the secondary school scientific education in Qabatia, Palestine. The sample was divided into two groups: an experimental group that studied the unit content (equations and inequalities) by using an interactive website based on life mathematics, and control group that studied the same content with a traditional method. The research and its tools were applied during the second semester of the 2023/2024 academic year. The results showed a significant difference between the two groups in the test of mathematical problem-solving skills in favor of the experimental group, which indicates the effectiveness of the interactive website based on life Mathematics of developing problem-solving skills. There is another result showed that there is no significant difference between the two groups in the post-test of mathematical anxiety scale, indicating that the interactive website was not able to reduce the level of anxiety in

Math for the experimental group comparing with control group. According to the results, some recommendations and suggestions were presented.

RESEARCH PROBLEM

The "Palestinian Ministry of Education" has been making efforts to develop Mathematics education, improve its outcomes, and increase its quality level. However, who see the educational indicators, will notice that the outcomes have not reached the required level. To clarify, results from the national assessment study in Mathematics indicated that more than 56% of fifth-grade students and 74% of ninth-grade students are still at a weak level of mathematical performance, which indicates a worrying situation regarding students' level in various mathematical skills (Research and Development Center, 2022; OECD, 2023). This is also confirmed by the results of the (2022) PIZA international study, which indicated that 48% of students scored below 358 in the mathematical proficiency test, which has a maximum score of 1000, and the average achievement of participating students in Mathematics was 366. Meanwhile, the highest average achievement was in Singapore (575), which indicated a significant gap between the results of Palestine and many other countries participating in this study, specifically in Mathematics education (Educational Research and Development Center, 2023).

Through the practical experience of the researchers and their field observations, they noted a weakness among students in many mathematical aspects, particularly in skills related to new mathematical situations (mathematical problems). This is shown in the inability of many students to identify accurately the nature of the mathematical problems presented to them, as well as their unwillingness to come up with efficient methods for resolving mathematical issues. Additionally, their capacity to confirm that their responses or the conclusions they draw are accurate is lacking, and students' solutions to mathematical problems are often limited to just one method without flexibility in inventing new methods and ideas to address the mathematical problems they learn. This way complicates the students' task of learning Mathematics correctly and leads to an increase in their anxiety levels. From their point of view, Mathematics is becoming a sophisticated language that is hard to comprehend and use in everyday situations.

Students' inability to handle mathematical issues is coupled with a number of emotional factors that impact their ability to study the subject properly, as many students get away from learning Mathematics and refuse to take any additional lessons outside the approved curriculum. Moreover, they are anxious as the tests are soon and they try very hard to reduce the mathematical content that required during daily and term exams. They avoid dealing with anything related to Mathematics outside the classroom and they constantly remark how hard math is and how they must do everything in their power to finish it.

Referring to a number of previous references and studies (Abu Sarah, 2020; Abu Sarah et al., 2019; Abu Zeina, 2017; Al-Munufi & Al-Mu'tham, 2018; Abdul Hamid, 2017; Eid, 2017; Abu Al-Rayat, 2014; Touba, 2014), they indicated a weakness in students regarding many mathematical problem-solving skills. These studies recommended the need to strive towards developing these skills through the use of modern learning tools, which contribute to help students think correctly and effectively. Conversely, many previous studies (Mohammed, 2018; Sawy, 2019; Ghali, 2021) indicated high levels of anxiety among students while learning Mathematics, and these studies recommended the need for innovation in teaching and learning methods in Mathematics. Besides, employing digital educational tools that are characterized by their ability to motivate students and stimulate the factors of excitement and interaction to attract students, and to reduce their levels of anxiety while learning the mathematical content inside and outside the classroom.

In an effort to identify potential remedies for the difficulties that pupils confront when learning Mathematics, the skills of solving mathematical problems require employing teaching methods that stimulate students' thinking and attract them towards learning Mathematics correctly. Additionally, any new teaching strategy needs to be able to inspire different learning styles, such as: cooperative, participatory, and constructive learning, and additional methods that can combine students' energies to improve their capacity to handle novel mathematical situations.

On the other hand, the methods employed in Mathematics should simplify the content presented to students and present it in a flexible and easy way to remove the negative perception that formed in the minds of students, considering Mathematics as a language of symbols and complexity. So from this perspective, an interactive website based on life Mathematics has been designed to create a digital learning environment that integrates teachers, students, and the curriculum in a comprehensive way, aim to solve the two main problems in this study, which are: solving mathematical problems and the high level of math anxious among students.

To confirm the validity of this new approach, some previous studies (Al-Saeedi, 2024; Nadzri et al., 2024; Herzallah & Abu Sarah, 2022; Abu Sarah, 2021; Herzallah et al., 2023; Shkokani, 2021; Abu Sarah & Yaseen, 2018) recommended the necessary of employing interactive digital tools and applications in Mathematics education in general, because of their characteristics and capabilities that contribute to increase student's interaction with the mathematical content presented to them. Likewise, these digital tools can convey goal Mathematics in a practical manner, avoiding the use of symbolic language and imaginative language that hinders students' understanding. Additionally, the ability to add two elements of excitement and enjoyment, which may contribute to reduce students' anxiety while learning

Mathematics. Based on the above, the problem of the current study is defined by the weak level of secondary school students in Palestine in problem-solving skills and the high level of anxiety when learning Mathematics. Therefore, an interactive website based on life Mathematics has been designed to solve this problem and answer the following main question:

What is the effectiveness of using an interactive website based on life Mathematics in developing problem-solving skills in Mathematics and reducing anxiety levels among secondary school students in Mathematics in Palestine?

The main question branches into the following sub-questions:

1-What is the effectiveness of using an interactive website based on life Mathematics in developing problem-solving skills in Mathematics secondary school students in Palestine?

2 -What is the effectiveness of using an interactive website based on life Mathematics in reducing the level of anxiety among secondary school students in Palestine?

THE TOOLS OF THE STUDY

1-Tests of the skills of solving mathematical problems.

2-The scale of mathematical anxiety.

THE RESULTS OF THE STUDY

1-There is a significant difference between the students of the experimental and control groups in the post-test of mathematical problem-solving skills in favor of the students of the experimental group, indicating the effectiveness of the interactive website based on life Mathematics in developing problem-solving skills among the students of the experimental group.

2-There is no significant difference between the students of the experimental and control groups in the post-test of mathematical anxiety level, indicating that the interactive website was not able to reduce the level of anxiety to the experimental groups compares with control group .

REFERENCES

Abdul Hamid, R. (2017). The effectiveness of using the WebQuest strategy in teaching geometry to develop mathematical proficiency among middle school female students. *Journal of Mathematics Education, 20*(3), 32–87.

Abu Al-Rayat, A. (2014). The effectiveness of using Marzano's Dimensions of Learning Model in teaching mathematics to develop mathematical competence among preparatory stage students. *Journal of Mathematics Education, 17*(4), 53–104.

Abu Sarah, A. (2020). Developing mathematical proficiency and spatial sense among sixth-grade students in Palestine using mathematical modeling based on interactive computer and augmented reality applications. *Unpublished Doctoral Dissertation*, Cairo University, Egypt.

Abu Sarah, A. (2021). Designing programming-based educational activities to develop procedural knowledge and motivation towards learning mathematics among first-year technological secondary school students in Palestine. *Al-Quds Open University Journal for Educational and Psychological Research and Studies, 12*(33), 40–53.

Abu Sarah, A., Kafafi, W., & Salha, S. (2019). Developing components of mathematical proficiency for sixth-grade students in Palestine using mathematical modeling based on interactive computer and augmented reality applications. *International Journal of Online Education, 18*(1), 65–128.

Abu Sarah, A & Yaseen, S. (2018). The effect of using three computer programs on the academic achievement of tenth-grade students in mathematics in the Qabatiya Directorate (A comparative study). *An-Najah National University Journal for Research - Humanities (B)*, 32(6), 1003-1023.

Abu Zeina, F. (2017). *School Mathematics Curricula and Their Teaching* (4th ed.). Al-Falah Library for Publishing and Distribution.

Al-Munufi, S., & Al-Mu'tham, K. (2018). The extent to which second intermediate grade students in the Qassim region master mathematical proficiency skills. *Journal of Mathematics Education, 21*(6), 59–105.

Al-Saeedi, A. (2024). Interactive mathematics learning through technology-enhanced resources: A review. *Journal of Mathematics Education, 1*(27), 205–246.

Educational Research and Development Center. (2023). *Published mathematical items from the PISA 2022 study*. Palestine.

Eid, N. (2017). The impact of employing two types of infographics in light of the visual approach to develop mathematical problem-solving skills among eighth-grade female students in Gaza. *Unpublished Master's Thesis*, Islamic University, Palestine.

Ghali, S. (2021). The effectiveness of a proposed program based on real-life mathematical applications in developing algebraic thinking skills and reducing math anxiety among first-grade preparatory students. *Journal of Mathematics Education*, *24*(9), 206–260.

Herzallah, H., Abu Jerad, O., Abu Sarahh, A., & Dweikat, L. (2023). The Reality of Employing the Microsoft Application Teams in Learning Mathematics Remotely During the COVID-19 Pandemic in Palestine. In Khlaif, Z., Sanmugam, M., & Itmazi, J. (Eds.), *Comparative Research on Diversity in Virtual Learning: Eastern vs. Western Perspectives* (pp. 62–87). IGI Global., DOI: 10.4018/978-1-6684-3595-3.ch004

Herzallah, H., & Abu Sarah, A. (2022). The effectiveness of using infographics through Web 2.0 tools in developing achievement and engineering drawing skills in the technology curriculum among first-year secondary school students in Palestine. *Paper presented at the 7th Virtual International Conference on ICT in Education and Training (TICET)*, Tripoli University, Libya; German Democratic Center for Strategic and Political Studies, Germany; and Human Resource Development Center, Germany, 89-110.

Mohammed, R. (2018). A proposed program in action research based on project-based learning via the web to develop research awareness and reduce teaching anxiety among female mathematics teachers-in-training. *Journal of Mathematics Education*, *21*(4), 168–223.

Nadzri, A., Ayub, A., Zulkifli, N., & Salim, N. (2024). Implications of AR Modules on Geometry Conceptual and Procedural Knowledge among Primary School Students. *Malaysian Journal of Mathematical Sciences*, *18*(1), 51–72. DOI: 10.47836/mjms.18.1.04

OECD. (2023). PISA 2022 Results: Vol. I. *The State of Learning and Equity in Education*. OECD Publishing.

Sawy, Y. (2019). The effect of using mathematical tales in teaching arithmetic operations to develop mathematical concepts and reduce math anxiety among primary school students. *Journal of Mathematics Education*, *22*(5), 117–141.

Shkokani, R. (2021). *Mathematics distance learning competencies among mathematics teachers in the upper basic stage in Nablus Governorate* (Published Master's thesis). An-Najah National University.

Touba, R. (2014). The impact of using a mathematical modeling strategy on understanding mathematical concepts and solving mathematical problems among seventh-grade students in the measurement unit. *Unpublished Master's Thesis*, An-Najah National University, Palestine.

Compilation of References

Abbas, N., Ali, I., Manzoor, R., Hussain, T., & Hussain, M. (2023). Role of artificial intelligence tools in enhancing students' educational performance at higher levels. *Journal of Artificial Intelligence Machine Learning and Neural Network*, (35), 36–49. DOI: 10.55529/jaimlnn.35.36.49

Abdou, E. D. (2023). *Education, civics, and citizenship in Egypt: Towards more inclusive curricular representations and teaching*. Springer Nature. DOI: 10.1007/978-3-031-33346-0

Abdul Hamid, R. (2017). The effectiveness of using the WebQuest strategy in teaching geometry to develop mathematical proficiency among middle school female students. *Journal of Mathematics Education, 20*(3), 32–87.

Abraham, M., Arficho, Z., & Habtemariam, T. (2022). Effects of training in ict-assisted english language teaching on secondary school english language teachers' knowledge, skills, and practice of using ict tools for teaching english. *Education Research International, 2022*, 1–10. DOI: 10.1155/2022/6233407

Abu Al-Rayat, A. (2014). The effectiveness of using Marzano's Dimensions of Learning Model in teaching mathematics to develop mathematical competence among preparatory stage students. *Journal of Mathematics Education, 17*(4), 53–104.

Abu Sarah, A & Yaseen, S. (2018). The effect of using three computer programs on the academic achievement of tenth-grade students in mathematics in the Qabatiya Directorate (A comparative study). *An-Najah National University Journal for Research - Humanities (B)*, 32(6), 1003-1023.

Abu Sarah, A. (2020). Developing mathematical proficiency and spatial sense among sixth-grade students in Palestine using mathematical modeling based on interactive computer and augmented reality applications. *Unpublished Doctoral Dissertation*, Cairo University, Egypt.

Abu Sarah, A. (2021). Designing programming-based educational activities to develop procedural knowledge and motivation towards learning mathematics among first-year technological secondary school students in Palestine. *Al-Quds Open University Journal for Educational and Psychological Research and Studies, 12*(33), 40–53.

Abu Sarah, A., Kafafi, W., & Salha, S. (2019). Developing components of mathematical proficiency for sixth-grade students in Palestine using mathematical modeling based on interactive computer and augmented reality applications. *International Journal of Online Education, 18*(1), 65–128.

Abu Zeina, F. (2017). *School Mathematics Curricula and Their Teaching* (4th ed.). Al-Falah Library for Publishing and Distribution.

Abubakari, M. S., Shafik, W., & Hidayatullah, A. F. (2024). Evaluating the Potential of Artificial Intelligence in Islamic Religious Education: A SWOT Analysis Overview. In *AI-Enhanced Teaching Methods* (pp. 216–239). IGI Global. DOI: 10.4018/979-8-3693-2728-9.ch010

Adekoya, C. O., Fasae, J. K., & Alade, A. V. (2024). Academic libraries, ICT use and sustainable higher education development. *Information Discovery and Delivery, 52*(1), 1–10. Advance online publication. DOI: 10.1108/IDD-01-2022-0002

Adeoye, M. A., & Otemuyiwa, B. I. (2024). Navigating the Future: Strategies of EdTech Companies in Driving Educational Transformation. JERIT. *Journal of Educational Research and Innovation Technology, 1*(1), 43–50. DOI: 10.34125/jerit.v1i1.10

Adipat, S. (2021). Developing technological pedagogical content knowledge (tpack) through technology-enhanced content and language-integrated learning (t-clil) instruction. *Education and Information Technologies, 26*(5), 6461–6477. DOI: 10.1007/s10639-021-10648-3 PMID: 34220281

Adnan, A. H. M., Rahmat, A. M., Mohtar, N. M., & Anuar, N. (2021). Industry 4.0 critical skills and career readiness of ASEAN TVET tertiary students in Malaysia, Indonesia and Brunei. *Journal of Physics: Conference Series, 1793*(1), 0–11. DOI: 10.1088/1742-6596/1793/1/012004

Affero, I., & Hassan, R. (2013). Issues and challenges of technical and vocational education & training in Malaysia for knowledge worker driven. *National Conference on Engineering Technology,* February 2015. https://doi.org/DOI: 10.13140/2.1.4555.2961

Afzal, A., Khan, S., Daud, S., Ahmed, Z., & Butt, A. (2023). Addressing the digital divide: Access and use of technology in education. *Journal of Social Sciences Review*, *3*(2), 883–895. DOI: 10.54183/jssr.v3i2.326

Agrawal, S. (2021). A survey on Recent Applications of Cloud Computing in Education: COVID-19 Perspective. *Journal of Physics: Conference Series*, *1828*(1), 012076. DOI: 10.1088/1742-6596/1828/1/012076

Ahmad, C. V. (2021). Causes of students' reluctance to participate in classroom discussions. *ASEAN Journal of Science and Engineering Education*, *1*(1), 47–62. DOI: 10.17509/ajsee.v1i1.32407

Ahmadi, S. (2024). Systematic Literature Review on Cloud Computing Security: Threats and Mitigation Strategies. *Journal of Information Security*, *15*(02), 148–167. DOI: 10.4236/jis.2024.152010

Aho, A. V. (2012). Computation and computational thinking. *The Computer Journal*, *55*(7), 833–835. DOI: 10.1093/comjnl/bxs074

Ajani, O. A., & Govender, S. (2023). Impact of ICT-driven teacher professional development for the enhancement of classroom practices in South Africa: A systematic review of literature. *Journal of Educational and Social Research*, *13*(5), 116. DOI: 10.36941/jesr-2023-0125

Akpen, C. N., Asaolu, S., Atobatele, S., Okagbue, H., & Sampson, S. (2024). Impact of online learning on students' performance and engagement: A systematic review. *Discover Education*, *3*(1), 205. DOI: 10.1007/s44217-024-00253-0

Alanazi, F., & Al Turki, Y. (2021). Knowledge and attitude of Attention-deficit and Hyperactivity Disorder (ADHD) among male primary school teachers. *Journal of Family Medicine and Primary Care*, *10*(3), 1218–1226. DOI: 10.4103/jfmpc.jfmpc_2194_20 PMID: 34041155

Al-Ansi, B. M., Jaboob, M., Garad, A., & Al-Ansi, A. (2023). Analyzing augmented reality (AR) and virtual reality (VR) recent development in education. *Social Sciences & Humanities Open*, *8*(1), 100532. DOI: 10.1016/j.ssaho.2023.100532

Aldhafeeri, F. M., & Alotaibi, A. A. (2023). Reimagining education for successful and sustainable digital shifting. *SAGE Open*, *13*(1), 21582440231154474. DOI: 10.1177/21582440231154474

Aleven, V., McLaren, B., & Sewall, J. (2009, April). Scaling Up Programming by Demonstration for Intelligent Tutoring Systems Development: An Open-Access Web Site for Middle School Mathematics Learning. *IEEE Transactions on Learning Technologies*, *2*(2), 64–78. DOI: 10.1109/TLT.2009.22

Alismaiel, O., Cifuentes-Faura, J., & Al-Rahmi, W. (2022). Social media technologies used for education: An empirical study on tam model during the covid-19 pandemic. *Frontiers in Education, 7*, 882831. Advance online publication. DOI: 10.3389/feduc.2022.882831

Ali, Z., Gongbing, B., & Mehreen, A. (2018). Understanding and predicting academic performance through cloud computing adoption: A perspective of technology acceptance model. *Journal of Computers in Education, 5*(3), 297–327. DOI: 10.1007/s40692-018-0114-0

Alkhasawneh, R., Cob, Z. C., & Latif, A. B. A. (2025). Cloud computing adoption in government organizations in developing countries. *International Journal of Public Administration in the Digital Age, 12*(1), 1–27. DOI: 10.4018/IJPADA.368246

Al-Munufi, S., & Al-Mu'tham, K. (2018). The extent to which second intermediate grade students in the Qassim region master mathematical proficiency skills. *Journal of Mathematics Education, 21*(6), 59–105.

Al-Mutairi, M. A. (2024). Unlocking Potential: Harnessing Language Learning Apps to Support Young Learners with ADHD. *International Journal of Scientific Research and Management, 12*(4), 3313–3319. DOI: 10.18535/ijsrm/v12i04.el02

Alokozay, W., Shadab, N., & Dauodzai, R. (2022). Utlilizing ict in english language teaching and learning from the perspective of afghan efl tearchers. *Randwick International of Education and Linguistics Science Journal, 3*(2), 256–271. DOI: 10.47175/rielsj.v3i2.451

Alrikabi, H. T. H. S., Al-Dujaili, M. J., Majeed, B. H., & Alrubeei, I. R. N.Mohammed Jawad Al-DujailiBan Hassan MajeedIbtihal Razaq Niama ALRubeei. (2024). Information and Communication Technology and its Impact on Improving the Quality of Engineering Education Systems. *International Journal of Engineering Pedagogy, 14*(1), 4–19. Advance online publication. DOI: 10.3991/ijep.v14i1.46943

Al-Saeedi, A. (2024). Interactive mathematics learning through technology-enhanced resources: A review. *Journal of Mathematics Education, 1*(27), 205–246.

Al-Samarraie, H., & Saeed, N. (2018). A systematic review of cloud computing tools for collaborative learning: Opportunities and challenges to the blended-learning environment. *Computers & Education, 124*, 77–91. DOI: 10.1016/j.compedu.2018.05.016

Alshammari, F., & Alhalafawy, W. (2023). Digital platforms and the improvement of learning outcomes: Evidence extracted from meta-analysis. *Sustainability (Basel), 15*(2), 1305. DOI: 10.3390/su15021305

Al-Sharhan, S., Al-Hunaiyyan, A., Alhajri, R., & Al-Huwail, N. (2020). Utilization of learning management system (LMS) among instructors and students. In Advances in electronics engineering (pp. 2). Lecture Notes in Electrical Engineering. DOI: 10.1007/978-981-15-1289-6_2

Anastasopoulou, E., Tsagri, A., Avramidi, E., Lourida, K., Mitroyanni, E., Tsogka, D., & Katsikis, I. (2024). The impact of ict on education. *Technium Social Sciences Journal, 58*, 48–55. DOI: 10.47577/tssj.v58i1.11144

Anderson, J. R., Corbett, A. T., Koedinger, K. R., & Pelletier, R. (1995, April). Cognitive Tutors: Lessons Learned. *Journal of the Learning Sciences*, *4*(2), 167–207. DOI: 10.1207/s15327809jls0402_2

Anderson, T., & Dron, J. (2011). Three generations of distance education pedagogy. *International Review of Research in Open and Distance Learning*, *12*(3), 80–97. DOI: 10.19173/irrodl.v12i3.890

Andreotti, V. (2014). Soft versus critical global citizenship education. In Abdi, A., Shultz, L., & Pillay, T. (Eds.), *Decolonizing global citizenship education* (pp. 21–31). Sense Publishers., DOI: 10.1007/978-94-6209-518-8_2

Anis, M. (2024). Teacher professional development in the digital age: Addressing the evolving needs post-COVID. *International Journal for Multidisciplinary Research*, *6*(1), 1–14. DOI: 10.36948/ijfmr.2024.v06i01.12386

Apple, M. W. (2006). Understanding and interrupting neoliberalism and neoconservatism in education. *Pedagogies*, *1*(1), 21–26. DOI: 10.1207/s15544818ped0101_4

Asqui, J. D. C., Quichimbo, D. M. D., Ajila, L. E. V., Molero, D. A. L., & Campos, M. E. V. (2024). Inclusive Education from Cultural Diversity and ICT. *Migration Letters : An International Journal of Migration Studies*, *21*(S2), 400–412. Advance online publication. DOI: 10.59670/ml.v20i8.6627

Atmowardoyo, H., Weda, S., & Sakkir, G. (2021). Learning strategies in english skills used by good language learners in millennial era: A positive case study in universitas negeri makassar. *Elt Worldwide Journal of English Language Teaching*, *8*(1), 28. DOI: 10.26858/eltww.v8i1.20339

Avisteva, R. (2020). Teachers' perspectives on the implementation of information and communication technology in language teaching.. DOI: 10.2991/assehr.k.200325.046

Ayanwale, O. R., Ayeni, F. O., & Adekunle, O. M. (2025). Generative artificial intelligence in pedagogical practices: A systematic review of the literature. *Cogent Education*, *12*(1), 2485499. DOI: 10.1080/2331186X.2025.2485499

Azid, N. W., Rawian, R., Sarimah, S.-A., & Kiong, T. T. (2019). The development of interactive case-based smart thinking and industrial problem-solving stimulator to enhance TVET students' thinking skills. *Journal of Engineering Science and Technology*, *14*(5), 2643–2656.

Bahari, A., Smith, M., & Scott, H. (2024). Examining the impact of chatbot-based language learning support, adaptive learning algorithms, and virtual reality language immersion on efl learners' language learning proficiency and self-regulated learning skills. DOI: 10.20944/preprints202403.1715.v1

Balalle, H. (2024). Exploring student engagement in technology-based education in relation to gamification, online/distance learning, and other factors: A systematic literature review. *Social Sciences & Humanities Open*, *9*, 100870. DOI: 10.1016/j.ssaho.2024.100870

BAMIDELE, E. F. (2024). EDUCATION FOR SOCIO-ECONOMIC AND POLITICAL RECONSTRUCTION IN THE 21ST CENTURY NIGERIA. *Journal of Education Innovation and Practice*, *8*(1), 1–11.

Baran, E., & Correia, A. P. (2014). A professional development framework for online teaching. *TechTrends*, *58*(5), 95–101. DOI: 10.1007/s11528-014-0791-0

BARON NS. (2023) Even kids are worried ChatGPT will make them lazy plagiarists, says a linguist who studies tech's effect on reading, writing and thinking. Fortune. https://fortune.com/2023/01/19/what-is-chatgpt-ai-effect-cheating-plagiarism-laziness-education-kids-students/

Basham, J., Gardner, J., & Smith, S. (2020). Measuring the implementation of udl in classrooms and schools: Initial field test results. *Remedial and Special Education*, *41*(4), 231–243. DOI: 10.1177/0741932520908015

Basri, M., & Paramma, M. (2019). Efl students' perspective on the usefulness of ict based learning in indonesian higher education. *Elt Worldwide Journal of English Language Teaching*, *6*(2), 105. DOI: 10.26858/eltww.v6i2.10515

Bates, A. W. (2019). *Teaching in a digital age* (2nd ed.). BCcampus. https://pressbooks.bccampus.ca/teachinginadigitalagev2/

Batool, H. (2022). Augmented reality applications as a digital learning innovation in response to the pandemic. *Frontiers in Education*, *7*, 937074. Advance online publication. DOI: 10.3389/feduc.2022.937074

Bayly-Castaneda, K., Ramirez-Montoya, M. S., & Morita-Alexander, A. (2024). Crafting personalized learning paths with AI for lifelong learning: A systematic literature review. *Frontiers in Education*, *9*, 1424386. DOI: 10.3389/feduc.2024.1424386

Belando-Montoro, M. R., Naranjo-Crespo, M., & Carrasco-Temino, M. A. (2022). Barriers and facilitators to the retention and participation of socially, economically, and culturally disadvantaged university students. An international systematic review. *International Journal of Educational Research*, *113*, 101968. DOI: 10.1016/j.ijer.2022.101968

Bell, L., & Bull, G. (2010). Digital video and teaching. *Contemporary Issues in Technology & Teacher Education*, *10*, 1–6. http://www.editlib.org/p/34120

Bernacki, M. L., Greene, M. J., & Lobczowski, N. G. (2021). A systematic review of research on personalized learning: Personalized by whom, to what, how, and for what purpose(s)? *Educational Psychology Review*, *33*(4), 1675–1715. DOI: 10.1007/s10648-021-09615-8

Bhadri, G., & Patil, L. (2022). Blended learning: An effective approach for online teaching and learning. *Journal of Engineering Education Transformations*, *35*(is1, S1), 53–60. DOI: 10.16920/jeet/2022/v35is1/22008

Bhandari, B. (2020). English teachers' perceptions and practices of integrating ict in elt. *Butwal Campus Journal*, *3*(1), 87–102. DOI: 10.3126/bcj.v3i1.36509

Bhuyan, N., & Chakraborty, A. (2024). The Problem of Efficiency: Redefining the Relation Between Success & Excellence in Business Ethics. *Philosophy of Management*, *23*(1), 17–39. DOI: 10.1007/s40926-023-00234-7

Biard, N., Cojean, S., & Jamet, E. (2018). Effects of segmentation and pacing on procedural learning by video. *Computers in Human Behavior*, *89*, 411–417. DOI: 10.1016/j.chb.2017.12.002

Bickerton, C. J., & Accetti, C. I. (2021). *Technopopulism: The new logic of democratic politics*. Oxford University Press. DOI: 10.1093/oso/9780198807766.001.0001

Bickmore, T., Schulman, D., & Yin, L. (2022). Compassionate computing: Design and evaluation of relational agents. *Journal of Human–Computer Interaction*, *38*(3), 201–227. DOI: 10.1080/07370024.2022.2035237

Biesta, G. (2021). *World-centred education: A view for the present*. Routledge., DOI: 10.4324/9781003098331

Bikmohammadi, M. (2023). *Incorporation of Google Docs in composition classes to Foster collaboration: An analysis of Google Docs' affordances.*

Bin Rashid, A., & Karim Kausik, M. A. (2024). AI revolutionizing industries worldwide: A comprehensive overview of its diverse applications. *Hybrid Advances*, *7*, 100277. DOI: 10.1016/j.hybadv.2024.100277

Bohara, D. (2024). Teachers' perceived value of integrating ict in transposing english language teaching in nepal. *KMC Journal*, *6*(1), 106–121. DOI: 10.3126/kmcj.v6i1.62332

Bondarchuk, O. I., Balakhtar, V. V., Pavlenok, K. S., Kyrichuk, V. O., Yakubovska, N. O., & Grubi, T. V. (2024). Monitoring the psychological component of higher education quality using ICT tools. *CTE Workshop Proceedings,* 11. DOI: 10.55056/cte.685

Bondoc, R. S.Jr. (2024). ICT-Driven instructional and assessment strategies for physical education in the new normal. *Environment and Social Psychology*, *9*(4). Advance online publication. DOI: 10.54517/esp.v9i4.2155

Boonmoh, A., Jumpakate, T., & Karpklon, S. (2021). Teachers' perceptions and experience in using technology for the classroom. *CALL-EJ*, *22*(1), 1–24.

Bora, U. J., & Ahmed, M. (2013). E-Learning using Cloud Computing. In Blue Eyes Intelligence Engineering & Sciences Publication [IJISME]. *International Journal of Science and Modern Engineering*, *1*(2), 9–10.

Borgos, J., Kinser, K., & Kline, L. (2023). The borderless market for open, distance, and digital education. In *Handbook of Open, Distance and Digital Education* (pp. 355–369). Springer Nature Singapore. DOI: 10.1007/978-981-19-2080-6_22

Borji, A., & Zhao, Y. (2024). *Generative AI and its impact on personalized intelligent tutoring systems: A roadmap for education*. arXiv preprint arXiv:2410.10650. https://arxiv.org/abs/2410.10650

Borup, J., Graham, C. R., West, R. E., Archambault, L., & Spring, K. (2012). The adolescent community of engagement framework. *Journal of Technology and Teacher Education*, *22*(1), 107–130.

Bose, S., Dey, S. K., & Bhattacharjee, S. (2022). Big data, data analytics and artificial intelligence in accounting: An overview. In Akter, S., & Wamba, S. F. (Eds.), *Handbook of big data methods* (pp. 1–34). Edward Elgar Publishing.

Bradford, S., & Cullen, F. (2013). Research and research methods for youth practitioners. In *Research and Research Methods for Youth Practitioners*. DOI: 10.4324/9780203802571

Brookfield, S. D. (2017). *Becoming a critically reflective teacher* (2nd ed.). Jossey-Bass.

Bruun, M. H. (2024). Algorithmic Governance, Public Participation and Trust: Citizen–State Relations in a Smart City Project. Social Anthropology. *Anthropologie et Sociétés, 32*(4), 13–30.

Bubandt, N., & Willerslev, R. (2015). The dark side of empathy. *Comparative Studies in Society and History, 57*(1), 5–34. DOI: 10.1017/S0010417514000589

Budiarto, M. K., Gunarhadi, G., & Rahman, A. (2024). Technology in education through mobile learning application (MLA) and its impact on learning outcomes: Literature review. *Journal of Education and Learning, 18*(2), 413–420. Advance online publication. DOI: 10.11591/edulearn.v18i2.20976

Bull, S., & Kay, J. (2016, January 12). SMILI: A Framework for Interfaces to Learning Data in Open Learner Models, Learning Analytics and Related Fields. *International Journal of Artificial Intelligence in Education, 26*(1), 293–331. DOI: 10.1007/s40593-015-0090-8

Buttar, S. (2016). Ict in higher education. *People International Journal of Social Sciences, 2*(1), 1686–1696. DOI: 10.20319/pijss.2016.s21.16861696

Bykov, V., Mikulowski, D., Moravcik, O., Svetsky, S., & Shyshkina, M. (2020). THE USE OF THE CLOUD-BASED OPEN LEARNING AND RESEARCH PLATFORM FOR COLLABORATION IN VIRTUAL TEAMS. In *Information Technologies and Learning Tools 76*(2), 304-305.

Cabellos, B., Siddiq, F., & Scherer, R. (2024). The moderating role of school facilitating conditions and attitudes towards ICT on teachers' ICT use and emphasis on developing students' digital skills. *Computers in Human Behavior, 150*, 107994. Advance online publication. DOI: 10.1016/j.chb.2023.107994

Cabrera, B. C. C., Leal, M. C., Martínez, J. S. A., Soto, A. L. R., Sepulveda, J., Valencia, J., Garcés-Giraldo, L. F., & Valencia-Arias, A. (2025). Artificial Intelligence (AI) and Learning Management Systems (LMS): A bibliometric analysis. *Journal of Infrastructure Policy and Development, 9*(1), 8029. DOI: 10.24294/jipd8029

Calloway-Thomas, C. (2018). Empathy in the global world: An intercultural perspective. *Sage (Atlanta, Ga.).*

Campos Soto, M. N. (2022). *Implementación y mejora de procesos lectoescritores en alumnos con trastorno por déficit de atención e hiperactividad (TDAH) a través de las Tecnologías de la Información y la Comunicación (TIC) basadas en la animación 3D*. Editorial Universidad de Granada.

Cancio-Méndez, M., Rodríguez-Fernández, J. E., López-García, S., & Rico-Díaz, J. (2024). The influence of information and communication technologies on physical activity in primary education: A systematic review. *Educar, 60*(1). Advance online publication. DOI: 10.5565/rev/educar.1854

Cartee, D. (2021). The pedagogy of empathy: Cultivating deeper learning. *Journal of Educational Research and Practice, 11*(1), 88–99. DOI: 10.5590/JERAP.2021.11.1.07

Carver, L. B., & Nouri, J. (2024). *Generative AI and agency in education: A critical scoping review and thematic analysis.* arXiv preprint arXiv:2411.00631. https://arxiv.org/abs/2411.00631

Casey, C. (2024). *Governing the Firm in the Social Interest: Corporate Governance Reimagined.* Taylor & Francis.

Castro-Alonso, J. C., Koning, B. B. De, Fiorella, L., & Paas, F. (2021). *Five strategies for optimizing instructional materials: Instructor- and learner-managed cognitive load.* 1379–1407.

Catlin, D., & Wollard, J. (2014). Educational robots and computational thinking. *Proceedings of 4th International Workshop Teaching Robotics, Teaching with Robotics & 5th International Conference Robotics in Education*, 144–151.

Charmaz, K. (2006). Constructing grounded theory: A practical guide through qualitative analysis. *Sage (Atlanta, Ga.).*

Chen, C. (2020). Ar videos as scaffolding to foster students' learning achievements and motivation in efl learning. *British Journal of Educational Technology, 51*(3), 657–672. DOI: 10.1111/bjet.12902

Chen, J., & Zhang, J. (2024). The nexus between green finance and energy consumption in regional comprehensive economic partnership countries. *Environmental Science and Pollution Research International, 31*(9), 14071–14087. Advance online publication. DOI: 10.1007/s11356-024-32003-y PMID: 38270766

Chittaro, L., & Buttussi, F. (2015, April 18). Assessing Knowledge Retention of an Immersive Serious Game vs. a Traditional Education Method in Aviation Safety. *IEEE Transactions on Visualization and Computer Graphics, 21*(4), 529–538. DOI: 10.1109/TVCG.2015.2391853

Christopoulos, A., & Sprangers, P. (2021). Integration of educational technology during the Covid-19 pandemic: An analysis of teacher and student receptions. *Cogent Education, 8*(1), 1964690. DOI: 10.1080/2331186X.2021.1964690

Ciarli, T., Kenney, M., Massini, S., & Piscitello, L. (2021). Digital technologies, innovation, and skills: Emerging trajectories and challenges. *Research Policy*, *50*(7), 104289. DOI: 10.1016/j.respol.2021.104289

Cocha Millingalle, M.C., Naveda Bonilla, D.J., & Tenelanda López, D.V. (2021). Estrategias metodológicas docentes y su asociación en el proceso de enseñanza-aprendizaje del idioma inglés de escolares con TDAH. *Journal of Science and Research, 6*(2), 132.145. https://doi.org/DOI: 10.5281/zenodo.5514319

Cortés Pascual, A., Moyano Muñoz, N., & Quilez Robres, A. (2019). The relationship between executive functions and academic performance: Review and meta-analysis. *Frontiers in Psychology*, *10*(1582), 1–18. DOI: 10.3389/fpsyg.2019.01582 PMID: 31354585

Dahal, A., & Bhat, N. (2024). Self-directed learning, its implementation, and challenges: A review. *Nepal Journal of Health Sciences*, *3*(1), 102–115. DOI: 10.3126/njhs.v3i1.63277

Das, A., Malaviya, S., & Singh, M. (2023). The impact of AI-driven personalization on learners' performance. *International Journal on Computer Science and Engineering*, *11*(8), 15–22. DOI: 10.26438/ijcse/v11i8.1522

De Araujo, C. S., Aragão, A. O., Boechat, G. P., de Carvalho, I. E., Costa, J. W., Lopes, J. M., Castro, T. S., & Honório, V. A. (2024). Technology as a support in teaching students with ADHD: A view from educators. *Contribuciones a aas. Ciencia y Sociedad*, *17*(2), 1–15. https://doi.org/https://doi.org/10.55905/revconv.17n.2-144

De Bruijn-Smolders, M., & Prinsen, F. R. (2024). Effective student engagement with blended learning: A systematic review. *Heliyon*, *10*(23), e39439. DOI: 10.1016/j.heliyon.2024.e39439 PMID: 39687123

De la Cruz, G., Ullauri Moreno, M. I., & Freire, J. (2020). Strategies of teaching English as a Foreign Language towards Attention Deficit Hyperactivity Disorder Students (ADHD). *Colombian Applied Linguistics Journal*, *22*(2), 169–182. DOI: 10.14483/22487085.16118

Del Cisne Saldarriaga, A., & Briones Palacios, Y. M. (2022). El aprendizaje de estudiantes con trastorno por déficit de atención e hiperactividad desde las TIC. *Episteme Koinonia*, *5*(1), 405–416. DOI: 10.35381/e.k.v5i1.1821

Denning, P. J. (2009). The profession of IT: Beyond computational thinking. *Communications of the ACM*, *52*(6), 28–30. DOI: 10.1145/1516046.1516054

Diab, H., Daher, W., Rayan, B., Issa, N., & Rayan, A. (2024). Transforming science education in elementary schools: The power of PhET simulations in enhancing student learning. *Multimodal Technologies and Interaction*, *8*(11), 105. DOI: 10.3390/mti8110105

Diakopoulos, N. (2016, January 25). Accountability in algorithmic decision making. *Communications of the ACM*, *59*(2), 56–62. DOI: 10.1145/2844110

Dianti, R. and Atmanegara, Y. (2019). Ict impact on efl secondary students' english achievements.. DOI: 10.2991/conaplin-18.2019.305

Dibrov, A. (2015). Innovation resistance: The main factors and ways to overcome them. *Procedia: Social and Behavioral Sciences*, *166*, 92–96. DOI: 10.1016/j.sbspro.2014.12.489

Domènech, J. (2023). Chatgpt in the classroom: friend or foe?., 339-347. DOI: 10.4995/HEAd23.2023.16179

Ebrahimzadeh, M., & Alavi, S. (2017). The effect of digital video games on efl students' language learning motivation. *Teaching English with Technology*, *17*(2), 87–112.

Economic Planning Unit. (2015). Strategy Paper 9: Transforming technical and vocational education and training to meet industry demand. In *Eleventh Malaysia Plan,2016-2020:Way Forward*. rmk11.epu.gov.my/pdf/strategy-paper/Strategy Paper 09.pdf

Eden, C. A., Chisom, O. N., & Adeniyi, I. S. (2024). Promoting digital literacy and social equity in education: Lessons from successful initiatives. *International Journal of Management & Entrepreneurship Research*, *6*(3), 687–696. DOI: 10.51594/ijmer.v6i3.880

Educational Research and Development Center. (2023). *Published mathematical items from the PISA 2022 study*. Palestine.

Eghrefuvwoma, O. G., Agada, S. A., Ighodaro, I. H., & Olasupo, S. F. (2024). ICT in Higher Education: A Tool for Sustainable Development in West Africa. *International Journal of Research and Innovation in Applied Science*, *IX*(I), 237–244. Advance online publication. DOI: 10.51584/IJRIAS.2024.90122

Eid, N. (2017). The impact of employing two types of infographics in light of the visual approach to develop mathematical problem-solving skills among eighth-grade female students in Gaza. *Unpublished Master's Thesis*, Islamic University, Palestine.

Elliott, A. (2022). *Algorithmic intimacy: The digital revolution in personal relationships*. John Wiley & Sons.

Ellis, V., Gatti, L., & Mansell, W. (2024). *The new political economy of teacher education: The Enterprise narrative and the shadow state*. Policy Press.

Ercan, T. (2010). Effective use of cloud computing in educational institutions. *Procedia: Social and Behavioral Sciences*, *2*(2), 938–942. DOI: 10.1016/j.sbspro.2010.03.130

Erenler, H. H. T. (2018). Heuristic evaluation of e-elearning. *International Journal of Organizational Leadership*, *7*(2), 195–210. DOI: 10.33844/ijol.2018.60235

Estrella Quintana, P. A., Guerrero Paredes, K. A., & Chiluisa Chiluisa, M. J. (2024). Attention deficit hyperactivity disorder and English Language teaching and learning. *Revista Científica de Innovación Educativa y Sociedad Actual*, *4*(1), 76–87. DOI: 10.62305/alcon.v4i1.66

Ewe, L. P. (2019). ADHD symptoms and the teacher–student relationship: A systematic literature review. *Emotional & Behavioural Difficulties*, *24*(2), 1–20. DOI: 10.1080/13632752.2019.1597562

Facer, K. (2021 forthcoming). *Futures in education: Towards an ethical practice. Paper commissioned for the UNESCO Futures of Education report*. UNESCO.

Fadillah, E., Widiastuti, S., Sulaeman, D., Sugiartini, P., & Setiawan, D. (2024). Efl learners' attitudes on the use of ict-based learning as efforts in improving english language achievement. *Journal of Languages and Language Teaching*, *12*(3), 1573. DOI: 10.33394/jollt.v12i3.11388

Fawns, T. (2022). An entangled pedagogy: Looking beyond the pedagogy—technology dichotomy. *Postdigital Science and Education*, *4*(3), 711–728. DOI: 10.1007/s42438-022-00302-7

Fawns, T., Mulherin, T., Hounsell, D., & Aitken, G. (2021). Seamful learning and professional education. *Studies in Continuing Education*, *43*(3), 360–376. DOI: 10.1080/0158037X.2021.1920383

Ferguson, R. (2012). Learning analytics: Drivers, developments and challenges. *International Journal of Technology Enhanced Learning*, *4*(5/6), 304. DOI: 10.1504/IJTEL.2012.051816

Fernández Gacho, L., Arias Gonzáles, V., Rodríguez Navarro, H., & Manzano Soto, N. (2020). Estudio e intervención en niños con Trastorno por Déficit de Atención e Hiperactividad de educación primaria. *Revista interuniversitaria de formación del profesorado, 34*(2), 247-274. https://doi.org/DOI: 10.47553/rifop.v34i2.77535

Fernández Menor, I. (2021). Evaluación de un estudiante con TDAH tras una intervención con recursos tecnológicos. *Páginas de Educación, 14*(2), 121–131. DOI: 10.22235/pe.v14i2.2565

Figueroa Pozo, D. C., Avilés Castellanos, N. F., Narváez Carvajal, L. I., Recalde Drouet, E. M., & Crespín Crespín, E. E. (2023). Recursos digitales para fortalecer el aprendizaje de niños con TDAH. *Ciencia Latina Revista Científica Multidisciplinar, 7*(2), 7648-7662. https://doi.org/https://doi.org/10.37811/cl_rcm.v7i2.5902

Fitria, T. (2023). Using information and communication technology (ict) in the education sector: Opportunities and challenges in english language teaching (elt). *Teaching English as Foreign Language Literature and Linguistics, 3*(1), 9–30. DOI: 10.33752/teflics.v3i1.4038

Fitri, N., Afrida, N., & Meutia, C. (2024). English foreign language teachers' perception on the use of information and communication technology in teaching english at smkn 3 langsa. *Jolcc, 2*(1), 1–17. DOI: 10.61320/jolcc.v2i1.1-17

Fitriyah, A. (2025). Inclusive Policy Design for Transformative Education: A Global Synthesis of Teacher Identity Formation in Resource-Constrained Contexts. *Sinergi International Journal of Education, 3*(2), 94–106. DOI: 10.61194/education.v3i2.703

Flores Nieves, C. H., Jiménez Guerrero, A. M., Vargas Galarza, M. V., Delgado Mora, L. E., Changoluisa Chicaiza, K. S., & Tapia Lincango, M. C. (2024). Relevancia del Uso de Recursos Tecnológicos en el Proceso de Aprendizaje de los Estudiantes con TDAH. *Ciencia Latina Revista Científica Multidisciplinar, 8*(4), 9958–9972. DOI: 10.37811/cl_rcm.v8i4.13143

Forte, A., & Guzdial, M. (2004). Computers for communication, not calculation: Media as a motivation and context for learning. *Proceedings of the Hawaii International Conference on System Sciences, 37*(January), 1513–1522. DOI: 10.1109/HICSS.2004.1265259

Freire, P. (1970). Pedagogy of the oppressed. *Continuum : an Interdisciplinary Journal on Continuity of Care.*

Fricker, M. (2007). *Epistemic injustice: Power and the ethics of knowing*. Oxford University Press., DOI: 10.1093/acprof:oso/9780198237907.001.0001

Gabriel, F., Marrone, R., Van Sebille, Y., Kovanovic, V., & de Laat, M. (2022). Digital education strategies around the world: Practices and policies. *Irish Educational Studies, 41*(1), 85–106. DOI: 10.1080/03323315.2021.2022513

García, J., Pérez, M., & López, A. (2021). Diagnóstico y manejo del TDAH en el entorno educativo. *Revista de Neurociencia Clínica, 34*(2), 123–134.

García-Martín, J., & García-Sánchez, J. N. (2022). The digital divide of know-how and use of digital technologies in higher education: The case of a college in Latin America in the COVID-19 era. *International Journal of Environmental Research and Public Health, 19*(6), 3358. DOI: 10.3390/ijerph19063358

Gardner, H. (2021). *Disciplined mind: what all students should understand.* Simon & Schuster.

Garrison, D. R., Anderson, T., & Archer, W. (1999). Critical inquiry in a text-based environment: Computer conferencing in higher education. *The Internet and Higher Education, 2*(2–3), 87–105. DOI: 10.1016/S1096-7516(00)00016-6

Gasevic, D., Siemens, G., & Rose, C. P. (2017, January 1). Guest Editorial: Special Section on Learning Analytics. *IEEE Transactions on Learning Technologies, 10*(1), 3–5. DOI: 10.1109/TLT.2017.2670999

Gautam, A., & Kotiyal, A. (2024). Connecting the future. In P. D. Singh & M. Angurala (Eds.), *Integration of Cloud Computing and IoT: Trends, Case Studies and Applications* (1st ed.), 274–308). Chapman and Hall/CRC eBooks. https://doi.org/ DOI: 10.1201/9781032656694-15

Gay, G. (2018). *Culturally responsive teaching: Theory, research, and practice* (3rd ed.). Teachers College Press.

Ghali, S. (2021). The effectiveness of a proposed program based on real-life mathematical applications in developing algebraic thinking skills and reducing math anxiety among first-grade preparatory students. *Journal of Mathematics Education, 24*(9), 206–260.

Gilliard, C. (2021). Tagged surveillance. The Journal of Interactive Technology and Pedagogy, 20.

Giroux, H. A. (2024). *Teachers as intellectuals: Toward a critical pedagogy of learning*. Bloomsbury Publishing. DOI: 10.5040/9781350458635

Gkora, V. (2024). Advancing ADHD Education: Autonomy, technology, and inclusive strategies. *GSC Advanced Research and Reviews, 18*(3), 101–121. DOI: 10.30574/gscarr.2024.18.3.0084

Gómez-Devís, M. B., & Herranz-Llácer, C. V. (2024). New Technologies: ICT and Education in future Early Childhood Education teachers. Characterization from lexical availability. *Ogigia, 35*. Advance online publication. DOI: 10.24197/ogigia.35.2024.7-30

Gruber, J., Hargittai, E., & Nguyen, M. H. (2022). The value of face-to-face communication in the digital world: What people miss about in-person interactions when those are limited. *Studies in Communication Sciences*, *22*(3), 1–19. DOI: 10.24434/j.scoms.2022.03.3340

Gümüş, M., Çakır, R., Korkmaz, Ö., & Erdoğmuş, F. (2021). Analysis of it acceptance levels, ict attitudes, and individual innovation levels of special education teachers and their opinions. *International Journal of Technology in Education*, *4*(3), 553–588. DOI: 10.46328/ijte.87

Gündüzalp, S. (2021). 21st century skills for sustainable education: Prediction level of teachers' information literacy skills on their digital literacy skills. *Discourse and Communication for Sustainable Education*, *12*(1), 85–101. DOI: 10.2478/dcse-2021-0007

Guo, B., Mondol, E., Karim, A., & Musallami, N. (2024). Impact of artificial intelligence on the enhancement of quality of teaching in the private sector tertiary education: International perspective. *International Journal of Academic Research in Progressive Education and Development*, *13*(4). Advance online publication. DOI: 10.6007/IJARPED/v13-i4/23546

Guo, M., Pang, D., & Dong, J. (2024). Promoting green growth: Resource management and reosurce efficiency in East Asian and Pacific ecnomies. *Resources Policy*, *90*, 104705. Advance online publication. DOI: 10.1016/j.resourpol.2024.104705

Habermas, J. (1987). *The theory of communicative action: Vol. 2. Lifeworld and system: A critique of functionalist reason*. Beacon Press.

Habók, A., & Nguyen, L. (2024). A comparative study of language learners' ict attitudes and behavior pre- and post-covid-19. *Scientific Reports*, *14*(1), 600. Advance online publication. DOI: 10.1038/s41598-023-50872-0 PMID: 38182683

Hafifah, G., & Sulistyo, G. (2020). Teachers' ict literacy and ict integration in elt in the indonesian higher education setting. *Turkish Online Journal of Distance Education*, •••, 186–198. DOI: 10.17718/tojde.762050

Haggard, S., & Kaufman, R. (2021). *Backsliding: Democratic regress in the contemporary world*. Cambridge University Press. DOI: 10.1017/9781108957809

Haleem, A., Javaid, M., Qadri, M. A., & Suman, R. (2022). Understanding the role of digital technologies in education: A review. *Sustainable Operations and Computers, 3*, 275–285. DOI: 10.1016/j.susoc.2022.05.004

Hamari, J., & Koivisto, J. (2014, November). Measuring flow in gamification: Dispositional Flow Scale-2. *Computers in Human Behavior, 40*, 133–143. DOI: 10.1016/j.chb.2014.07.048

Hamzah Nasution, B., Siregar, M., Agustina, R., & Sitompu, U. (2024). *Green Banking Concept Implementation in Banking Credit Governance in Indonesia: Comparison Between Indonesia and China*. KnE Social Sciences., DOI: 10.18502/kss.v8i21.14761

Hart, S. G., & Staveland, L. E. (1988). Development of NASA-TLX (Task Load Index): Results of empirical and theoretical research. In Hancock, P. A., & Meshkati, N. (Eds.), *Human Mental Workload* (pp. 139–183). North Holland Press., DOI: 10.1007/s10749-010-0111-6

Hase, A., & Kuhl, P. (2024). Teachers' use of data from digital learning platforms for instructional design: A systematic review. *Educational Technology Research and Development, 72*(4), 1925–1945. DOI: 10.1007/s11423-024-10356-y

Hasler, B. S., Kersten, B., & Sweller, J. (2007, September). Learner control, cognitive load and instructional animation. *Applied Cognitive Psychology, 21*(6), 713–729. DOI: 10.1002/acp.1345

Hassan, N., Yau, K. A., & Wu, C. (2019). Edge Computing in 5G: A review. *IEEE Access : Practical Innovations, Open Solutions, 7*, 127276–127289. DOI: 10.1109/ACCESS.2019.2938534

Hatsidimitris, G., & Kalyuga, S. (2013). Guided self-management of transient information in animations through pacing and sequencing strategies. *Educational Technology Research and Development, 61*(1), 91–105. DOI: 10.1007/s11423-012-9276-z

Hersi, M. (2024). Gamification-cum-motivational strategies in english language learning. *International Journal of Linguistics, 16*(3), 1. DOI: 10.5296/ijl.v16i3.21797

Herzallah, H., & Abu Sarah, A. (2022). The effectiveness of using infographics through Web 2.0 tools in developing achievement and engineering drawing skills in the technology curriculum among first-year secondary school students in Palestine. *Paper presented at the 7th Virtual International Conference on ICT in Education and Training (TICET)*, Tripoli University, Libya; German Democratic Center for Strategic and Political Studies, Germany; and Human Resource Development Center, Germany, 89-110.

Herzallah, H., Abu Jerad, O., Abu Sarahh, A., & Dweikat, L. (2023). The Reality of Employing the Microsoft Application Teams in Learning Mathematics Remotely During the COVID-19 Pandemic in Palestine. In Khlaif, Z., Sanmugam, M., & Itmazi, J. (Eds.), *Comparative Research on Diversity in Virtual Learning: Eastern vs. Western Perspectives* (pp. 62–87). IGI Global., DOI: 10.4018/978-1-6684-3595-3.ch004

Hidayat, M. (2019). The analysis of teacher's belief on digital literacy in forign language teaching. [Eeal Journal]. *English Education and Applied Linguistics Journal, 2*(1), 36–41. DOI: 10.31980/eealjournal.v2i1.1080

Hidayatulloh, S. (2024). Book review: Integrating information and communication technologies in english for specific purposes. *LLT Journal a Journal on Language and Language Teaching*, *27*(1), 557–560. DOI: 10.24071/llt.v27i1.6750

Hinostroza, J. E., Armstrong-Gallegos, S., & Villafaena, M. (2024). Roles of digital technologies in the implementation of inquiry-based learning (IBL): A systematic literature review. *Social Sciences & Humanities Open*, *9*, 100874. DOI: 10.1016/j.ssaho.2024.100874

Hirsh-Pasek, K., Zosh, J. M., Hadani, H. S., Golinkoff, R. M., Clark, K., Donohue, C., & Wartella, E. (2022). *A Whole New World: Education Meets the Metaverse. Policy Brief.* Center for Universal Education at The Brookings Institution.

Hodges, C., Moore, S., Lockee, B., Trust, T., & Bond, A. (2020). The difference between emergency remote teaching and online learning. *EDUCAUSE Review*, *27*, •••. https://er.educause.edu/articles/2020/3/the-difference-between-emergency-remote-teaching-and-online-learning

hooks, b. (1994). *Teaching to transgress: Education as the practice of freedom.* Routledge.

Horn, M. (2024). The European green deal, retail investors and sustainable investments: A perspective article covering economic, behavioral, and regulatory insights. *Current Research in Environmental Sustainability*, *7*, 100241. Advance online publication. DOI: 10.1016/j.crsust.2024.100241

Huang, R., Tlili, A., Yang, J., & Chang, T.-W. (2020). The Chinese experience in maintaining undisrupted learning in COVID-19 outbreak. In *Handbook on Facilitating Flexible Learning During Educational Disruption* (Issue March). Beijing: Smart Learning Institute of Beijing Normal University. https://www.researchgate.net/publication/339939064

Huang, J., Pinmanee, S., & Chaveesuk, S. (2025). Acceptance and Use of Digital Education Platform: An Application of UTAUT2. *Operational Research in Engineering Sciences: Theory and Applications*, *8*(1), 237–264.

Huang, L. (2023). Ethics of artificial intelligence in education: Student privacy and data protection. *Science Insights Education Frontiers*, *16*(2), 2577–2587. DOI: 10.15354/sief.23.re202

Huang, X., Zou, D., Cheng, G., & Xie, H. (2021). A systematic review of ar and vr enhanced language learning. *Sustainability (Basel)*, *13*(9), 4639. DOI: 10.3390/su13094639

Hunter, A. (2023). Transformative learning in international education. In *Developing intercultural competence and transformation* (pp. 92–107). Routledge. DOI: 10.4324/9781003444169-7

Hussain, I., Angaiz, D., & Jan, S. (2021). Relationship between prospective teachers' attitude towards information and communication technology (ict) and its use for educational purposes in gilgit-baltistan. *Pakistan Journal of Social Research*, *03*(04), 398–411. DOI: 10.52567/pjsr.v3i4.302

Hussain, S. M. A. (2023). Actualising soft power through cultural diplomacy: A media, education, and communication perspective. *Dogo Rangsang Research Journal*, *13*(6), 196–205.

Ibrahim, U. (2024). Assessing the Impact of Mobile Applications on Student Engagement in ICT and Computer Science Education. [IJAER]. *International Journal of Applied Educational Research*, *2*(1), 51–62. Advance online publication. DOI: 10.59890/ijaer.v2i1.1457

Isaias, P., Miranda, P., & Pífano, S. (2017). Emotional presence in e-learning systems. *Information Systems Frontiers*, *19*(4), 747–763. DOI: 10.1007/s10796-015-9614-x

Iskrenović-Momčilović, O., & Momčilović, A. (2021). Ict in teaching for children with intellectual disabilities. *Journal of Intellectual Disabilities*, *26*(2), 365–374. DOI: 10.1177/1744629521995376 PMID: 33827310

Islam, M. (2015). Impact of ict on women empowerment in south asia. *Journal of Economic & Financial Studies*, *3*(03), 80. DOI: 10.18533/jefs.v3i03.166

Islamov, R., Greenwald, O., & Tunyova, N. (2020). Information and communication technologies in foreign language teaching in multinational higher education environment of mining region. E3s Web of Conferences, 174, 04042. DOI: 10.1051/e3sconf/202017404042

Ismail, M., Ismail, I., & Razali, N. (2017). Video animation as teaching aid for engineering drawing course in Malaysia vocational college. *International Journal of Trend in Research and Development*, *4*(5), 475–479.

Jeyaseelan, S. (2025). Vendor Lock-In Issues In Cloud Computing and How to Neutralize Them. In M. Muchnick, *A Dissertation Presented in Fulfillment of the Requirements for the Degree Doctor of Philosophy*.

Joseph, O. B., Onwuzulike, O. C., & Shitu, K. (2024). Digital transformation in education: Strategies for effective implementation. *World Journal of Advanced Research and Reviews*, *23*(2), 2785–2799. DOI: 10.30574/wjarr.2024.23.2.2668

Joshipura, M., Mathur, S., & Kedia, N. (2024). Sustainable investing and financing for sustainable development: A hybrid review. In *Sustainable Development*. DOI: 10.1002/sd.2912

Josué, A., Bedoya-Flores, M. C., Mosquera-Quiñonez, E. F., Bautista-Sánchez, J. V., & Bautista-Sánchez, J. V. (2023). Educational platforms: Digital tools for the teaching-learning process in education. *Ibero-American Journal of Education & Society Research*, *3*(1), 259–263. DOI: 10.56183/iberoeds.v3i1.626

Juškevičiene, A., & Dagiene, V. (2018). Computational thinking relationship with digital competence. *Informatics in Education*, *17*(2), 265–284. DOI: 10.15388/infedu.2018.14

Kabudi, T., Pappas, I., & Olsen, D. H. (2021). AI-enabled adaptive learning systems: A systematic mapping of the literature. *Computers and Education: Artificial Intelligence*, *2*, 100017. DOI: 10.1016/j.caeai.2021.100017

Kaewsaeng-on, R., & Mehmood, S. (2024). Quantile modeling for environmental risk: SAARC's journey with green finance, policies, and regulations. *Journal of Cleaner Production*, *434*, 140234. Advance online publication. DOI: 10.1016/j.jclepro.2023.140234

Kasturi, D., Lakshmi, D. V. G., Naidu, D. M. P., & Sheela, P. (2024). Evolution of the Green Investment: Prospectus to the Companies. *Migration Letters : An International Journal of Migration Studies*, *21*(4), 581–590. Advance online publication. DOI: 10.59670/ml.v21i4.7534

Kayyali, M. (2021). Positive impact of high technology on higher education.

Kayyali, M. (2022). Post-COVID-19 Pandemic Main Challenges with e-Learning.

Kayyali, M. (2025 a). E-Learning Platforms and Their Influence on Lifelong Learning. In Impact of Digitalization on Education and Social Sustainability (pp. 155-176). IGI Global.

Kayyali, M. (2025 b). E-Learning Platforms and Their Influence on Lifelong Learning. In Impact of Digitalization on Education and Social Sustainability (pp. 155-176). IGI Global.

Kayyali, M. (2020). Post COVID-19: New era for higher education systems. *International Journal of Applied Science and Engineering*, *8*(2), 131–145. DOI: 10.30954/2322-0465.2.2020.6

Kayyali, M. (2022 a). *Universities of the Future: HEIs that are Oriented to Accreditation*. Quality Assurance, Research, and Ranking.

Kayyali, M. (2022b). Equity, Equality, Diversity, and Inclusion as Key Performance Indicators in Higher Education. *Online Submission*, *3*(5), 1–3.

Kayyali, M. (2023). Importance of Rankings in Academia. *Online Submission*, *4*(1), 1–3.

Kayyali, M. (2025 a). Leadership's Role in Fostering a Resilient Organizational Culture. In *Enhancing Resilience in Business Continuity Management* (pp. 267–294). IGI Global Scientific Publishing. DOI: 10.4018/979-8-3693-8809-9.ch009

Kayyali, M. (2025 b). The Impact of Personalized Learning on Student Engagement and Achievement in STEAM. In *Integrating Personalized Learning Methods Into STEAM Education* (pp. 51–78). IGI Global Scientific Publishing. DOI: 10.4018/979-8-3693-7718-5.ch003

Kayyali, M. (Ed.). (2024). *Building Resiliency in Higher Education: Globalization, Digital Skills, and Student Wellness: Globalization, Digital Skills, and Student Wellness*. IGI Global.

Kennedy, L. J., Richdale, A. L., & Lawson, L. P. (2025). A comparison of neurodivergent and non-neurodivergent students' characteristics and higher education experiences in Australia and New Zealand. *Autism in Adulthood: Challenges and Management*, aut.2024.0159. DOI: 10.1089/aut.2024.0159

Khadka, K. (2021). Prospects and challenges faced by secondary teachers while integrating ict in nepalese esl classrooms. *Dristikon a Multidisciplinary Journal*, *11*(1), 89–99. DOI: 10.3126/dristikon.v11i1.39141

Khan, K. M., & Malluhi, Q. (2010). Establishing trust in cloud computing. *IT Professional, 12*(5), 20–27. DOI: 10.1109/MITP.2010.128

Khan, M. A. (2016). A survey of security issues for cloud computing. *Journal of Network and Computer Applications, 71*, 11–29. DOI: 10.1016/j.jnca.2016.05.010

Khayer, A., Bao, Y., & Nguyen, B. (2020). Understanding cloud computing success and its impact on firm performance: An integrated approach. *Industrial Management & Data Systems, 120*(5), 963–985. DOI: 10.1108/IMDS-06-2019-0327

Kincheloe, J. L. (2005). *Critical constructivism primer*. Peter Lang.

Kincheloe, J. L., & McLaren, P. (2005). Rethinking critical theory and qualitative research. In Denzin, N. K., & Lincoln, Y. S. (Eds.), *The SAGE handbook of qualitative research* (3rd ed., pp. 303–342). SAGE.

Kithungu, R., Gakunga, D., Nungu, M., & Nungu, M. (2021). A teaching strategy based upon a model of agentic learning. *Journal of Studies in Education, 11*(1), 58. DOI: 10.5296/jse.v11i1.17285

Kiwonde, F. (2023). E-learning training as a strategy for teachers' professional development in tanzania: A study of selected primary schools in musoma district, tanzania. *Asian Research Journal of Arts & Social Sciences, 21*(4), 43–53. DOI: 10.9734/arjass/2023/v21i4491

Kjellsdotter, A. (2020). What matter(s)? a didactical analysis of primary school teachers' ict integration. *Journal of Curriculum Studies, 52*(6), 823–839. DOI: 10.1080/00220272.2020.1759144

Klosi, I. (2024). THE TEACHER OF THE FUTURE: VISION AND SKILLS. *KNOWLEDGE-International Journal, 64*(5), 609–613.

Knight, J., Dooly, M., & Barberà, E. (2023). Getting smart: Towards critical digital literacy pedagogies. *Social Semiotics, 33*(2), 326–349. DOI: 10.1080/10350330.2020.1836815

Koedinger, K. R., Corbett, A. T., & Perfetti, C. (2012, April 9). The Knowledge-Learning-Instruction Framework: Bridging the Science-Practice Chasm to Enhance Robust Student Learning. *Cognitive Science, 36*(5), 757–798. DOI: 10.1111/j.1551-6709.2012.01245.x

Kojayan, A., Statti, A. L., & Torres, K. (2021). The Effects of Technology Integration in the Classroom for Students With ADHD. *International Journal of Curriculum Development and Learning Measurement, 11*(1), 1–10. DOI: 10.4018/IJCDLM.2021010101

Konrad, V. (2021). New directions at the post-globalization border. *Journal of Borderlands Studies*, *36*(5), 713–726. DOI: 10.1080/08865655.2021.1980733

Konyrova, L. (2024). The evolution of language learning: Exploring ai's impact on teaching english as a second language. *Eurasian Science Review*, *2*(2), 133–138. DOI: 10.63034/esr-42

Kshetri, N. (2010, January). Cloud computing in developing economies: drivers, effects, and policy measures. In *Proceedings of PTC 10*, 1-22.

Kumar, S., & Priyanka. (2024). The Effects of Information and Communication Technology (ICT) on Pedagogy and Student Learning Outcome in Higher Education. *EAI Endorsed Transactions on Scalable Information Systems, 11*(2). DOI: 10.4108/eetsis.4629

Kurniasari, N., Widodo, J., & Yappi, S. (2022). The effectiveness of youtube on elementary school students' reading skills during the covid-19 pandemic. *Jurnal Basicedu*, *6*(3), 3840–3846. DOI: 10.31004/basicedu.v6i3.2756

Kurti, E., Ferati, M., & Kalonaityte, V. (2024). Closing the gender gap in ICT higher education: Exploring women's motivations in pursuing ICT education. *Frontiers in Education*, *9*, 1352029. Advance online publication. DOI: 10.3389/feduc.2024.1352029

Lacelle-Webster, A. (2023). A Democratic Theory of Hope: Collective Agency in Uncertain Times (Doctoral dissertation, UNIVERSITY OF BRITISH COLUMBIA (Vancouver).

Lakshminarayanan, R., Kumar, B., & Raju, M. (2013). Cloud computing benefits for educational institutions. *arXiv (Cornell University)*. https://doi.org//arxiv.1305 .2616DOI: 10.48550

Lalwani, V. (2024). Incorporating green assets in equity portfolios. *Finance Research Letters*, *59*, 104815. Advance online publication. DOI: 10.1016/j.frl.2023.104815

Lamb, A. J., & Weiner, J. M. (2021). Technology as infrastructure for change: district leader understandings of 1: 1 educational technology initiatives and educational change. *Journal of Educational Administration*, *59*(3), 335–351. DOI: 10.1108/JEA-10-2020-0220

Lane, J. N., Leonardi, P. M., Contractor, N. S., & DeChurch, L. A. (2024). Teams in the digital workplace: Technology's role for communication, collaboration, and performance. *Small Group Research*, *55*(1), 139–183. DOI: 10.1177/10464964231200015

Latorre Cosculluela, C., Liesa-Orús, M., & Vázquez-Toledo, S. (2018). Escuelas inclusivas: Aprendizaje cooperativo y TAC con alumnado con TDAH. *MAGIS Revista Internacional de Investigación en Educación, 10*(21), 137–152. DOI: 10.11144/Javeriana.m10-21.eatt

Latorre Cosculluela, C., & Quintas Hijós, A. (2022). *Inclusión educativa y tecnologías para el aprendizaje*. Ediciones Octaedro. DOI: 10.36006/16298-4

Laubheimer, P. (2018). *Beyond the NPS: Measuring perceived usability with the SUS, NASA-TLX, and the single ease question after tasks and usability tests experience*. Nielsen Norman Group: World Leaders in Research-Based User. https://www.nngroup.com/articles/measuring-perceived-usability/

Ledgerwood, A., Lawson, K. M., Kraus, M. W., Vollhardt, J. R., Remedios, J. D., Westberg, D. W., Uskul, A. K., Adetula, A., Leach, C. W., Martinez, J. E., Naumann, L. P., Reddy, G., Chucky Tate, C., Todd, A. R., Weltzien, K., Buchanan, N. C., González, R., Montilla Doble, L. J., Romero-Canyas, R., & Zou, L. X. (2024). Disrupting racism and global exclusion in academic publishing: Recommendations and resources for authors, reviewers, and editors. *Collabra. Psychology, 10*(1), 121394. DOI: 10.1525/collabra.121394

Le, H., Janssen, J., & Wubbels, T. (2017). Collaborative learning practices: Teacher and student perceived obstacles to effective student collaboration. *Cambridge Journal of Education, 48*(1), 103–122. DOI: 10.1080/0305764X.2016.1259389

Lemay, D. J., Bazelais, P., & Doleck, T. (2021). Transition to online learning during the COVID-19 pandemic. *Computers in Human Behavior Reports, 4*, 100130. DOI: 10.1016/j.chbr.2021.100130 PMID: 34568640

Lens. Lens. Retrieved October, 2025, from https://www.lens.org/

Lera, F. J. R., González, D. F., Rico, F. M., Guerrero-Higueras, Á. M., & Conde, M. Á. (2021). Measuring students acceptance and usability of a cloud virtual desktop solution for a programming course. *Applied Sciences (Basel, Switzerland), 11*(15), 7157. DOI: 10.3390/app11157157

Liang, M., Lim, C., Park, J., & Mendoza, N. (2022). A review of ict-enabled learning for schoolgirls in asia and its impacts on education equity. *Educational Technology Research and Development, 71*(2), 267–293. DOI: 10.1007/s11423-022-10178-w PMID: 36533220

Liao, C. W., Chen, C. H., & Shih, S. J. (2019). The interactivity of video and collaboration for learning achievement, intrinsic motivation, cognitive load, and behavior patterns in a digital game-based learning environment. *Computers & Education, 133*, 43–55. DOI: 10.1016/j.compedu.2019.01.013

Lin, Y. L., & Wang, W. T. (2024). Enhancing students' online collaborative PBL learning performance in the context of co-authoring-based technologies: A case of wiki technologies. *Education and Information Technologies, 29*(2), 2303–2328. DOI: 10.1007/s10639-023-11907-1 PMID: 37361760

Liu, C. K. (2019). A holistic approach to flipped classroom: A conceptual framework using e-platform. *International Journal of Engineering Business Management, 11*, 1847979019855205. DOI: 10.1177/1847979019855205

Liu, G., & Liang, K. (2024). The role of technological innovation in enhancing resource sustainability to achieve green recovery. *Resources Policy, 89*, 104659. Advance online publication. DOI: 10.1016/j.resourpol.2024.104659

Li, W., Yu, S., Wang, M., Li, X., Ma, G., Ju, X., & Ling, C. (2024). How do life course events affect the accumulation of digital literacy?: Based on qualitative research of 16 medical university teachers of china. *Medicine, 103*(27), e38755. DOI: 10.1097/MD.0000000000038755 PMID: 38968540

Lockwood, J., & Mooney, A. (2017). *Computational thinking in education: Where does it fit? A systematic literary review. March.*

Lu, J. J., & Fletcher, G. H. L. (2009). Thinking about computational thinking categories and subject descriptors. *Sigcse*, 260–264.

Lubuva, E. E., Ndibalema, P., & Mbwambo, E. (2024). Designing ICT integrated lesson activities: An intervention to bolster tutors' pedagogical use of ICT competences in Tanzania teacher education. *Arab Gulf Journal of Scientific Research, 42*(4), 1917–1936. Advance online publication. DOI: 10.1108/AGJSR-05-2023-0191

Lu, D., & Xie, Y. N. (2024). The application of educational technology to develop problem-solving skills: A systematic review. *Thinking Skills and Creativity, 51*, 101454. DOI: 10.1016/j.tsc.2023.101454

Lusk, D. L., Evans, A. D., Jeffrey, T. R., Palmer, K. R., Wikstrom, C. S., & Doolittle, P. E. (2009). Multimedia learning and individual differences: Mediating the effects of working memory capacity with segmentation. *British Journal of Educational Technology, 40*(4), 636–651. DOI: 10.1111/j.1467-8535.2008.00848.x

LUTSIV, B., MAYOROVA, T., & LUTSIV, P. (2024). "GREEN FINANCE" IN THE PARADIGM OF SUSTAINABLE BANKING DEVELOPMENT OF THE ECONOMY OF UKRAINE. *WORLD OF FINANCE, 3(76).* DOI: 10.35774/sf2023.03.064

Ma, H. (2024). Role of green innovation and green finance on green recovery: Analysis of natural resources rents. *Resources Policy, 89*, 104571. Advance online publication. DOI: 10.1016/j.resourpol.2023.104571

Mahdum, M., Hadriana, H., & Safriyanti, M. (2019). Exploring teacher perceptions and motivations to ict use in learning activities in indonesia. *Journal of Information Technology Education, 18*, 293–317. DOI: 10.28945/4366

Malik, A., & Om, H. (2017). Cloud computing and Internet of Things Integration: architecture, applications, issues, and challenges. In *Springer eBooks.* 1–24. https://doi.org/DOI: 10.1007/978-3-319-62238-5_1

Martínez-Pérez, A., Lezcano-Barbero, F., Casado-Muñoz, R., & Zabaleta-González, R. (2024). ICT training in Spanish non-formal education: A revolution in the making. *European Journal of Social Work, 27*(1), 177–191. Advance online publication. DOI: 10.1080/13691457.2022.2162006

Maryanti, R., Nandiyanto, A. B., Hufad, A., & Sunardi, S. (2021). Science education for students with special needs in Indonesia: From definition, systematic review, education system, to curriculum. *Indonesian Journal of Community and Special Needs Education, 1*(1), 1–8. DOI: 10.17509/ijcsne.v1i1.32653

Maryuni, Y., Ribawati, E., Prasadana, M. A. F., Nurhasanah, A., Wibowo, T. U. S. H., Nashar, N., Fauzan, R., Putra, A. P., & Sari, I. J. (2024). The Effect of STEM-based History Podcast Media on Literacy in Digital Era through ICT Courses for History Education Students. *International Journal of STEM Education for Sustainability, 4*(1), 70–78. Advance online publication. DOI: 10.53889/ijses.v4i1.310

Mashelkar, R. A. (2018). Exponential technology, industry 4.0 and future of jobs in India. *Review of Market Integration, 10*(2), 1–20. DOI: 10.1177/0974929218774408

Mayer, R. E., & Chandler, P. (2001). When learning is just a click away: Does simple user interaction foster deeper understanding of multimedia messages? *Journal of Educational Psychology, 93*(2), 390–397. DOI: 10.1037/0022-0663.93.2.390

Mayer, R. E., Wells, A., Parong, J., & Howarth, J. T. (2019). Learner control of the pacing of an online slideshow lesson: Does segmenting help? *Applied Cognitive Psychology, 33*(5), 930–935. DOI: 10.1002/acp.3560

Mazin, K. A., Norman, H., Nordin, N., & Ibrahim, R. (2020). Student self-recording videos for TVET competency in MOOCs. *Journal of Physics: Conference Series, 1529*(4), 042061. Advance online publication. DOI: 10.1088/1742-6596/1529/4/042061

McCarthy, A. M., Maor, D., McConney, A., & Cavanaugh, C. (2023). Digital transformation in education: Critical components for leaders of system change. *Social Sciences & Humanities Open, 8*(1), 100479. DOI: 10.1016/j.ssaho.2023.100479

McCaslin, S. E., & Young, M. (2015). Increasing student motivation and knowledge in mechanical engineering by using action cameras and video productions. *Advances in Production Engineering & Management, 10*(2), 87–96. DOI: 10.14743/apem2015.2.194

McCosker, A., & Wilken, R. (2012, May 2). Café Space, Communication, Creativity, and Materialism. *M/C Journal, 15*(2). https://doi.org/DOI: 10.5204/mcj.459

Means, B., Toyama, Y., Murphy, R., Bakia, M., & Jones, K. (2009). Evaluation of evidence-based practices in online learning: A meta-analysis and review of online learning studies. In *U.S. Department of Education*. www.ed.gov/about/offices/list/opepd/ppss/reports.html

Mehta, P. S., Shedge, P. P., Vyavahare, S., Soundarrajan, S., & Chillarge, G. (2024). Next-Gen Learning cloud integration for academic and industrial synergy. In Mahalle, P. N. (Ed.), *Advances in educational technologies and instructional design book series* (pp. 73–108)., DOI: 10.4018/979-8-3693-6705-6.ch004

Mendonça dos Santos, R. (2024). Teachers' narratives of resisting deprofessionalization (Doctoral dissertation, University of British Columbia).

Menon, P., Samaddar, K., & Aparna, K. H. (2023). Towards globalisation of B-schools: Best practices for sustainable higher education. *International Journal of Business and Globalisation, 33*(1-2), 130–146. DOI: 10.1504/IJBG.2023.128326

Mezirow, J. (1991). *Transformative dimensions of adult learning*. Jossey-Bass.

Mhlongo, S., Mbatha, K., Ramatsetse, B., & Dlamini, R. (2023). Challenges, opportunities, and prospects of adopting and using smart digital technologies in learning environments: An iterative review. *Heliyon, 9*(6), e16348. DOI: 10.1016/j.heliyon.2023.e16348 PMID: 37274691

Michael, E., & Shah, P. (2020). Students' perception on youtube usage in rural esl classroom. *International Journal of Academic Research in Business & Social Sciences, 10*(3). Advance online publication. DOI: 10.6007/IJARBSS/v10-i3/7059

Miles, J. (2024). Guilt, complicity, and responsibility for historical injustice: Towards a pedagogy of complex implication. *Pedagogy, Culture & Society*, *32*(3), 619–635. DOI: 10.1080/14681366.2022.2064537

Millefiorini, A. (Ed.). (2025). *Democracies on the Move: Contributions to a Social Theory of Contemporary Democracies*. Taylor & Francis. DOI: 10.4324/9781032722528

Min, A. (2023). Artificial intelligence and bias: Challenges, implications, and remedies. *Journal of Social Research*, *2*(11), 3808–3817. DOI: 10.55324/josr.v2i11.1477

Ministry of Higher Education. (2017). TVET for human resource development. In *Technical Vocational Education & Training (Tvet) in Malaysia : Selected Works*. http://mycc.my/document/files/PDF Dokumen/Technical Vocational Education %26 Training (TVET) in Malaysia Selected Works.pdf

Mohammed, R. (2018). A proposed program in action research based on project-based learning via the web to develop research awareness and reduce teaching anxiety among female mathematics teachers-in-training. *Journal of Mathematics Education*, *21*(4), 168–223.

Mondal, S., & Goswami, S. S. (2024). A narrative literature review on the economic impact of cloud computing: Opportunities and challenges. *Computing and AI.*, *1934*, 1934. Advance online publication. DOI: 10.59400/cai1934

Moreno, R., & Mayer, R. (2007). Interactive multimodal learning environments: Special issue on interactive learning environments: Contemporary issues and trends. *Educational Psychology Review*, *19*(3), 309–326. DOI: 10.1007/s10648-007-9047-2

Moro, S. (2021). Educational Intervention Strategies in the English Language Classroom for Students With Attention-Deficit/Hyperactivity Disorder in Primary Education. *Revista Electrónica Educare*, *25*(3), 1–20. DOI: 10.15359/ree.25-3.20

Morrison-Smith, S., & Ruiz, J. (2020). Challenges and barriers in virtual teams: A literature review. *SN Applied Sciences*, *2*(6), 1096. DOI: 10.1007/s42452-020-2801-5

Mukul, E., & Büyüközkan, G. (2023). Digital transformation in education: A systematic review of Education 4.0. *Technological Forecasting and Social Change*, *194*, 122664. DOI: 10.1016/j.techfore.2023.122664

Mulaudzi, L. (2024). Evaluating impact: South Africa's e-education white paper on digital equity and post-COVID education leadership. *Cogent Education*, *11*(1), 2415749. Advance online publication. DOI: 10.1080/2331186X.2024.2415749

Munje, P., & Jita, T. (2020). The impact of the lack of ICT resources on teaching and learning in selected South African primary schools. International Journal of Learning. *Teaching and Educational Research, 19*(7), 263–279. DOI: 10.26803/ijlter.19.7.15

Murphy, C., Martin, M., & Parmar, N. (2022). Designing for emotion in online adult education. *Adult Education Quarterly, 72*(1), 34–56. DOI: 10.1177/07417136221080742

Murtaza, M., Ahmed, Y., Shamsi, J., Sherwani, F., & Usman, M. (2022). Ai-based personalized e-learning systems: Issues, challenges, and solutions. *IEEE Access : Practical Innovations, Open Solutions, 10*, 81323–81342. DOI: 10.1109/ACCESS.2022.3193938

Mustafa, A., Taha, N., Al-Shboul, O., Alsalem, M., & Malki, M. (2020). Using youtube to learn anatomy: perspectives of jordanian medical students. Biomed Research International, 2020(1). DOI: 10.1155/2020/6861416

Mustofa, M., & Sari, A. (2020). Video subtitle to teach listening skill of junior high school students. [Journal of English Educators Society]. *Jees, 5*(2), 149–153. DOI: 10.21070/jees.v5i2.874

Nadiyah, R. S., & Faaizah, S. (2015). The development of online project based collaborative learning using ADDIE model. *Procedia: Social and Behavioral Sciences, 195*, 1803–1812. DOI: 10.1016/j.sbspro.2015.06.392

Nadzri, A., Ayub, A., Zulkifli, N., & Salim, N. (2024). Implications of AR Modules on Geometry Conceptual and Procedural Knowledge among Primary School Students. *Malaysian Journal of Mathematical Sciences, 18*(1), 51–72. DOI: 10.47836/mjms.18.1.04

Naveed, Q. N., Alam, M. M., Qahmash, A. I., & Quadri, K. M. (2021). Exploring the determinants of service quality of cloud E-Learning system for active system usage. *Applied Sciences (Basel, Switzerland), 11*(9), 4176. DOI: 10.3390/app11094176

Ndlovu-Gatsheni, S. J. (2018). *Epistemic freedom in Africa: Deprovincialization and decolonization*. Routledge., DOI: 10.4324/9780429492204

Neenan, M., & Dryden, W. (2021). Understanding and developing resilience. In *Cognitive Behavioural Coaching in Practice* (pp. 199–220). Routledge. DOI: 10.4324/9781003027164-9

Nhat Hanh, T. (1998). *Interbeing: Fourteen guidelines for engaged Buddhism*. Parallax Press.

Nielsen, J. (2012). *How many test users in a usability study*. https://www.nngroup.com/articles/how-many-test-users/

Nikita (2023) Advantages and Disadvantages of Artificial Intelligence. Simplilearn. https://www.simplilearn.com/advantages-and-disadvantages-of-artificial-intelligence-article

Nishnianidze, A. (2024). Surveillance in the digital age. *European Scientific Journal*, *20*(37), 1–25. DOI: 10.19044/esj.2024.v20n37p1

Noddings, N. (2013). *Caring: A relational approach to ethics and moral education* (2nd ed.). University of California Press.

Nohong, M., Sobarsyah, M., Sohilauw, M. I., & Herman, B. (2024). Nexus Between Strategic Green Finance and Green Competitive Advantage: Study at Indonesia Corporate. *Journal of Environmental Assessment Policy and Management*, *26*(2), 2450003. Advance online publication. DOI: 10.1142/S1464333224500030

Novak, E., McDaniel, K., & Li, J. (2023). Factors that impact student frustration in digital learning environments. *Computers and Education Open*, *5*, 100153. DOI: 10.1016/j.caeo.2023.100153

Nst, A., Daulay, S., & Dewi, U. (2023). Using rosetta stone application as a mobile assited language learning (mall) in efl: documentary study. Scope Journal of English Language Teaching, 8(1), 08. DOI: 10.30998/scope.v8i1.17653

Odhiambo, M. O. (2024). Availability of academic resources presents opportunities and challenges. In *Advances in linguistics and communication studies*. 435–458. https://doi.org/DOI: 10.4018/979-8-3693-3579-6.ch017

OECD. (2023). PISA 2022 Results: Vol. I. *The State of Learning and Equity in Education*. OECD Publishing.

Ofosu-Asare, Y. (2024). Developing classroom ict teaching techniques, principles and practice for teachers in rural ghana without access to computers or internet: A framework based on literature review. *International Journal of Information and Learning Technology*, *41*(3), 262–279. DOI: 10.1108/IJILT-04-2023-0045

Ok, E., Grace, J., & John, M. (2024). *Collaborative Learning Tools*. ResearchGate.

Omboto, C., Kanga, A., & Njageh, A. (2022). Myth or reality: Digital literacy programme implementation in primary special schools in nairobi, kenya. *European Journal of Education*, *5*(2), 51–66. DOI: 10.2478/ejed-2022-0014

Othman, W., Makrakis, V., Kostoulas-Makrakis, N., Hamidon, Z., Keat, O. C., Abdullah, M. L., Shafie, N., & Mat, H. (2024). Predictors of Motivation and Barriers to ICT-Enabling Education for Sustainability. *Sustainability (Basel), 16*(2), 749. Advance online publication. DOI: 10.3390/su16020749

Owusu-Boampong, A., & Holmberg, C. (2015). *Distance education in European higher education – The state of play.* UNESCO Institute for Lifelong Learning.

Oyedotun, T. D. (2024). Digital pedagogy: The role of e-learning tools in facilitating 21st century skills. *International Journal of Innovation in Education, 9*(1), 32–46. DOI: 10.1504/IJIIE.2024.144155

Ozalp, H., Ozcan, P., Dinckol, D., Zachariadis, M., & Gawer, A. (2022). "Digital colonization" of highly regulated industries: An analysis of big tech platforms' entry into health care and education. *California Management Review, 64*(4), 78–107. DOI: 10.1177/00081256221094307

Özmacun, T. E. (2023). A critical perspective on intellectual property rights through the lens of the free and open-source software experience (Doctoral dissertation, Middle East Technical University (Turkey)).

Padhy, R. P., Patra, M. R., & Satapathy, S. C. (2011). Cloud computing: security issues and research challenges. In *IRACST - International Journal of Computer Science and Information Technology & Security*. IJCSITS.

Padmanaban, H. (2024). *Quantum computing and AI in the cloud.* Journal of Computational Intelligence and Robotics.

Papert, S. (1980). MINDSTORMS: Children, Computers, and Powerful Ideas. In *Basic Books* (Vol. 66).

Parmaxi, A., & Demetriou, A. (2020). Augmented reality in language learning: A state-of-the-art review of 2014–2019. *Journal of Computer Assisted Learning, 36*(6), 861–875. DOI: 10.1111/jcal.12486

Paskevicius, M. (2021). Educators as content creators in a diverse digital media landscape.

Paudel, P. (2020). Information and communication technology in foreign language classes in english: Roles and practices. *International Journal of Technology in Education and Science, 5*(1), 37–55. DOI: 10.46328/ijtes.179

Paudel, P. (2022). Challenges of using information and communication technology in teaching english in nepal. *KMC Journal, 4*(2), 1–18. DOI: 10.3126/kmcj.v4i2.47725

Paul, S., & Das, D. (2025). Future Trends and Innovations in Cloud-Driven Education. In *Advances in educational technologies and instructional design book series*. 321–346. https://doi.org/DOI: 10.4018/979-8-3693-6705-6.ch013

Peixoto, B., Pinto, D., Krassmann, A., Melo, M., Cabral, L., & Bessa, M. (2019). Using virtual reality tools for teaching foreign languages., 581-588. DOI: 10.1007/978-3-030-16187-3_56

Pendy, B. (2023). Artificial intelligence: The future of education. *Jurnal Indonesia Sosial Sains*, *2*(11). Advance online publication. DOI: 10.59141/jiss.v2i11.801

Peng, R., Razak, R., & Halili, S. (2024). Integration of information and communication technology in teaching english among in-service teachers. *International Journal of Evaluation and Research in Education (Ijere)*, *13*(2), 1235. DOI: 10.11591/ijere.v13i2.26347

Perales, F., & Ballespí, S. (2023). Prevalencia y manifestaciones clínicas del TDAH en población infantil. *Journal of Child Psychology and Psychiatry, and Allied Disciplines*, *45*(1), 12–24.

Pérez-Jorge, D., González-Herrera, A. I., Alonso-Rodríguez, I., & Rodríguez-Jiménez, M. del C. (2024). Challenges and Opportunities in Inclusive Education with ICT: Teachers' Perspectives in the Canary Islands during the COVID-19 Pandemic. *Education Sciences*, *14*(3), 283. Advance online publication. DOI: 10.3390/educsci14030283

Pérez, M. P., Pesek, I., Zmazek, B., & Lipovec, A. (2020). Video explanations as a useful digital source of education in the COVID 19 situation. *Journal of Elementary Education*, *13*(4), 395–412. DOI: 10.18690/rei.13.4.395-412.2020

Pezo, E. (2024). Adoption Of ICT By Basic Education Teachers: A Systematic Review. *Journal of Namibian Studies : History Politics Culture*, *40*, 314–331. Advance online publication. DOI: 10.59670/3tn2an17

Pilamunga Valla, E. V., Jiménez Pacheco, N. J., Mendiburu Rojas, A. F., & Medina Gonzáles, R. H. (2021). Las TICS y su influencia en el proceso de aprendizaje en los estudiantes con TDAH. *Journal of Science and Research*, *7*, 1163–1180.

Pillow, W. (2003). Confession, catharsis, or cure? *International Journal of Qualitative Studies in Education : QSE*, *16*(2), 175–196. DOI: 10.1080/0951839032000060635

Polonetsky, J., & Tene, O. (2014, July 1). The Ethics of Student Privacy: Building Trust for Ed Tech. *International Journal of Information Ethics*, *21*, 25–34. DOI: 10.29173/irie367

Ponde, P. (2024). Security and privacy considerations in Cloud-Based education. In *Advances in educational technologies and instructional design book series*. 175–194. https://doi.org/DOI: 10.4018/979-8-3693-6705-6.ch007

Porozovs, J., Dudkina, A., & Valdemiers, A. (2019). The use of information and communication technologies of pedagogical specialities students in the study process., 104-110. DOI: 10.22616/REEP.2019.013

Posner, T., & Fei-Fei, L. (2020). AI will change the world, so it's time to change A. *Nature*, *588*(7837), S118–S118. DOI: 10.1038/d41586-020-03412-z

Postholm, M. B. (2012). Teachers' professional development: A theoretical review. *Educational Research*, *54*(4), 405–429. DOI: 10.1080/00131881.2012.734725

Poudel, A. (2022). Information and communication technology in english language teaching: Some opportunities and challenges. *Journal of Comparative & International Higher Education*, *14*(4). Advance online publication. DOI: 10.32674/jcihe.v14i4.3874

Power, D. J., Heavin, C., & O'Connor, Y. (2021). Balancing privacy rights and surveillance analytics: A decision process guide. *Journal of Business Analytics*, *4*(2), 155–170. DOI: 10.1080/2573234X.2021.1920856

Pratama, A., Faroqi, A., & Mandyartha, E. (2023). Design of integrated school learning information system using codeigniter framework.. DOI: 10.11594/nstp.2023.3326

Pregowska, A., Masztalerz, K., Garlińska, M., & Osial, M. (2021). A worldwide journey through distance education—From the post office to virtual, augmented and mixed realities, and education during the COVID-19 pandemic. *Education Sciences*, *11*(3), 118. DOI: 10.3390/educsci11030118

Prensky, M. (2009). H . Sapiens digital: From digital immigrants and digital natives to digital wisdom. *Innovate: Journal of Online Education.*, *5*(3).

Preuveneers, D., Garofalo, G., & Joosen, W. (2021). Cloud and edge based data analytics for privacy-preserving multi-modal engagement monitoring in the classroom. *Information Systems Frontiers*, *23*(1), 151–164. DOI: 10.1007/s10796-020-09993-4

Qasem, Y. M., Abdullah, R., Yaha, Y., & Atana, R. (2020). Continuance Use of Cloud computing in Higher Education Institutions: A Conceptual model. *Applied Sciences (Basel, Switzerland)*, *10*(19), 6628. DOI: 10.3390/app10196628

Qashou, A. M. A., Bahar, N., & Mohamed, H. (2025). Qualitative exploration of data security risks in mobile cloud computing for higher education. *Security and Privacy*, *8*(2), e70001. Advance online publication. DOI: 10.1002/spy2.70001

Quintero Gutiérrez del Álamo, F. J. (2019). *Actualización en el manejo del TDAH*. Lúa Ediciones.

Raab, C. D. (2020). Information privacy, impact assessment, and the place of ethics. *Computer Law & Security Review*, *37*, 105404. DOI: 10.1016/j.clsr.2020.105404

Ragi, T. (2024). Unveiling influence: Assessing the effect of ict tools on student activities through knowledge management processes. *CRR*, *1*(2), 50–57. DOI: 10.21844/crr.v1i02.1112

Rahimi, N., Rashidi, A., Hakimi, M., & Shirpoor, M. (2024). Navigating E-Governance and ICT Dynamics in Afghanistan Higher Education. *International Journal of Scientific Multidisciplinary Research*, *3*(2), 137–152. Advance online publication. DOI: 10.55927/ijsmr.v3i2.8011

Rahimi, R. A., & Oh, G. S. (2024). Rethinking the role of educators in the 21st century: Navigating globalization, technology, and pandemics. *Journal of Marketing Analytics*, *12*(2), 182–197. DOI: 10.1057/s41270-024-00303-4

Raj, S., Ann, W., Subramaniam, P., & Yunus, M. (2019). Using youtube as a platform to learn social expression. *Creative Education*, *10*(02), 288–296. DOI: 10.4236/ce.2019.102023

Rantyuk, I. I., & Vakaliuk, T. A. (2024). Selecting project management ICT tools for informal education of IT professionals. *Educational Dimension*, *10*, 167–197. Advance online publication. DOI: 10.55056/ed.630

Rashid, S., Yasin, M., & Ashaari, N. (2019). Undergraduate students of special education's readiness towards the use of information and technology (ict) in teaching and learning the sign language. *Creative Education*, *10*(11), 2374–2385. DOI: 10.4236/ce.2019.1011169

Raza, S. A., & Khan, K. A. (2022). Knowledge and innovative factors: How cloud computing improves students' academic performance. *Interactive Technology and Smart Education*, *19*(2), 161–183. DOI: 10.1108/ITSE-04-2020-0047

Reddy, K. S., & Babu, N. S. (2024). Navigating the digital divide: Assessing faculty ICT skills and challenges in shaping the future of higher education in India. *Brazilian Journal of Development*, *10*(1), 355–378. Advance online publication. DOI: 10.34117/bjdv10n1-023

Riga, A., Ifanti, T., & Papayiannis, N. (2019). Recommended educational approaches for teaching children with Attention Deficit Hyperactivity Disorder (ADHD) coordination and organization skills. *IJRDO-Journal of Educational Research*, *2*(12), 1–14.

Rimal, B. P., Jukan, A., Katsaros, D., & Goeleven, Y. (2011). Architectural Requirements for cloud computing Systems: An Enterprise Cloud approach. *Journal of Grid Computing, 9*(1), 3–26. DOI: 10.1007/s10723-010-9171-y

Robles Villalba, F. J., & Reyes Moreno, E. R. (2024). Evaluando el impacto de las TIC's en estudiantes con TDAH. *Una Ciencia Revista de Estudios e Investigaciones, 17*(32), 68–83. DOI: 10.35997/unaciencia.v17i32.771

Rodgers, C., & LaBoskey, V. K. (2018). Digital video creation in the LSP classroom. *The EUROCALL Review, 26*(1), 71–104. DOI: 10.4995/eurocall.2018.9666

Rodrigo Yanguas, M., González Tardón, C., Bella Fernández, M., & Blasco Fontecilla, H. (2022). Serious Video Games: Angels or Demons in Patients with Attention-Deficit Hyperactivity Disorder. *Frontiers in Psychiatry, 13*, 1–19. DOI: 10.3389/fpsyt.2022.798480 PMID: 35573357

Rodríguez-Jiménez, M., Pérez-Jorge, D., Contreras, A., & Larumbe-Zabala, E. (2023). Development and validation of edu-cov-nee-neae questionnaire to measure educational response, ict usage and impact of covid-19 pandemic on students with special educational needs. *Heliyon, 9*(9), e19231. DOI: 10.1016/j.heliyon.2023.e19231 PMID: 37674819

Rodriguez, M. (2022). Artificial affect: A critique of empathy in algorithmic education. *Learning, Media and Technology, 47*(2), 210–227. DOI: 10.1080/17439884.2022.2029280

Roudometof, V. (2024). How should we think about globalization in a post-globalization era?. Dialogues in Sociology, 29768667241293053.

Roy, K., & Swargiary, K. (2023). *ICT Revolutionalizing Education*. Scholar's Press.

Rubiano Caballero, M. L., & Lozano Zárate, Y. Y. (2024). Educación inclusiva y TIC: Prácticas pedagógicas para el rendimiento escolar de estudiantes con TDAH en instituciones públicas. *Ciencia Latina Revista Científica Multidisciplinar, 8*(5), 9857–9878. DOI: 10.37811/cl_rcm.v8i5.14364

Runiasih, M., Rugaiyah, R., & Ika, L. (2025). Systematic Literature Review (SLR): Implementation of Flip Classroom using Learning Management System (LMS) to improve students' learning independence. *International Education Trend Issues, 3*(1), 1–8. DOI: 10.56442/ieti.v3i1.963

Ryn, A., & Sandaran, S. (2020). Teachers' practices and perceptions of the use of ict in elt classrooms in the pre-covid 19 pandemic era and suggestions for the 'new normal'. *LSP International Journal, 7*(1). Advance online publication. DOI: 10.11113/lspi.v7n1.100

Sabi, H. M., Uzoka, F. E., Langmia, K., & Njeh, F. N. (2016). Conceptualizing a model for adoption of cloud computing in education. *International Journal of Information Management*, *36*(2), 183–191. DOI: 10.1016/j.ijinfomgt.2015.11.010

Salas-Vallina, A., Rofcanin, Y., & Las Heras, M. (2022). Building resilience and performance in turbulent times: The influence of shared leadership and passion at work across levels. *Business Research Quarterly*, *25*(1), 8–27. DOI: 10.1177/23409444211035138

Salinas, I., Fernández, M. B., Johnson, D., & Bastías, N. (2023). Freire's hope in radically changing times: A dialogue for curriculum integration from science education to face the climate crisis. *Cultural Studies of Science Education*, *18*(1), 21–39. DOI: 10.1007/s11422-023-10157-4 PMID: 36845560

Samier, E. A. (2022). To surveil or not to surveil: Educational surveillance in populist and nationalist times. In Populism and Educational Leadership, Administration and Policy (pp. 188-205). Routledge.

Samyan, N., & St Flour, P. O. (2021). The impact of cloud computing on e-Learning during COVID-19 pandemic. [IJSES]. *International Journal of Studies in Education and Science*, *2*(2), 146–172.

Sancar, R., Atal, D., & Deryakulu, D. (2021). A new framework for teachers' professional development. *Teaching and Teacher Education*, *101*, 103305. DOI: 10.1016/j.tate.2021.103305

Sarva, E., & Purina-Bieza, K. E. (2023). Educators' perspectives on the main challenges and opportunities for implementing digital solutions in learning and teaching. International Journal of Emerging Technologies in Learning (iJET, 18(24), 62-80. https://doi.org/DOI: 10.3991/ijet.v18i24.44745

Sato, M., & Loewen, S. (2022). The research–practice dialogue in second language learning and teaching: Past, present, and future. *Modern Language Journal*, *106*(3), 509–527. DOI: 10.1111/modl.12791

Saud, D. (2023). Teachers and students' perceptions towards using ict in elt in model schools. Journal of Tikapur Multiple Campus, 78-93. DOI: 10.3126/jotmc.v6i01.56347

Sauro, J. (2011). *Measuring usability with the system usability scale (SUS)*. https://measuringu.com/sus/

Sawy, Y. (2019). The effect of using mathematical tales in teaching arithmetic operations to develop mathematical concepts and reduce math anxiety among primary school students. *Journal of Mathematics Education*, *22*(5), 117–141.

Sayeed Salih, O., Husain, M., Hamdan, M., Abdelsalam, S., Elshafie, H., & Motwakel, A. (2025). Transforming education with AI: A systematic review of ChatGPT's role in learning, academic practices, and institutional adoption. *Results in Engineering*, *25*, 103837. DOI: 10.1016/j.rineng.2024.103837

Scalera, M., Gentile, E., Plantamura, P., & Dimauro, G. (2020). A Systematic mapping study in cloud for Educational innovation. *Applied Sciences (Basel, Switzerland)*, *10*(13), 4531. DOI: 10.3390/app10134531

Schmid, R. F., Borokhovski, E., Bernard, R. M., Pickup, D. I., & Abrami, P. C. (2023). A meta-analysis of online learning, blended learning, the flipped classroom, and classroom instruction for pre-service and in-service teachers. *Computers and Education Open*, *5*, 100142. DOI: 10.1016/j.caeo.2023.100142

Sein-Echaluce, M. L., Fidalgo-Blanco, A., García-Peñalvo, F. J., & Fonseca, D. (2021). Impact of Transparency in the Teamwork Development through Cloud Computing. *Applied Sciences (Basel, Switzerland)*, *11*(9), 3887. DOI: 10.3390/app11093887

Selbst, A. D., & Barocas, S. (2018). The Intuitive Appeal of Explainable Machines. SSRN *Electronic Journal*. https://doi.org/DOI: 10.2139/ssrn.3126971

Selwyn, N. (2016). *Education and technology: Key issues and debates* (2nd ed.). Bloomsbury.

Shafik, W. (2024a). Enhancing AI-Enabled Education Through Mobile Learning and Bring Your Own Device (BYOD) Integration. In *Exploring Youth Studies in the Age of AI* (pp. 140–173). IGI Global.

Shafik, W. (2024b). Generative AI for Social Good and Sustainable Development. In *Generative AI: Current Trends and Applications* (pp. 185–217). Springer. DOI: 10.1007/978-981-97-8460-8_10

Shafik, W. (2024c). Mobile Learning and Bring Your Own Device (BYOD): Enhancing Education in the Digital Age. In *Integrating Cutting-Edge Technology Into the Classroom* (pp. 240–267). IGI Global.

Shafik, W. (2024d). Towards Trustworthy and Explainable AI Educational Systems. In *Explainable AI for Education: Recent Trends and Challenges* (pp. 17–41). Springer. DOI: 10.1007/978-3-031-72410-7_2

Sharma, P., Joshi, S., Gautam, S., Maharjan, S., Khanal, S. R., Reis, M. C., . . . de Jesus Filipe, V. M. (2022, August). Student engagement detection using emotion analysis, eye tracking and head movement with machine learning. In International conference on technology and innovation in learning, teaching and education (pp. 52-68). Cham: Springer Nature Switzerland. DOI: 10.1007/978-3-031-22918-3_5

Sharma, S. U. (2025). Using Cloud Technology to Maximise Resources for Education. *Journal of Scientific research and Technology, 3*(2), 33–36.

Sharma, C., & Sharma, C. (2024). Cloud Computing Security: Threats and Mitigation Strategies. In *2024 International Conference on Signal Processing and Advance Research in Computing (SPARC).* 1–6. IEEE. https://doi.org/DOI: 10.1109/SPARC61891.2024.10829335

Sharma, M., Vadalkar, S., Singh, A., Chavan, G., & Tsagarakis, K. P. (2025). How does cloud computing adoption accelerate education? – Enhancing academic results by integrating knowledge management pre and post pandemic. *International Journal of Human-Computer Interaction*, •••, 1–21. DOI: 10.1080/10447318.2025.2558030

Shin, M.-H. (2018). Effects of project-based learning on students' motivation and self-efficacy. *English Teaching*, *73*(1), 95–115. DOI: 10.15858/engtea.73.1.201803.95

Shkokani, R. (2021). *Mathematics distance learning competencies among mathematics teachers in the upper basic stage in Nablus Governorate* (Published Master's thesis). An-Najah National University.

Shrestha, S., Haque, S., Dawadi, S., & Giri, R. (2021). Preparations for and practices of online education during the covid-19 pandemic: A study of bangladesh and nepal. *Education and Information Technologies*, *27*(1), 243–265. DOI: 10.1007/s10639-021-10659-0 PMID: 34341654

Singh, A. (2024). The Future of Learning: AI-Driven Personalized Education. Available at *SSRN* 5076438. DOI: 10.56557/ajocr/2024/v9i49018

Singh, J., Steele, K., & Singh, L. (2021). Combining the best of online and face-to-face learning: Hybrid and blended learning approach for COVID-19, post vaccine, & post-pandemic world. *Journal of Educational Technology Systems*, *50*(2), 140–171. DOI: 10.1177/00472395211047865

Sintema, E., & Jita, T. (2022). Pre-service teachers' self-concept and views toward using ict for teaching science. *Eurasia Journal of Mathematics, Science and Technology Education, 18*(9), em2154. Advance online publication. DOI: 10.29333/ejmste/12396

Sitzmann, T., Kraiger, K., Stewart, D., & Wisher, R.SITZMANN. (2006, September). THE COMPARATIVE EFFECTIVENESS OF WEB-BASED AND CLASSROOM INSTRUCTION: A META-ANALYSIS. *Personnel Psychology*, *59*(3), 623–664. DOI: 10.1111/j.1744-6570.2006.00049.x

Smeplass, E. (2023). Nurturing inclusivity and professional growth among vocational teachers through communities of practice. *Pedagogy, Culture & Society*, , 1–20.

Snelson, C., Yang, D., & Temple, T. (2021). Addressing the challenges of online video analysis in qualitative studies: A worked example from computational thinking research. *The Qualitative Report*, *26*(6), 1974–1988. DOI: 10.46743/2160-3715/2021.4734

Sobkowiak, M., Senn, J., & Vollmer, H. (2023). Rethinking planetary boundaries: Accounting for ecological limits. *Social and Environmental Accountability Journal*, *43*(3), 259–272. DOI: 10.1080/0969160X.2023.2283019

Song, A., Wu, K., & Ding, J. (2024). Developing an immersive game-based learning platform with generative artificial intelligence and virtual reality technologies – "LearningverseVR." Computers & Education: X Reality, 4, 100069. DOI: 10.1016/j.cexr.2024.100069

Stanley, P. D., & Clinton, J. R. (2024). *Connecting: The mentoring relationships you need to succeed in life*. NavPress.

Stiller, K. D., Freitag, A., Zinnbauer, P., & Freitag, C. (2009). How pacing of multimedia instructions can influence modality effects: A case of superiority of visual texts. *Australasian Journal of Educational Technology*, *25*(2), 184–203. DOI: 10.14742/ajet.1149

Stiller, K. D., Petzold, K., & Zinnbauer, P. (2011). Presentation time concerning system-paced multimedia instructions and the superiority of learner pacing. *Australasian Journal of Educational Technology*, *27*(4), 693–708. DOI: 10.14742/ajet.945

Sugeng, S., Fitria, A., & Bastomi, S. (2024). Epistemic Injustice and Digital Disinformation: Addressing Knowledge Inequities in the Digital Age. *DISKURSUS-JURNAL FILSAFAT DAN TEOLOGI STF DRIYARKARA*, *20*(1), 135–169. DOI: 10.36383/diskursus.v20i1.483

Sun, Y., & Yan, Z. (2025). Factors influencing teachers' technology adoption in technology-rich classrooms: Model development and test. *Educational Technology Research and Development*, *73*(3), 1481–1496. Advance online publication. DOI: 10.1007/s11423-025-10454-5

Tan, C. Y., Jang, S. T., Lam, S. M., An, A. Q., & Lo, U. K. V. (2025). Teaching and learning challenges due to the COVID-19 pandemic: A systematic review. *Educational Research Review*, *100*, 100667. DOI: 10.1016/j.edurev.2025.100667

Tapalova, O., & Zhiyenbayeva, N. (2022). Artificial intelligence in education: Aied for personalised learning pathways. *Electronic Journal of e-Learning*, *20*(5), 639–653. DOI: 10.34190/ejel.20.5.2597

Tari, M., Ghobaei-Arani, M., Pouramini, J., & Ghorbian, M. (2024). Auto-scaling mechanisms in serverless computing: A comprehensive review. *Computer Science Review, 53*, 100650. DOI: 10.1016/j.cosrev.2024.100650

Tariq, G., Sun, H., & Ali, S. (2024). Environmental footprint impacts of green energies, green energy finance and green governance in G7 countries. *Carbon Footprints, 3*(1). Advance online publication. DOI: 10.20517/cf.2023.48

Tariq, M. U. (2024). Navigating the personalization pathway: Implementing adaptive learning technologies in higher education. In *Adaptive Learning Technologies for Higher Education* (pp. 265–291). IGI Global. DOI: 10.4018/979-8-3693-3641-0.ch012

Thomas, T. (2024). The role of artificial intelligence in formal and informal education for students. *International Journal for Research in Applied Science and Engineering Technology, 12*(3), 69–71. DOI: 10.22214/ijraset.2024.58738

Touba, R. (2014). The impact of using a mathematical modeling strategy on understanding mathematical concepts and solving mathematical problems among seventh-grade students in the measurement unit. *Unpublished Master's Thesis*, An-Najah National University, Palestine.

Trabelsi, Z., Alnajjar, F., Parambil, M. M. A., Gochoo, M., & Ali, L. (2023). Real-time attention monitoring system for classroom: A deep learning approach for student's behavior recognition. *Big Data and Cognitive Computing, 7*(1), 48. DOI: 10.3390/bdcc7010048

Trust, T., Krutka, D. G., & Carpenter, J. P. (2016). "Together we are better": Professional learning networks for teachers. *Computers & Education, 102*, 15–34. DOI: 10.1016/j.compedu.2016.06.007

Tsamados, A., Aggarwal, N., Cowls, J., Morley, J., Roberts, H., Taddeo, M., & Floridi, L. (2021). The ethics of algorithms: key problems and solutions. Ethics, governance, and policies in artificial intelligence, 97-123.

Tsourli, E. B., Dalakoura, Z. D., & Fragkaki, M. (2022). Developing 21st-century skills in university distance education. *International Conference on Open & Distance Education, 11*(7A), 154–169. https://doi.org/DOI: 10.12681/icodl.3316

Tuhuteru, L. (2023). The role of citizenship education in efforts to instill democratic values. International journal of humanities education and social sciences, 2(4).

Turnbull, D., Chugh, R., & Luck, J. (2020). Learning management systems, an overview. In *Encyclopedia of education and information technologies* (p. 248). Springer Nature., DOI: 10.1007/978-3-030-10576-1_248

Uleanya, C., & Prinsloo, P. (2024). Mapping the imaginary of Information Communication Technologies (ICTs) in education: The case of Botswana. *Learning, Media and Technology*. Advance online publication. DOI: 10.1080/17439884.2024.2306553

Umar, I., U, I., J. A, I., B, S., & I. M, M. (. (2024). Advancing Teachers' Professional Development: Exploring Models and Impact Assessment on Ict Integration in Colleges of Education. *International Journal of Integrative Research, 2*(1). Advance online publication. DOI: 10.59890/ijir.v2i1.792

UNESCO. (2021). Technical Vocational and Education Training (TVET) 1. *Sub-Education Policy Review Report: Technical and Vocational and Training (TVET)*, 74. https://en.unesco.org/sites/default/files/tvet_final_-_january_2021.pdf

Uyaguari Paredes, D.R., Uyaguari Paredes, M.J., Ríos Chichande, D.A. & Vásquez Villacis, M.I. (2024). Desarrollo de un entorno educativo inclusivo para niños con trastorno por déficit de atención e hiperactividad mediante el uso de tecnologías de la información y comunicación. *South Florida Journal of Development, 5*(3), 1-17. https://doi.org/ 5n3-017DOI: 10.46932/sfjdv

van de Werfhorst, H. G., Kessenich, E., & Geven, S. (2022). The digital divide in online education: Inequality in digital readiness of students and schools. *Computers and Education Open, 3*, 100100. DOI: 10.1016/j.caeo.2022.100100 PMID: 40478103

Vanlehn, K. (2011, October). The Relative Effectiveness of Human Tutoring, Intelligent Tutoring Systems, and Other Tutoring Systems. *Educational Psychologist, 46*(4), 197–221. DOI: 10.1080/00461520.2011.611369

Villafuerte Holguín, J. S., & Alonzo Rezabala, M. E. (2020). Touchscreens and teaching English to children with attention deficit disorder: language practices and recreational games. *Apertura, 12*(2), 52.73. https://doi.org/DOI: 10.32870/Ap.v12n2.1908

Vindigni, G. (2024). Strategies for overcoming barriers to inclusive and equitable education: A global analysis towards achieving sustainable development goal 4. European Journal of Arts. *Humanities and Social Sciences, 1*(5), 3–47. DOI: 10.59324/ejahss.2024.1(5).01

Voogt, J., Fisser, P., Good, J., Mishra, P., & Yadav, A. (2015). Computational thinking in compulsory education: Towards an agenda for research and practice. *Education and Information Technologies, 20*(4), 715–728. DOI: 10.1007/s10639-015-9412-6

Voronkova, V., Vasyl'chuk, G., Nikitenko, V., Kaganov, Y., & Metelenko, N. (2023). Transformation of digital education in the era of the fourth industrial revolution and globalization.

VOSviewer - Visualizing scientific landscapes. VOSviewer. Retrieved October, 2025, from https://www.vosviewer.com/

Vouglanis, T., & Driga, A. M. (2024). The use of ICT in the education of students with alexia/acquired dyslexia. *International Journal of Science and Research Archive*, *11*(1), 116–123. Advance online publication. DOI: 10.30574/ijsra.2024.11.1.0004

Wagner, P., & Kössler, T. (2022). Moulding democratic citizens: democracy and education in modern European history–an introduction. European Review of History: Revue européenne d'histoire, 29(6), 859-883.

Wagner, M., Ley, T., Kammerer, L., & Helm, C. (2024). Exploring teacher educators' challenges in the context of digital transformation and their self-reported TPACK: A mixed methods study. *European Journal of Teacher Education*, **•••**, 1–19. DOI: 10.1080/02619768.2024.2340689

Wang, C., Chen, X., Yu, T., Liu, Y., & Jing, Y. (2024). Education reform and change driven by digital technology: A bibliometric study from a global perspective. *Humanities & Social Sciences Communications*, *11*(1), 1–17. DOI: 10.1057/s41599-024-02717-y

Wang, R., Destek, M. A., Weimei, C., Albahooth, B., & Khan, Z. (2024). Drivers of Sustainable Green Finance: Country's Level Risk and Trade Perspective for OECD Countries. *Journal of Environment & Development*, *33*(1), 125–148. Advance online publication. DOI: 10.1177/10704965231217046

Wang, Y.-H. (2017). Integrating self-paced mobile learning into language instruction: Impact on reading comprehension and learner satisfaction. *Interactive Learning Environments*, *25*(3), 397–411. DOI: 10.1080/10494820.2015.1131170

Wang, Y., & Zhang, H. (2024). Generative AI in education and research: A systematic mapping review. *Review of Education*, *12*(2), e3489. DOI: 10.1002/rev3.3489

Williamson, B. (2017). Big data in education: The digital future of learning, policy and practice. *Sage (Atlanta, Ga.)*. Advance online publication. DOI: 10.4135/9781529714920

Wilson, A., & Brezicha, K. F. (2023). Good citizenship and the "True and Inspiring Story of America": A critical policy analysis of the South Dakota civics and history initiative. *Peabody Journal of Education*, *98*(5), 500–515. DOI: 10.1080/0161956X.2023.2261311

Wing, J. M. (2008). Computational thinking and thinking about computing. *Philosophical Transactions of the Royal Society A: Mathematical, Physical and Engineering Sciences, 366*(1881), 3717–3725. DOI: 10.1098/rsta.2008.0118

Wing, J. (2006). Computational thinking. *Communications of the ACM, 49*(3), 22–24. DOI: 10.1145/1118178.1118215

Wojewska, A. N., Staritz, C., Tröster, B., & Leisenheimer, L. (2024). The criticality of lithium and the finance-sustainability nexus: Supply-demand perceptions, state policies, production networks, and financial actors. *The Extractive Industries and Society, 17*, 101393. Advance online publication. DOI: 10.1016/j.exis.2023.101393

Woodman, T. C., Whatley, M., & Glass, C. R. (Eds.). (2023). *Digital internationalization in higher education: Beyond virtual exchange*. Taylor & Francis. DOI: 10.4324/9781003444237

Wu, T. (2024). Virtual research laboratory in foreign language teaching: Strategies and role. Scientific Bulletin of Mukachevo State University Series. *Pedagogy and Psychology, 10*(1), 107–115. DOI: 10.52534/msu-pp1.2024.107

Xiao, Z., Duritan, M. J. M., & Jia, R. (2024). Resourceful futures: Integrating responsible mining and green education for sustainable development in developing and emerging economies. *Resources Policy, 88*, 104377. Advance online publication. DOI: 10.1016/j.resourpol.2023.104377

Xu, E., Wang, W., & Wang, Q. (2023). The effectiveness of collaborative problem solving in promoting students' critical thinking: A meta-analysis based on empirical literature. *Humanities & Social Sciences Communications, 10*(1), 16. DOI: 10.1057/s41599-023-01508-1

Yaacob, Z., & Saad, N. (2020). Acceptance of youtube as a learning platform during the covid-19 pandemic: the moderating effect of subscription status. Tem Journal, 1732-1739. DOI: 10.18421/tem94-54

Yang, H., Zhu, S., & MacLeod, J. (2018). Promoting education equity in rural and underdeveloped areas: Cases on computer-supported collaborative teaching in china. *Eurasia Journal of Mathematics, Science and Technology Education, 14*(6). Advance online publication. DOI: 10.29333/ejmste/89841

Yang, S., Tian, H., Sun, L., & Yu, X. (2019). From one-size-fits-all teaching to adaptive learning: The crisis and solution of education in the era of AI. *Journal of Physics: Conference Series, 1237*(4), 042039. DOI: 10.1088/1742-6596/1237/4/042039

Yasser, I., El-Khalek, A. A., Khalid, H. E., Salama, A. A., & Essa, A. K. (2025). *Cloud Computing and neutrosophic Logic: Navigating uncertainty in Educational administration*. UNM Digital Repository.

Young, H. (2018). *Digital humanism: Technology and the future of humanity*. Routledge.

Youssef, A. B., Jain, P., & Al-Fadhli, S. (2024). Generative AI in higher education: A global perspective of adoption strategies. *Discover Education*, *2*, 15. Advance online publication. DOI: 10.1016/j.discove.2024.1001516

Zamfir, A. M., Davidescu, A. A., & Năstasă, A. (2023, June). Implications of Educational Poverty in a Post-Globalization World. In Griffiths School of Management and IT Annual Conference on Business, Entrepreneurship and Ethics (pp. 265-284). Cham: Springer Nature Switzerland.

Zavgorodniaia, A., Tilanterä, A., Korhonen, A., Seppälä, O., Hellas, A., & Sorva, J. (2021). Algorithm visualization and the elusive modality effect. *ICER 2021 - Proceedings of the 17th*

Zembylas, M. (2013). The "emotional turn" in education. In Wong, P. T. P. (Ed.), *The human quest for meaning* (2nd ed., pp. 387–400). Routledge.

Zembylas, M. (2018). Reinventing critical pedagogy as decolonising pedagogy. *Review of Education, Pedagogy & Cultural Studies*, *40*(5), 404–421. DOI: 10.1080/10714413.2019.1570794

Zeng, W. (2022). An empirical research on china's policy for ict integration in basic education from 1988 to 2021. *Educational Technology Research and Development*, *70*(3), 1059–1082. DOI: 10.1007/s11423-022-10079-y PMID: 35221627

Zhang, C., Zhang, L., Liu, L., & Du, C. (2024). The study of the relationship between green finance and resource efficiency in east asian economies. *Resources Policy*, *89*, 104658. Advance online publication. DOI: 10.1016/j.resourpol.2024.104658

Zhang, Y., Adams, D., & Lee, K. C. S. (2024). Assessing academic staff's ICT integration in higher education: Instrument development and validation. *Journal of Computers in Education*, *11*(1), 75–94. Advance online publication. DOI: 10.1007/s40692-022-00248-6

Zhou, Q., Lee, C., Sin, S., Lin, S., Hu, H., & Ismail, M. (2020). Understanding the use of youtube as a learning resource: A social cognitive perspective. *Aslib Journal of Information Management*, *72*(3), 339–359. DOI: 10.1108/AJIM-10-2019-0290

Zubković, B., Pahljina-Reinić, R., & Kolić-Vehovec, S. (2022). Predictors of ict use in teaching in different educational domains. *Humanities Today Proceedings*, *1*(1), 75–91. DOI: 10.2478/htpr-2022-0006

About the Contributors

Faiz Abhelhafid is preparing a Ph.D. in teaching and learning at An-Najah National University and working for Al-Istiqlal University since 2016, Faiz Abdelhafid is working as a lecturer at the faculty of intermediate studies and head of the planification and development department at the Continuous Education Center. He actively planned and coordinated many innovative international projects evolving around educational technologies. He has a Master degree in social use of technologies for education and training from a French university.

Sameh Alqubaj earned a Ph.D. in international relations. In addition to his MA in Regional Studies, he holds a bachelor's degree in computer information systems and an intermediate diploma in computer programming and system analysis. Al-Qubaj taught computers in several Ramallah schools. He has taught many courses at al-Quds Open University and Al-Istiqlal University. Al-Qubaj prepared and developed a large number of course syllabuses during his academic career in teaching. Al-Qubaj has worked in information technology since 1991, holding positions such as programmer, database developer, consultant, and Head of Technical Department. From 2007 to 2017, he was the director of Al-Istiqlal University's computer center. Al-Qubaj is a member of the Academic Quality Team at Al-Istiqlal University.

***

Amra Ariyani, S.Pd., M.Pd., is a lecturer in the English Language Education Program at the Faculty of Languages and Literature, Universitas Negeri Makassar (UNM). She earned her Bachelor's, Master's, and Doctoral degrees in English Language Education from the same institution, demonstrating a strong and continuous academic commitment to her field. Dr. Amra specializes in courses related to English language teaching and learning, including instructional media, language teaching methodologies, and classroom-based language practices. Her teaching is characterized by a strong emphasis on practical application, innovation, and student-centered learning. Her research interests encompass English language education, language teaching methodology, learner characteristics, and innovative strategies in teaching English as a Foreign Language (EFL). Through her scholarly work and pedagogical practice, Dr. Amra is dedicated to advancing the quality of English education and fostering the professional growth of future English language educators.

Muralidhar B is a professionally qualified dental surgeon. He holds Ph.D. from Canara Bank School of Management Studies, Bangalore University after his MBA from Christ University. He has 18 years of total teaching experience spanning across dental education, healthcare management and general management and 4 years of corporate experience. He has worked as research associate in Institute for Social and Economic Change, Bangalore in a project entitled as "Management By and For Common Man: An Exploration of the Concept and Direction for Future"post Ph.D. He had secured 28th rank in S.S.L.C examination for Karnataka State in 1993. His 25 patents have been published and 2 design patents have been granted. He has 36 research publications to his credit out of which 8 are Scopus-indexed and 5 are ABDC indexed. He has co-authored 11 books. He is currently working as Assistant Professor in Department of Business Management Studies, School of Commerce, Jain Deemed to be University, Bangalore.

Kashvi Chaturvedi is studying B.Tech, specialisation in Artificial Intelligence and Data Science, at Ajeenkya DY Patil University. She started solidifying her programming skills by completing various certification courses like Data Structures in C, Tableau functions and Operating Systems along with doing few projects using languages like Python, C, C++ and creating Web Page using HTML, CSS and Java Script. She had also attended an Arduino-Based Automation course at Dolphin Labs to enhance her software knowledge. There, she learned how to work with sensors, microcontrollers, and circuit design. She has also written multiple review papers on topics such as Geographic Information Systems, Comparative Analysis of BJT Configurations, Swiss Bank vs. European Central Bank: Data Analysis. She is also researching on and learning many topics such as Small Modular Reactors, Bibliometric Analysis, UNSDGs etc. Because of her interest in latest technologies, she also participated in Global Hyperloop Competition hosted by IIT Madras representing her college as team Vac-Vectoor where she learnt more about various subtopics from the mentors.

Susanta Das is a seasoned educator with a proven record of teaching and mentoring diverse students, working at multiple institutions, and providing service to the various initiatives of organizations. He received his Ph.D. and M.A. degrees from Western Michigan University (WMU), USA, and M.Sc. degree from Banaras Hindu University (BHU), India all in Physics. He continued his research as a Marie-Curie post-doctoral fellow at Stockholm University, Stockholm, Sweden on beam diagnostics for the DESIREE (Double ElectroStatic Ion Ring ExpEriment) in the project DITANET (Diagnostic Techniques for particle Accelerators – a Marie-Curie initial training NETwork), at the Indian Institute of Science Education and Research-Kolkata, India on high-pressure physics as a Project Scientist-B, and at the University of Electro-Communications, Tokyo, Japan on ion-surface interactions as a Post-doctoral fellow. Throughout his career, Dr. Das worked and collaborated with many researchers, Ph.D., master, and visiting students from different countries. He further visited many countries to discuss research and an international conference participant (UK, Italy, Germany, Greece, Belgium, Bulgaria, Romania, Brazil etc.). He worked at several institutions before assuming his current role at Ajeenkya D Y Patil University. He received the Marie-Curie post-doctoral fellowship in Sweden, Science Academies' Summer Research Fellowship in India, Gowen Frostic Doctoral Fellowship, Department Graduate Research and Creative Scholar Award by WMU, and Leo R. Parpart Doctoral Fellowship by Dept. of Physics, WMU, among many others, throughout his academic journey. He co-authored several research articles in WoS/Scopus indexed international journals, conference proceedings, and scholarly book chapters and co-edited several books published by renowned international publishing houses (IGI-Global, Springer Nature, Nova Science Publishers, Wiley, Scrivener-Wiley, Notion Press). At present, Dr. Das continues to delve into cutting-edge fields, with a keen interest in nanotechnology, UN SDGs, and data science. His multifaceted contributions, spanning teaching, research, and administrative leadership, showcase a dedicated professional who is instrumental in advancing the vision and mission of the institutions he serves.

Pooja Dehankar is working in Department of Artificial Intelligence and Data Science, School of Engineering, Ajeenkya DY Patil, Pune. She has completed BE in Computer Engineering from Priyadarshini College of Engineering, Nagpur and ME in Information Technology from Sinhgad College of Engineering, Pune. She has done Post Graduate Diploma in Advance Computing. She is currently pursuing Ph.D. from Symbiosis International(Deemed University). She has 16 years of experience in academics. Her area of interest are Artificial Intelligence, Cyber Security and Data Mining. She has published papers in Scopus Journals. She actively participates in workshops, faculty development programs, short term training programs, conferences and webinars.

Y Fathima, BBA.,MFC.,MBA.,M.Phil.,DGT.,Ph.D.,is an Professor in the Department of Commerce, School of Commerce at Jain (Deemed-to-be University) Bangalore. An accomplished educator with demonstrated ability to teach, motivate, and direct students while maintaining high interest and achievement. Articulate communicator able to effectively interact with diverse population of students at a variety of academic levels. She has 18 years of Teaching Experience and 10 years of Research experience. She has published papers in reputed journal and authored 4 books.

Maria Fragkaki is a senior researcher, academic coordinator, and professor with extensive expertise in ICT in education, critical pedagogy, and the design of digitally mediated learning environments. She currently holds the position of Academic Coordinator and Professor-Counsellor for four postgraduate modules within the Master's Programme in Educational Studies at the Open University of Cyprus. In parallel, she serves as a Professor-Counsellor at the Hellenic Open University and as an Academic Reading and Writing Tutor for international postgraduate students in the United Kingdom, providing specialised support in academic literacy, critical analysis, and research writing. Prof. Fragkaki holds a PhD in ICT in Education, a Master's degree in Informatics in Education, a Master's in Computer Science in Education, and a Bachelor's degree in Pedagogical Studies. She is also a Fellow of the UK Higher Education Academy (FHEA), a distinction that reflects her sustained commitment to excellence in university teaching, academic mentoring, and inclusive pedagogical practice. Over the course of her career, she has served as a senior researcher and project manager at the University of Patras, Greece, and has held visiting academic positions at the University of Surrey (UK) and Al Quds Open University (Palestine). Her research interrogates the intersections of empathy, digital pedagogy, transformative learning, and decolonial education, particularly within online and distance education contexts. Prof. Fragkaki has an extensive record of academic publication, including peer-reviewed journal articles, scholarly book chapters, and conference contributions. She has led and participated in numerous European and international research initiatives, with a focus on digital inclusion, pedagogical innovation, and social justice in education. Her current work introduces the Empathic Anthropological Framework for Distance Education (EAF@UDE), which conceptualises empathy as both an epistemological and pedagogical imperative. Through this lens, she explores how emotional intelligence, cultural responsiveness, and ethically grounded design can rehumanise technology-mediated learning environments.

Isabel García has a degree in English Studies from the University of Alicante and a PhD from the National University of Distance Education (UNED). She is currently working as a full time lecturer at the Centro Universitario de la Defensa in San Javier (Spain). She has been awarded a scholarship by the Franklin Institute (University of Alcala de Henares, Spain) and the Radcliffe Institute for Advanced Study (Harvard University, USA), where she conducted a pre-doctoral research stay in the year 2012. Among her main lines of research, we can highlight the role of different women in literature and culture of the United States in contrast to Francophone writers with publication in Spanish journals like Revista Estudios Humanísticos (University of León), Tonos Digital (University of Murcia), Dossiers Feministes (University Jaime I), Prisma Social, Raudem (University of Almeria), Camino Real (University of Alcala de Henres), or Nomadas (University Complutense of Madrid). She also focuses on the study of the history of the teaching of English and gender studies with publications in journals such as Revista Feminismo (University of Alicante) or even Quaderns Digitals.

Muliaty Ibrahim is a dedicated educator with over 10 years of experience as a lecturer in the Bachelor's Program of English Education at Universitas Megarezky. She holds a Bachelor's, Master's, and Doctoral degree from Universitas Negeri Makassar, showcasing her deep commitment to academic excellence. She is an active member of various professional English language organizations, Muliaty is also an avid writer, contributing to national and international journals and authoring several book chapters. Her passion for teaching, research, and professional development continues to inspire and empower her students and colleagues in the field of English education.

Antonio Juan Rubio has a degree in English Studies from the University of Murcia and a PhD from the National University of Distance Education (UNED) with the positive accreditation by the ANECA body, being given the Extraordinary Doctorate Award. He is currently working as a professor at the University of Granada. He has been awarded a scholarship by University College (Cork, Ireland), the Franklin Institute (University of Alcala de Henares, Spain) and the Radcliffe Institute for Advanced Study (Harvard University, USA), where he conducted a pre-doctoral research visit in the year 2012. He is currently a member of the scientific committee of several national and international journals as well as a member of the editorial board of several international journals. He also belongs to the organizing and scientific committee of several conferences organized by the Athens Institute for Education and Research (Greece). Among his main lines of research we can emphasize the following aspects: cultural studies in the United States; gender issues associated with the role of women in the Anglo-American literature; or the teaching practice and process of English.

Mustafa Kayyali is an ardent advocate for excellence in higher education, driven by a relentless pursuit of quality, recognition, and innovation. Their journey in academia started with a Master's in Quality Management and Evaluation in Higher Education from Universitat Oberta de Catalunya, followed by a Ph.D. in Quality Management from Azteca University. These academic pursuits have laid a strong foundation for their expertise in Accreditation, Quality Assurance, and Higher Education Rankings. As an entrepreneur, researcher, translator, and publisher, they are deeply involved in various facets of the academic world. Their diverse interests encompass Management, Translation, Interpretation, and Academic consulting, contributing to a well-rounded understanding of the industry. Throughout their career, they have made significant contributions to academic literature, with more than 40 published papers and 30 book chapters to my name. Additionally, they take pride in having translated 6 books, bridging language gaps, and promoting knowledge exchange on a global scale. They are committed to fostering positive change in higher education and contributing to its continuous improvement.

Manju Priya R, currently an Associate Professor at the School of Commerce, JAIN (Deemed-to-be) University in Bangalore, holds a M.Com., M.Phil, MBA, PGDCA, and Ph.D. She is a distinguished educator known for her ability to inspire, engage and guide students towards high academic achievement. With 19 years of teaching experience and 10 years of research expertise, she excels in maintaining students' interest and fostering a dynamic learning environment. She has guided numerous undergraduate, postgraduate and M.Phil students and is presently mentoring Ph.D candidates. Her academic interests are focused on Marketing and Human Resources(HR). She has served as a resource person for various colleges, sharing her insights with both students and faculty members. She has actively participated in over 50 National and International seminars and conferences, and has published more than 45 research papers in UGC care listed journals, Scopus, WOS, ABDC. Furthermore, she has authored 3 books and she has also contributed 10 book chapters to various academic publications.

Varanasi Rahul, an Assistant Professor at JAIN University, Bangalore, holds three postgraduate degrees (M.Com in Accounting & Taxation, M.Com in Management, and MA in History) and is a semi-qualified Cost Accountant. He has cleared UGC NET in Commerce twice, AP SET in Commerce, History, and Management, and TS SET in Commerce. With 3 years of teaching experience and 6 months at PwC, he is also a freelance archaeologist, epigraphist, and numismatist. He has authored 3 books, published 6 research articles, contributed 9 book chapters, and written 30+ newspaper articles. Rahul has applied for 11 patents, with 1 granted and 09 published.

Vivit Rosmayanti, S.Pd.I., M.Pd was born in Botta on June 26, 1987. She has over 10 years of experience teaching at the University. She obtained a bachelor's degree in English Education from Alauddin State Islamic University. She completed her master's and doctoral degrees in English Education from Makassar State University. She is currently a lecturer in the Post Graduate Program, Department of English Education at Makassar State University, South Sulawesi, Indonesia. She is also actively writing research articles to be published in nationally and internationally accredited journals. She has written in some chapter books, monographs, and reference books. She is interested in teaching methods for teaching English. Her research interests include language teaching methodology, ICT instruction, and ESP teaching.

N Sathyanarayana has over 15 years of teaching experience at the UG and PG levels. He holds a Master's in Business Administration and Commerce and is pursuing a Ph.D. He currently holds the position of Assistant Professor in the School of Commerce at Jain (Deemed-to-be University), Bangalore. He has published 22 research papers in international journals, 7 research papers in national journals, and presented 14 research papers at international conferences and 11 research papers at national conferences. He has co-authored 2 books at national level and 7 book chapters (Scopus indexed).

Wasswa Shafik (Member, IEEE) received a Bachelor of Science degree in information technology from Ndejje University, Uganda, a Master of Engineering degree in Information Technology (Computer and Communication Networks) from Yazd University, Iran, and a PhD in Computer Science with the School of Digital Science, Universiti Brunei Darussalam, Brunei Darussalam. He is also the Founder and Research Director of the Dig Connectivity Research Laboratory (DCRLab) after serving as a Research Associate at the Network Interconnectivity Research Laboratory at Yazd University. Prior to this, he worked as a Community Data Analyst at Population Services International (PSI-Uganda), a Community Data Officer at the Programme for Accessible Health Communication (PACE-Uganda), a Research Assistant at the Socio-Economic Data Centre (SEDC-Uganda), Uganda, an Assistant Data Officer at TechnoServe, Kampala, Uganda, and Asmaah Charity Organization as Cheif Executive Officer. He has hundreds of publications with renowned publishers. His research interests include Artificial Intelligence, Neural Networks, Computer Vision, Smart Agriculture, Digital Health, and Green Computing.

G Srividhya, M.Com, M.Phil, Ph.D., is currently an Assistant Professor in the Department of Commerce at the School of Commerce, Jain (Deemed-to-be University), Bangalore. With a career spanning since 2005, she has accumulated

19 years of teaching experience and 10 years of research experience across various institutions. She has published papers in reputed journals and contributed book chapters to various national and international conferences. She is a life member of professional bodies such as IARA and IAA. She also possesses expertise in accounting and finance.

Wan Ahmad Jaafar Wan Yahaya has been Director at the Centre for Instructional Technology and Multimedia (PTPM), Universiti Sains Malaysia (USM), Penang, Malaysia since 2016. He is also the former President of the Malaysian Educational Technology Association (META)He is the Chairman of the Technical Committee on Audio, Video, Information Technology, and Communication Equipment at the Department of Standard Malaysia, Ministry of International Trade and Industry Malaysia. He holds a PhD in Multimedia Education from the University of Leeds and a Master of Education (Computer in Education) from the University of Manchester, United Kingdom. He has published more than 170 articles/papers in indexed journals like Scopus, Web of Science (ISI), SCSI, SSCI, conference proceedings, and book chapters, primarily in the field of multimedia education. He is the recipient of the Thailand Association International Honor Award for Outstanding Contributions to the field of Educational Communications and Technology for the year 2023.

Mohd Noorulfakhri Yaacob, a Senior Lecturer at Universiti Malaysia Perlis (UniMAP), is a renowned expert in information technology, system development, and biometric research. Born on September 1, 1981, in Malaysia, he holds a Ph.D. in Communication and Information Technology from UniMAP (2021), a Master's in Information Technology from Universiti Utara Malaysia (2009), and a Bachelor's in Computer Science from Universiti Kebangsaan Malaysia (2003). With over 20 years of experience, Dr. Noorulfakhri has contributed significantly to system development, data center management, and national-level ICT projects. He has led the development of systems for the Ministry of Higher Education, including SUKIPT and ASEAN University Games. His research focuses on biometric authentication, particularly keystroke dynamics, with numerous publications in prestigious journals. A recipient of the I-RIA 2019 Silver Medal, Dr. Noorulfakhri is also a certified Professional Technologist (Ts.) and holds multiple IT certifications. Residing in Arau, Perlis, he is a dedicated academic, innovator, and family man committed to advancing technology and education.

Index

A

Academic Institutions 55
Access to Learning 274
AI in Education 87, 161, 164, 165, 166, 167, 168, 179, 180, 186, 205
AI-Powered Simulation 162, 163
Algorithmic Control 17
Algorithmic Thinking 220, 222, 223, 224, 225, 226, 235
Anthropological Framework 1, 2, 6, 7, 11, 12, 16, 17, 20, 23, 24, 26, 29, 32, 34, 44, 47
Anthropology of Education 65, 72, 76
Artificial Intelligence 48, 57, 72, 75, 78, 85, 86, 88, 89, 91, 99, 105, 109, 110, 111, 113, 135, 159, 160, 161, 162, 164, 169, 178, 181, 185, 186, 190, 212, 269, 275, 279, 285, 287, 290, 291
Attention Deficit Disorder 243, 244, 247, 248, 268
Attention Deficit Hyperactivity Disorder 243, 244, 265, 267

C

Challenges 2, 15, 16, 17, 21, 23, 30, 31, 44, 45, 63, 69, 71, 81, 82, 86, 87, 89, 90, 93, 96, 98, 99, 100, 101, 103, 104, 107, 109, 111, 112, 113, 114, 123, 126, 138, 155, 165, 181, 185, 193, 197, 212, 213, 214, 222, 224, 225, 238, 241, 243, 245, 248, 249, 255, 260, 262, 263, 264, 280, 281, 286, 288, 289, 290
Chat GPT 174, 175
Civic Disengagement 117
Civic Engagement 6, 10, 44, 49, 116, 119, 120
Cloud Computing 85, 86, 91, 105, 112, 187, 188, 189, 191, 192, 193, 194, 195, 196, 197, 198, 199, 200, 201, 202, 203, 204, 205, 206, 207, 208, 209, 210, 211, 212, 213, 214, 215, 216, 217, 294
Cognitive Engagement 4, 26, 178, 181
Cognitive Load 163, 219, 220, 221, 223, 237, 238, 239, 240
Collaboration 34, 36, 63, 68, 74, 92, 94, 96, 97, 98, 99, 100, 102, 103, 104, 105, 106, 107, 110, 111, 124, 130, 166, 168, 187, 188, 191, 192, 195, 196, 208, 210, 211, 212, 216, 220, 240, 272, 273, 275, 277, 284
Communications 72, 114, 157, 185, 200, 238, 242
Creativity 60, 68, 94, 106, 111, 120, 121, 124, 125, 126, 127, 159, 160, 161, 164, 165, 167, 168, 169, 170, 171, 172, 174, 175, 176, 177, 178, 179, 180, 181, 182, 186, 187, 193, 196, 208, 258, 259, 270
Cultural Responsiveness 28

D

Decision Making 185
Digital Divide 61, 62, 63, 69, 71, 81, 87, 90, 93, 100, 101, 103, 104, 107, 108, 113, 140, 143, 154, 269, 281
Digital Education 3, 6, 7, 8, 10, 11, 14, 15, 18, 23, 32, 35, 48, 63, 87, 88, 90, 103, 104, 106, 107, 110, 124, 125, 132, 139, 145, 156, 198, 205, 213, 235
Digital Inclusion 104, 138
Digital Learning 1, 2, 5, 7, 9, 11, 12, 14, 18, 20, 22, 30, 35, 41, 42, 46, 47, 53, 88, 91, 96, 101, 102, 104, 110, 112, 125, 141, 143, 189, 195, 202, 208, 209, 286, 296
Digital Pedagogy 6, 7, 9, 10, 16, 28, 47, 49, 124, 125, 126, 127, 134
Digital Transformation 1, 85, 86, 87, 88, 89, 90, 91, 92, 93, 94, 96, 97, 98, 99, 100, 101, 102, 103, 104, 105, 106, 107, 110, 111, 114, 115
Digital Video Courseware 219, 224, 226, 227, 229, 230, 231, 232, 234, 235, 236, 237

E

Educational Equity 138, 198, 274
Educational Technology 48, 53, 54, 68, 70, 72, 73, 74, 104, 110, 111, 113, 126, 131, 139, 141, 143, 144, 149, 150, 152, 154, 155, 196, 208, 220, 221, 222, 223, 239, 240, 242, 275, 286, 289, 292
EFL 239, 243, 245, 278, 285, 286, 289
Emotional Intelligence 37, 126, 166, 167
Empathy 1, 2, 3, 4, 5, 6, 7, 8, 9, 10, 11, 12, 13, 14, 15, 16, 17, 18, 20, 21, 22, 23, 24, 25, 26, 27, 28, 29, 30, 32, 33, 34, 35, 36, 37, 38, 39, 41, 42, 43, 44, 45, 46, 47, 48, 49, 50, 52, 120, 142, 145, 150, 165, 167, 250, 253
Equity in Education 154, 274, 299

F

Future of Learning 52, 134

H

Human-Centered Design 12, 150

I

ICT 37, 53, 54, 55, 56, 57, 58, 59, 60, 61, 62, 63, 64, 66, 67, 68, 69, 70, 71, 72, 73, 74, 75, 76, 78, 79, 80, 81, 82, 83, 108, 112, 115, 149, 150, 151, 156, 188, 195, 243, 244, 245, 246, 248, 252, 254, 255, 256, 257, 258, 259, 260, 261, 262, 263, 264, 270, 271, 272, 273, 274, 275, 276, 277, 281, 282, 283, 285, 286, 287, 288, 289, 290, 291, 292, 299
ICT in Education 53, 56, 59, 60, 61, 62, 63, 70, 72, 76, 255, 299
ICT in ELT 270, 271, 282, 286, 291
Instructional Design 9, 11, 14, 90, 110, 170, 214, 223, 225, 227
Interactive Website 293, 294, 296, 297

L

Learner-Paced Segmentation 224
Life Mathematics 293, 294, 296, 297

M

Mathematical Problems 295, 296, 297, 300

O

Online Learning 7, 8, 11, 14, 15, 16, 18, 20, 22, 23, 26, 41, 42, 51, 63, 71, 75, 87, 92, 95, 101, 108, 111, 113, 139, 152, 205, 207, 221, 223, 224, 225, 226, 241, 270
Outcomes 25, 58, 59, 60, 68, 75, 78, 85, 86, 88, 90, 91, 93, 94, 97, 99, 101, 108, 121, 148, 159, 168, 169, 178, 179, 180, 181, 182, 183, 187, 188, 192, 219, 223, 226, 228, 232, 234, 236, 237, 271, 273, 277, 282, 295

P

Post-Globalization 137, 138, 139, 140, 141, 142, 149, 150, 151, 152, 153, 155, 156, 157

S

Segmentation Techniques 219
Students 9, 14, 22, 23, 24, 25, 26, 27, 28, 32, 33, 34, 35, 36, 37, 38, 39, 41, 42, 43, 44, 48, 54, 55, 58, 59, 60, 62, 63, 64, 68, 69, 70, 72, 73, 74, 76, 78, 81, 82, 86, 87, 88, 89, 90, 91, 92, 93, 94, 95, 97, 98, 99, 100, 101, 102, 103, 104, 105, 106, 107, 108, 111, 113, 114, 117, 118, 119, 120, 121, 122, 123, 124, 125, 126, 128, 129, 130, 132, 138, 139, 140, 141, 142, 143, 144, 145, 146, 147, 148, 149, 150, 151, 155, 159, 160, 161, 162, 163, 164, 165, 166, 167, 168, 169, 170, 171, 172, 173, 174, 175, 176, 177, 178, 179, 180, 181, 182, 183, 184,

185, 187, 188, 189, 191, 194, 195, 196, 197, 201, 210, 211, 213, 214, 215, 216, 217, 220, 221, 222, 223, 224, 225, 226, 227, 230, 231, 232, 233, 234, 235, 236, 237, 238, 239, 241, 243, 244, 245, 246, 247, 249, 250, 251, 252, 253, 254, 255, 256, 257, 258, 259, 260, 261, 262, 263, 264, 265, 266, 267, 271, 272, 273, 274, 275, 276, 278, 279, 280, 281, 282, 283, 284, 285, 286, 288, 289, 290, 291, 294, 295, 296, 297, 298, 299, 300

Surveillance 7, 16, 22, 27, 37, 60, 74, 88, 102, 112, 115, 119, 122, 124, 126, 128, 131, 133, 139, 144, 147, 148, 156, 200

T

Teaching Practices 72, 75, 93, 94, 96, 97, 101, 106, 165, 169, 171, 174, 275, 282

Technology Integration 67, 76, 221, 266

www.ingramcontent.com/pod-product-compliance
Lightning Source LLC
LaVergne TN
LVHW080903300926
849162LV00003B/534